Digitisation

In recent years, digital technologies have become pervasive in academic and everyday life. This comprehensive volume covers a wide range of concepts for studying the new cultural dynamics that are evident as a result of digitisation. It considers how the cultural changes triggered by digitisation processes can be approached empirically. The chapters include carefully chosen examples and help readers from disciplines such as Anthropology, Sociology, Media Studies, and Science and Technology Studies to grasp digitisation theoretically as well as methodologically.

Gertraud Koch is a Professor and Head of the Institute of European Ethnology/ Cultural Anthropology at the University of Hamburg, Germany.

Digitisation
Theories and Concepts for Empirical Cultural Research

Edited by Gertraud Koch

LONDON AND NEW YORK

First published 2017
by Routledge
2 Park Square, Milton Park, Abingdon, Oxon OX14 4RN

and by Routledge
711 Third Avenue, New York, NY 10017

Routledge is an imprint of the Taylor & Francis Group, an informa business

© 2017 selection and editorial matter, Gertraud Koch; individual chapters, the contributors

The right of Gertraud Koch to be identified as the author of the editorial material, and of the authors for their individual chapters, has been asserted in accordance with sections 77 and 78 of the Copyright, Designs and Patents Act 1988.

All rights reserved. No part of this book may be reprinted or reproduced or utilised in any form or by any electronic, mechanical, or other means, now known or hereafter invented, including photocopying and recording, or in any information storage or retrieval system, without permission in writing from the publishers.

Trademark notice: Product or corporate names may be trademarks or registered trademarks, and are used only for identification and explanation without intent to infringe.

British Library Cataloguing-in-Publication Data
A catalogue record for this book is available from the British Library

Library of Congress Cataloging-in-Publication Data
A catalog record for this book has been requested

ISBN: 978-1-138-64610-0 (hbk)
ISBN: 978-1-315-62773-1 (ebk)

Typeset in Times New Roman
by codeMantra

Printed and bound in Great Britain by
TJ International Ltd, Padstow, Cornwall

Contents

List of illustrations vii
Notes on contributors viii

Introduction: digitisation as challenge for empirical cultural research 1
GERTRAUD KOCH

PART I
Coded culture 11

1 **Cultural techniques, practices, programmes: how to study the anthropo-logic of digitisation** 13
CARSTEN OCHS

2 **Archive** 40
ISTO HUVILA

3 **Imperfect imaginaries: digitisation, mundanisation, and the ungraspable** 53
ROBERT WILLIM

4 **Ethnography of digital infrastructures** 78
GERTRAUD KOCH

PART II
Doing digital culture 93

5 **Hackers and hacking** 95
LUIS FELIPE R. MURILLO AND CHRISTOPHER KELTY

6 'A brilliant copy every time!': aspects of a cultural proportion 117
CHRISTIAN SCHÖNHOLZ

7 The manifestation of mash-up categories 132
JOAN KRISTIN BLEICHER

8 Big Data 158
KATHARINA E. KINDER-KURLANDA

PART III
Approaching the world digitally 177

9 From GUI to No-UI: locating the interface for the
Internet of Things 179
NISHANT SHAH

10 Ubiquitous computing and the Internet of Things 197
KATHARINA E. KINDER-KURLANDA AND DANIEL BOOS

11 Calculating spaces: digital encounters with maps and geodata 209
INA DIETZSCH AND DANIEL KUNZELMANN

12 Augmented realities 230
GERTRAUD KOCH

PART IV
Concepts of culture revisited 249

13 The political economy of digital technologies: outlining
an emerging field of research 251
ANDREAS WITTEL

14 Ludification of culture: the significance of play and games
in everyday practices of the digital era 276
ANNE DIPPEL AND SONIA FIZEK

15 Media genealogy: back to the present of digital cultures 293
CLEMENS APPRICH AND GÖTZ BACHMANN

Index 307

Illustrations

Figures

3.1	Servers at Google's Douglas County, Georgia, data centre	66
3.2	Steel works, Pittsburgh, PA	68
3.3	Pipelines at Google's Douglas County, Georgia, data centre	69
3.4	Servers at Google's Hamina, Finland, data centre	70
5.1	'Hackers are Doers' Funders and Founders website	96
5.2	Stack of power: an analytic decomposition of hacking practices and how they might relate	112
11.1	Network plans of the London Underground	211
11.2	The 'clotted' grid of the children's map of Basel	219
11.3	Narrative elements of the plot as visualised data layers	219
11.4	Screenshots of the geodata-based navigational software Waze	223
12.1	Reality-Virtuality Continuum according to Milgram et al.	232
12.2	WiFi box, London, Summer 2015	234
12.3	Piccadilly Circus advertising screens	237
12.4	MoSoho, interactive screen at the Museum of Soho, London	238

Box

5.1	Promotional language for the Vitra 'Hack' desk	104

Contributors

Clemens Apprich is a Research Associate at the Centre for Digital Cultures at Leuphana University of Lüneburg. His research ranges from the history to the theory to the techno-political economy of digital cultures. His upcoming book on the *Media Genealogy of Net Cultures* will be published by Rowman & Littlefield International.

Götz Bachmann is Professor for Digital Cultures at Leuphana University of Lüneburg. His work explores the interstices of ethnography, media and social theory, and workplace studies. He has authored a monograph on collegiality and is currently working on an ethnography of engineering elites in the Bay Area.

Joan Kristin Bleicher is, since 2001, Professor for Media Studies at the Department for Media and Communication at the University of Hamburg. Since 2009, she is Member of the Research Centre for Media and Communication at the University of Hamburg. Her research interests are borderlines between fact and fiction in the media, German quality TV, online television, and current media developments. Her most influential publications are *Television as Mythology* (1999), *Forcing Community – Filmfunding by Television* (2014), *Sexy Media* with Skadi Loist and Sigrid Kannengießer (2013), and *Internet* (2010).

Daniel Boos investigates future work experiences at Swisscom. He received a Ph.D. at the ETH Zurich. The topic of his dissertation was *Accountability and Internet of Things Applications*. His research interests include the relation between ubiquitous computing, work and organisation, digital workplace, and virtual collaboration.

Ina Dietzsch is a Cultural Anthropologist at the University of Basel and acting Chair of Microsociology at the Technical University of Dresden. Her research interests are in the field of the anthropology of knowledge, the mathematisation of society, and particularly in the relation between images, texts, and numbers.

Anne Dippel is an Anthropologist and Historian. In 2015, she published *Dichten und Denken in Österreich*, an ethnography based on two years

Contributors ix

of fieldwork in Austria. She is a Lecturer at the Department of Socio-Cultural Anthropology and Cultural History at Friedrich-Schiller University Jena. Her current research focuses on practices and conditions of knowledge production at CERN, focussing on the impact of playfulness and games in science.

Sonia Fizek is a Lecturer in game theory and design at Abertay University. In 2013–2015, she was a postdoctoral researcher at Leuphana University of Lüneburg, where she co-edited *Rethinking Gamification* (2014). Her current research focuses on work and play, data games for science, and design. She is also an associate editor of the *Journal of Gaming and Virtual Worlds*.

Isto Huvila holds the Chair in Information Studies at the Department of ALM (Archives, Library, and Information, Museum and Cultural Heritage Studies) at Uppsala University, Sweden. His primary areas of research include information and knowledge management, information work, knowledge organisation, documentation, and participatory information practices.

Christopher Kelty is Professor in the departments of Anthropology and Information Studies, and the Institute for Society and Genetics at UCLA.

Katharina E. Kinder-Kurlanda is a Cultural Anthropologist and Web Scientist working as a team leader at GESIS. She obtained her Ph.D. in an interdisciplinary ubiquitous computing project at Lancaster University. Her current research interests are Big Data, algorithmic public spheres, social media data, and Internet research ethics.

Gertraud Koch is Professor and Head of the Institute of European Ethnology / Cultural Anthropology at the University of Hamburg. She is vice-chairwoman of the Expert Committee 'Intangible Cultural Heritage' of the German Commission for UNESCO. Her research fields include knowledge anthropology, digital anthropology, working cultures, and urban and regional diversities.

Daniel Kunzelmann is doing a binational doctorate in Cultural Anthropology and Political Science at the LMU Munich and the University of Basel. Being interested in political anthropology, social media, and algorithmic power, his research focuses on digital transformations of local politics.

Luis Felipe R. Murillo is an Anthropologist whose research interests include science and technology studies, computing, language, and politics. He is currently working as a postdoctoral researcher at the Conservatoire National des Arts et Métiers (CNAM) in the interdisciplinary laboratory of economic sociology (LISE/CNRS) on questions of 'open science' with support from the Institute for Research and Innovation in Society (IFRIS) in Paris.

x   Contributors

Carsten Ochs is Postdoc at the Sociological Theory Department, University of Kassel. He deals with digitisation issues from an STS perspective. Current projects include the digital transformation of the public/private distinction. Publications include *Digitale Glokalisierung: Das Paradox von weltweiter Sozialität und lokaler Kultur* (2013).

Nishant Shah is a Guest Professor at Leuphana University of Lüneburg, Co-founder of the Centre for Internet & Society India, and Dean-Research at ArtEZ University of the Arts, The Netherlands. His works lie at the intersection of digital cultures, identity politics, and affective accounts of social justice.

Christian Schönholz lives and works in Marburg, Germany, where he studied European Ethnology/Cultural Studies, Art History, and Sociology. Since his Magister exam in 2009, he is a research assistant at the Institute for European Ethnology/Cultural Studies at Philipps-University Marburg. His Ph.D. thesis on the German anthropologist Rudolf Virchow is a contribution to the history of anthropological sciences. His current research interests are digitisation in everyday life, the making of originality and discourses on value, and the history of science. Publications include *Rudolf Virchow und die Wissenschaften vom Menschen. Wissensgenerierung und Anthropologie im 19. Jahrhundert* (2013).

Robert Willim is an Ethnologist, Artist, and Associate Professor of European Ethnology, Lund University. His research deals with themes such as digital culture, imaginaries, and materiality; his artworks are positioned close to his practices as an ethnologist. More information at www.robertwillim.com.

Andreas Wittel has worked at University of Bochum (Institut für Arbeitswissenschaft), University of London (Goldsmiths College, Centre for Cultural Studies), and Nottingham Trent University (School of Arts and Humanities). He has published two monographs, *Belegschaftskultur im Schatten der Firmenideologie: Eine ethnographische Fallstudie* (1997) and *Digital Transitions* (2016). His research interests lie at the intersection of political economy, social interaction, and digital media and technologies.

Introduction
Digitisation as challenge for empirical cultural research

Gertraud Koch

Digital culture – analogue culture

Independently from the relevance with respect to its contribution to the culturality of computerisation, digitisation is a technical term that signifies the transformation from analogue into discrete data, i.e., values in a stepped value system or value stock clearly to be distinguished from each other. The principle of digitisation is thus not necessarily tied to the computer but has been implemented in other alphanumeric formats such as the Morse alphabet, the teletypewriter, or further forms of sign systems. Reciprocally, the computer is not necessarily a digital device, as it has been created with analogue technologies, operating with radio valves, before it was then imagined and designed as a digital machine.

There is no consensus on the relevance of digitisation for culture and society. Not everyone would agree that digitisation is of epochal relevance, demarcating a new era in cultural history and demanding new approaches for theorising and studying culture, which is here understood in a broad sense as the practices, meanings, and materiality constitutive for everyday life. Often, digitisation is seen as a part of computerisation and thus as just another step in the advancement of media development, which then does not demand new epistemological instruments but rather can be studied with the instruments cultural analysis already has – albeit, advanced and adapted to the digital (Horst & Miller 2012). Others challenge the relevance of digitisation with relation to further technological features. They claim that computation as we know it today is the sum of many elements and that the algorithms and computability are the most significant features of the computer with respect to its effects on cultural and social change. However, both digitisation and the algorithm are the results of particular ideational traditions that developed over long periods of time and go back far into the history of mankind (Heintz 1993).

Even from the perspective of computer engineers, the potentials and relevance of digitisation are far from being uncontroversial. During the foundation phase of computer science in the 1940s and 1950s, cybernetics, the basic discipline for computer development, debated on the range and scope of digitisation as a technical principle for computer systems, and whether digitisation was the principle on which the human brain (as a biological system) ran. How else – as per reasoning of John von Neumann, the inventor of today's still relevant digital computer architecture – should the brain be able to gain all its capability for coping with complexity (Pias 2004)? The idea of digital in the cybernetics community is generally conceptualised as the opposite to analogue, and both are understood as transferable into each other. However, more detailed observations and the debates of computer scientists in the foundation phase in the United States show that the analogue does not work seamlessly as an acronym. The dichotomy of analogue/digital shows moments of a continuum and a relational connection, rather than a strict opposition. This is seen, for example, when the digital Morse signals are transmitted via analogue communication channels or when the analogue is entirely unable to undergo a transformation to the digital, as the philosopher John Haugeland emphasises with respect to death as something that cannot be transferred into a digital state (Haugeland 1981, Schröter 2004).

The discussion about the distinguishability of analogue and digital is a debate that comes to its end not through scientifically gained knowledge but through a pragmatic argument, which is the triumph of the computer architecture invented by John von Neumann, working digitally, and disseminating the digital computer technology most efficiently. Home and personal computers became standardised products that contributed to the circulation of millionfold digital texts, images, and sounds. This raises other theoretical questions about the digital than those debated for the large-scale computers in the cybernetic foundation phase, as the German media studies scholar Claus Pias reflects these debates (Pias 2004: 309). Thus, von Neumann's computer architecture becomes a standard and is decisive for the dissemination and multiplication of the digital as a basic element of computer technology. The IT discipline, involved in these developments in the 1960s and beyond, thus intensely work on the development of techniques for transforming analogue to digital and digital to analogue data (Brennan & Linebarger 1964, Forgacs & Warnick 1967, Hoeschele 1968, Nguyen et al. 1996).[1] Without digitisation, the computer in its current form would not be conceivable and vice versa; only the computer and its binary codes enabled the rapid dissemination and application of the digital as a means of computability. The recording and processing of digital data is the basis upon which the computer works today. Computer development and the digital have thus become synonymous in the everyday perception (Schröter & Böhnke 2004). Moreover, digitisation has become the allegory for the bandwidth of the diverse developments related to information technology since the image of the computer, as a device and a machine, does not have enough symbolic power to comprise the variety of the intruding information technologies, devices, and paradigms, as well as the velocity of the intrusion into people's lives by becoming increasingly invisible.[2]

It becomes evident how prominent these cybernetic ideas have become regarding common understanding in the context of computation when having a look at the reflections about analogue and digital data that relate directly to the research on steering principles in technological, biological, and social (human) systems and the ways in which they are controlled through information and communication.[3] This early focus of scientific research on information indicates that cybernetics somehow acts as a stoke for the self-understanding of current modern societies, who understand themselves principally as information societies and – by the information paradigm perpetuating – as knowledge societies. Working with the information paradigm, these disciplines have outlined those concepts and coined those terms that later are introduced as guiding ideas for societal and cultural development. Thus, cybernetic knowledge production has been obviously inspirational and powerful enough to play a guiding role also in the social and cultural fields.

It is thus partially a technological argument and partially a cultural-historical argument why digitisation has emerged as a concept relevant in social and cultural theory. It is possible that in particular the (often criticised) fuzziness of the concept *digital* facilitated its migration across disciplines and inspired cultural research, which is either way permanently concerned with multifaceted, multidimensional, paradoxical, and contradictory phenomena. Similar to the debates of engineers about the role of the digital within computer technology, as outlined previously, the ontological dimensions and the meaning of the digital are negotiated in cultural research. Media researcher Wolfgang Ernst, for example, defines replication as a crucial element of the digital; the option to create identical copies emerges from a given, fixed set of discrete values, allowing an exact transformation into other digital formats, or to put it differently, it leads to the loss of the original (Ernst 2004). Such attempts to define the nature of the digital must remain eclectic at the current state of cultural research, when the exploration of digital formats as cultural articulations and objects has just begun. Empirical cultural research has just scratched on the surface and the dynamics and variety of the development are so multifaceted that more empirical research is necessary to gain a comprehensive understanding and to come to a sound theorising of the cultural activity of the digital. It is the goal of this volume to inspire empirical cultural research and thus create a stock of knowledge allowing advanced theoretical thinking about the nature of the digital in cultural theory. To this theoretical endeavour, most chapters in this book contribute conceptual approaches for empirical cultural research from a disciplinary background of European ethnology and cultural anthropology, inspired and enriched with international perspectives of science and technology studies, media studies, sociology, cultural studies, and cultural philosophy.

The digital and digitisation – the contributions

The digital is used as a conceptual approach to empirical cultural analysis that is borrowed from cybernetics and computer science and enhanced towards a relational concept entangling the cultural, social, biological, and technological dimensions of the digital as an intellectual idea. Although the intentions of cybernetics and empirical cultural analysis widely differ, this relational idea is in some respect already included in the cybernetic approach to information and data. However, cybernetics is interested in mechanics of control, while cultural analysis is in search of the various facets, connections, relations, and hierarchies where the cultural and the digital merge and constitute each other in this process. Conversely, these points of interaction and merging constitute the point of departure for empirical cultural research and the search for understanding how the digital has become a fundamental cultural technology of today's lifeworlds and in which way it contributes to doing culture in everyday life.

The contributions in this volume are selected and arranged according to this idea of the digital as entangled with social practices and as a multifaceted phenomenon that demands approaches from more than one perspective. The analytical perspectives suggested here are: (a) the 'coding' of culture through digital, informational objects such as hard- and software, (b) the social and cultural practices of doing culture that emerge with the digital and are specific for it, but may also affect and inspire analogue practices, (c) the approaches and perceptions of the world enabled by technology and only possible when (media) technologies mediate between humans and the world and bring original perspectives into being which would not exist otherwise, and (d) the revisiting of well-established, relevant theories and concepts of cultural research, such as historical materialism with the question of how the digital affects them and which revisions might become apparent.

The conceptual approaches in this volume are not assembled with the intention of comprehensiveness but rather guided by the aim to give examples for how each of the four analytical foci can be translated into empirical approaches to the culturality of the digital in its relations. All contributions inquire digitisation from a European knowledge background.

The digital is, to a certain extent, materialised culture or, more precisely, coded culture, as suggested earlier. During the design process, the cultural is inscribed in the devices, programmes, and infrastructures, which – while they do not determine the use – set up a particular framework, although it is not immediately visible but works unnoticed and insistently. These cultural inscriptions in digital technologies can be made visible through analyses of the materiality of hard- and software. Moreover, the imaginations and world views associated discursively to digital technologies work as second order codings. These images guide the uses of, the attitudes towards, and the meanings given to the digital and thus inform people's handling of digital technologies.

In the first contribution of this part, Carsten Ochs suggests to understand digitisation not as a development with epochal character but rather to think about it in the plural, because of its particular character in diverse contexts, and to study it related to social, biological, and technological dimensions. Further, he reflects digitisation methodologically and theoretically as a cultural technique and develops a vocabulary for empirical cultural research. In view of this and in reference to an 'empirical naturalism' or 'naturalist empirism' of John Dewey and the reflections of the anthropologist of technology André Leroi-Gourhan, a cultural analysis of the digital needs to differentiate between diverse forms of inscriptions. The terms cultural technique, practice, and programme entangle cultural, technological, and biological knowledge to a conceptual repertoire for cultural analysis.

Robert Willim's contribution puts a spotlight on another form of how culture is coded, when he develops his approach of the incomplete imaginary of the digital. These are the 'programmes in the mind' that are outlined as crucial for the perception and the use of the digital technologies. The evocative imaginaries stand beside the more mundane everyday interaction with computers. How they come together and merge is informed by the 'grand stories' of the digital, which remain necessarily incomplete and thus are a source of constantly ongoing negotiations about the digital.

A further mode of coding culture is outlined by Isto Huvila in his elaborations on archives. Initially, he outlines diverse concepts of the archive and moves on to the archival as a more metaphorical and abstract concept. The archival can be used analytically for studying social practices and experiences, as Huvila outlines, and thus inspires cultural research in a particular way, most relevant for digital practices.

The fourth and last contribution in this part is written by Gertraud Koch on the ethnography of digital infrastructures, which is a conceptual approach on the entanglement of technology, infrastructure, and social practice and the options to disentangle these relations on an analytical level. The chapter elaborates on the grounds of the Ethnography of Infrastructure – a concept developed from the 1990s on by Susan Leigh Star and Geoffrey Bowker – as a methodological point of departure for studying digital and retrospectively digitised infrastructures, with an emphasis on overlapping of and simultaneous digital infrastructures.

The digital offers a particular repertoire for cultural production or – in practice theoretical terms (Beck 2015) – for doing digital culture, which is the approach to digitisation suggested in Part II of this book. Hacking, copy-paste, mash-up, Big Data – not occasionally these terms have migrated as anglicisms into other languages for naming rather common social practices of the digital world. Computer science operates internationally and this is consequentially mirrored in the language of digital culture. Mostly, the 'new' digital practices are not at all restricted to the digital, but refer to practices in the 'old' analogue world at the same time. Obviously, these social practices benefit particularly from the discrete values of the digital, their reproducibility, convertibility, and 'link-ability' across networks.

Over the years, hackers have moved up the social ladder from marginal, criminal outsiders to heroes of computerisation, impersonating positive values such as freedom, autonomy, and privacy, as Luis Felipe R. Murillo and Christopher Kelty point out. The two authors, working in California – the epicentre of computer science – develop in their contribution a basic framework for empirical cultural studies on hacker communities and the social practice of hacking.

In Christian Schönholz's reflections on the cultural remapping of originality and, similarly, the formerly remote and problematic – the copy – has moved into the centre and has become part of a more or less common standard repertoire of social practice. Although still contested and partially restricted, practices of copy and paste gain acceptance and the idea of originality is reconfigured digitally, which raises questions on exploitation and exploitation right in a new and interesting way, on democratisation of cultural objects and forms, on resistance, and on irony or power relations.

The mash-up is an example of a new-old form of cultural production, long known but newly blossoming due to the convergence of media, with its options to combine and integrate content relatively easily from one medium to another. Particularly inspiring seem to be the rather uncomplicated options for making new, creative arrangements similar to the bricolage. The classification, Joan Kristin Bleicher suggests, provides rich insights and, moreover, a source of inspiration for empirical cultural analysis. The contribution demonstrates the gain of interdisciplinary approaches between media studies and empirical cultural research.

Last but not least, the forth contribution in this part, from Katharina E. Kinder-Kurlanda, deals with Big Data, which has gained evocative, imaginative charisma in the current debate on the digital. The supposed effects for social and cultural analytics are enormous and demonstrate the expectations and the enthusiasm about the new options of computability, which are expected to intrude into all areas of private and social life. The calculating analysis of data seems to lay bare daily practices on a large scale. However, the contribution raises questions with respect to the qualities relevant for understanding what data means.

The digital offers new modes for experiencing the world, by creating information technologies that set up new relations between people, things, and information and thus reconfigure approaches digitally, as the contributions outline in Part III. All chapters take up specific applications of information technology as a point of departure for their analysis of how digital technologies mediate between humans and the world, how they affect perception, and also how this differs from the ways mass media – such as TV, radio, or newspaper – work in this respect. When cultural techniques like cartography are brought to the digital realm, new representations of the world emerge that are technology-driven and create new links between representation and social practices. Moreover, entirely new constructions of reality emerge by fading and aligning physical and virtual objects. Often, the technological quality of spaces, informationally enhanced in such a way, is not readily visible. The range of perspectives introduced here shows how complex and diverging the initiated social and cultural changes are.

Nishant Shah outlines in his contribution the development of the user interface towards an interface without a user, which is currently emerging due to a paradigm shift towards the Internet of Things and will lead to the computer gradually disappearing out of the field of view while in fact being more present in social spaces than ever. Shah emphasises the necessity to analyse these developments with respect to power and gender relations as they are hidden in particular ways in these new interfaces.

The Internet of Things is the subject matter of Katharina E. Kinder-Kurlanda and Daniel Boos, but approached from an entirely different analytical perspective, which is the ubiquity of computing and of computer networks in work environments. The various computer networks operating today and the Internet of Things, in combination with mobile computing, literally build a second, third, or fourth skin over social and cultural spaces, permeate them, and have the potential to deeply affect the lifeworlds in these spaces. Thus, the authors provide a vocabulary for research and heavily emphasise the need of further inquiry.

The third contribution in this part, written by Ina Dietzsch and Daniel Kunzelmann, problematises plotting as a cultural technique. In its digital version, 'calculating spaces' emerge on a massive scale, in which maps are interlinked and enhanced with individually created information brought in by users from their mobile devices. Maps as an academic medium are thus entirely newly interpreted, and the rather metaphysical idea of 'calculating spaces' – brought up by computer pioneer Konrad Zuse – is furnished with a new meaning. For empirical cultural analysis recombination, new social orderings of visibility and invisibility, as well as sociotechnical arrangements of plotting, become relevant points of departure.

The part ends with a conceptual approach, written by Gertraud Koch, on augmented realities (ARs), which shows that information technology takes up cultural ideas and social practices of enriching spaces with information when developing new technological solutions for augmenting spaces. Taking the technical as a starting point, the alignment of physical and virtual objects is put into a broader context of social and cultural practices of enriching physical spaces. ARs are explored in realms such as museums, education, entertainment, medical applications, and the economy, which means that there are already interesting fields for empirical cultural research. This is relevant for an empirically informed theory building on how ARs are inspiring changes and reconfigurations of how the world is perceived.

The digital and corresponding changes in both the social and cultural realm do not mean inevitably the obsolescence or far-reaching revision of approaches in cultural theory, which refer to earlier periods. On the contrary, a more detailed look on the changes and effects of digitisation and an asking ductus are helpful to gain an understanding of if and how a comprehensive review is necessary and in which way they then shall be organised. In the fourth and last part of this edited volume, distinguished theoretical approaches on culture are discussed with respect to how they contribute to the current developments of digitisation and in which way they are relevant for empirical cultural research.

In his chapter on the political economy of digital technologies, Andreas Wittel outlines the contours of this young and fast-growing field of research, which is related to the political economy of media and communication, in need of change and adaptation because of the technological character of the digital. Moreover, in this context, political activism has become another issue. Wittel emphasises the need of further empirical studies that carve out, in ethnographic or auto-ethnographic approaches, the principles of the digital economy in its interrelation with the material and immaterial aspects.

Anne Dippel and Sonia Fizek turn towards play as a social practice closely tied to the history of humankind. They follow the current assessments of an increased relevance of games in the digital and speak about a ludification of culture. They thus put another emphasis on their approach and distinguish it from the intensely debated 'gamification', which they locate on the level of a particular design approach in information technology. The concept of ludification reversely takes the manifestations of the game as the starting point for the analysis and turns from there to the digital and the ways in which the game is revived, modified, integrated, and used there, including the effects that emerge when these elements of gamification meet people, everyday contexts, and cultures.

The volume closes with a look towards the future, as the authors mention in their chapter. This puts an emphasis on the necessity to think about possible futures in context and with respect to their references in the past. The study of media genealogies provides a range of methodological procedures – as Clemens Apprich and Götz Bachmann outline – and is thus an important approach for cultural analysis, which also differs from the media archaeology in some respects. The handling of the genealogical multilinearity is a major challenge and a significant potential at the same time.

Acknowledgements

I am very grateful for being able to edit this book, as many others contributed to this project with their expertise, their time, and their dedication. First to be mentioned is Alejandra Tijerina García. She has done a tremendous job in copy-editing this edition. Stefanie Everke Buchanan has been of major help, always readily available to do translations. Stella Bandemer and Bianka Schaffus have given great support with proofreading and checking citations. Gratitude is owed to Gisela Welz, who has commented on the introduction, and the three peer-reviewers for their helpful comments. Last but not least, I want to give thanks to the authors for their willingness to support the conceptual approach of this book with their expertise, their creativity, and their patience in discussing changes and revisions.

Notes

1 Before computer science was founded as an academic discipline, the scholars developing information processing technologies were educated and working in physics, mathematics, electronic engineering, and other disciplines.
2 Computer science itself seems to have a rather skeptical estimation of the role of digitisation for the progress of information technologies. At least, digitisation is not mentioned in the self-reflection of the scientific community in Germany on the most important paradigms in computer science, but is only mentioned indirectly in the context of the von Neumann computer architecture (Hellige 1994).
3 Cybernetics refers to information as one of the main physical variables beside matter and energy.

References

Beck, S., 2015. *Von Praxistheorie 1.0 zu 3.0 - oder: wie analoge und digitale Praxen relationiert werden sollten*. Available at: www.academia.edu/10952692/Von_ Praxistheorie_1.0_zu_3.0_oder_wie_analoge_und_digitale_Praxen_elationiert_ werden_sollten (accessed 19 July 2016).
Brennan, R.D. and Linebarger, R.N., 1964. A survey of digital simulation: digital analog simulator programs. *Simulation*, 3 (6), pp. 22–36.
Ernst, W., 2004. Den A/D Umbruch aktiv denken -medienarchäologisch, kulturtechnisch. In J. Schröter and A. Böhnke (Eds.), *Analog/digital-opposition oder Kontinuum. Zur Theorie und Geschichte einer Unterscheidung*. Medienumbrüche, 2. Bielefeld: Transcript, pp. 49–66.
Forgacs, R.L. and Warnick, A., 1967. Digital-analog magnetometer utilizing superconducting sensor. *Review of Scientific Instruments*, 38 (2), pp. 214–220.
Haugeland, J., 1981. Analog and analog. *Philosophical Topics*, 12 (1), pp. 213–225.
Heintz, B., 1993. *Die Herrschaft der Regel: Zur Grundlagengeschichte des Computers*. Frankfurt am Main: Campus.
Hellige, H.-D. (Ed.), 1994. Leitbilder der Informatik und Computer-Entwicklung. *A Paper Presented at the Conference "Historische Aspekte von Informatik und Gesellschaft" of the GI and the Deutsches Museum*. Munich, Germany, 4–6 October 1993. Artec Paper, 33. Link (accessed 12 July 2016).
Hoeschele, D.F., 1968. *Analog-to-digital/digital-to-analog conversion techniques*. New York: Wiley.
Horst, H.A. and Miller, D. (Eds.), 2012. *Digital anthropology*. London: Bloomsbury.
Nguyen, T.M., Rana, D., Ruiz, A., and Willner, B.E., 1996. Hybrid digital/analog multimedia hub with dynamically allocated/released channels for video processing and distribution. Google Patents. US5515511 A.
Pias, C., 2004. Elektronengehirn und verbotene Zone. Zur kybernetischen Ökonomie des Digitalen. In J. Schröter and A. Böhnke (Eds.), *Analog/digital-opposition oder*

Kontinuum. Zur Theorie und Geschichte einer Unterscheidung. Medienumbrüche, 2. Bielefeld: Transcript, pp. 295–309.

Schröter, J., 2004. Analog/digital-opposition oder Kontinuum. In J. Schröter and A. Böhnke (Eds.), *Analog/digital-opposition oder Kontinuum. Zur Theorie und Geschichte einer Unterscheidung.* Medienumbrüche, 2. Bielefeld: Transcript, pp. 7–30.

Schröter, J. and Böhnke, A. (Eds.), 2004. *Analog/digital-opposition oder Kontinuum. Zur Theorie und Geschichte einer Unterscheidung.* Bielefeld: Transcript.

Part I
Coded culture

1 Cultural techniques, practices, programmes
How to study the anthropo-logic of digitisation

Carsten Ochs

Introduction

It has become somewhat of a truism by now that each and everything is transformed by digitisation, be it politics, the economy, or society (at least, if the latter is not anyway conceived as being in the process of dissolution; Faßler 2009).[1] We may presume, then, that culture and cultural analysis will not remain unaffected, too. While the question of how to come to analytical terms with processes of digitisation has been on anthropology's agenda for quite a while now, recent sociotechnical developments have further increased the urgency of finding fruitful answers. What are the premises to set out from? Are we to act on the assumption that digitisation marks an epochal rupture, the repercussions of which are now to be analysed? Or should we rather come from the myriad local micro-practices, e.g., from digitisations in the plural? Or perhaps the distinction between micro- and macro-perspectives implicitly alluded to here is mistaken right from the outset?

Whereas digitisation obviously raises old questions in new terms, the general public's discourse as well as the social sciences' and humanities' diagnoses in recent decades have tended to rather take an 'epochal rupture' view. Numerous otherwise excellent analyses stemming from this era aim at spelling out the actual characteristics of the digital age (see Castells 2001, Lash 2002, Galloway 2004, Baecker 2007, Faßler 2009, 2014). However, it is precisely the 'epochal view' that has raised criticism recently. Ruppert, Savage, and Law, for instance, hold that

> [o]ne influential approach imparts intrinsic properties to the digital, which is imagined to grow and unfold so that its qualities become more widely disseminated. The suggestion that the digital marks a profound, epochal, rupture in social change is familiar. (...) However, a re-reading of many of these seminal texts a decade later suggests that they treat information technologies and the digital in a derivative way. Rather than offering novel arguments about its revolutionary capacities, reflections on the innovatory character of the digital tend to reflect concerns with epochal change originally developed in the context of other kinds of claims.
> (2013: 26)

Just like in other fields of research, also in the one of digitisation, sustained scrutiny on the researchers' side seems to result in a differentiation of aspects relevant to the phenomenon, and, subsequently and consequently, in the pluralisation of the research object.[2] Such differentiation obviously is accompanied by a shift of research perspectives towards empirical microanalyses and diagnoses (Savage et al. 2010, Ochs 2012, 2013, Ruppert et al. 2013).

This chapter aims at contributing to the development of a theoretically sound anthropological research perspective on digitisation processes that can be put into operation empirically. Thus, it pursues methodological intentions by seizing the recent relativisation of epochal diagnoses as a suggestion to provide a conceptual equipment that is able to account for both, the relativised view ('epochal rupture') and the relativisation (digitisations in the plural). My hope is that it eventually becomes clear that the analytical frame to be developed allows for shifting from micro- to macro-perspectives on digitisation, while preventing breakage between these seemingly incommensurable views. By so doing, I attempt to theoretically elaborate 'glocal' anthropo-logic of digitisation that I identified in prior work (Ochs 2013).

Now, what exactly is the term 'anthropo-logic' meant to designate in this methodological context? To begin with, the approach presented here is situated mainly within two research traditions. The first is the one that J. Dewey (1995: 15) has called 'empirical naturalism' or 'natural empiricism', turning explicitly against the separation of humans and their experience of nature. The second tradition is that of André Leroi-Gourhan's Anthropology of Technology, an approach that can be easily tied to empirical naturalism insofar as it strives to develop a *biology of technology*, which conceives of the social as being based upon, but still independent from, the zoological realm (Leroi-Gourhan 1993: 188). The perspective established here thus forms part of a broader movement within the social sciences and humanities, which tends to conceive of culture as something that emerges by interacting with processes traditionally dubbed 'biological' and 'technical' – however, without reducing culture to a mere function of these processes. What is at stake, consequently, is the development of a research pragmatics allowing to fruitfully take up and combine multifaceted knowledge (Beck 2008), such as biological and technical, without necessarily ending up in biological reductionism nor technological determinism. In other words, while the task of cultural analysis is to harness those overlapping areas of expertise that may be shared with other disciplines, it must not misconceive culture as a function of biology or technology, if it is to prevent conceptual inaccuracies.

The general thrust of the argument unfolded below is thus clarified. The methodological equipment for the analysis of the anthropo-logic of digitisation, in the sense just specified, will be proposed and explained in detail. From the discussion so far follow three conditions this equipment has to fulfil so as to count as robust and coherent: first, it shall be able to integrate 'epochal' and 'micro' perspectives; second, it shall allow for fruitfully taking up technological and biological knowledge; and third, to pass as methodology, it must be possible to empirically put the equipment into operation. To achieve these three goals, I will proceed as follows:

In the section *Digitisation as a cultural technique*, digitisation is conceptualised as a cultural technique and thus analysed in an utmost abstract way. Doing so means to theorise the formal, general, and comprehensive characteristics of the *modus operandi*, which can be identified throughout the great variety of all the empirically observable digitisation processes. This can be done, or so I will claim, via conducting a cultural analysis of the techno-mathematical principles of the cultural technique of digitisation.

As was mentioned in passing, the virtual *modus operandi* of the cultural technique of digitisation is empirically only existent insofar as it is actualised as practice. Hence, in the section *Actualising cultural techniques: operational chains/practices*, the perspective is reversed and the lens turned from the abstract towards its concrete practical instantiation. Accordingly, it will be argued that digital practices may fruitfully be analysed as the translation of abstract cultural techniques into concrete operational chains 'here and now'.

Having said this, the question arises, how does such translation exactly occur? *Programming practices* deals with this question and attempts to provide a precise theoretical answer, thus approaching the core of the methodology proposed here. As will be explicated, the translation of the digital *modus operandi* is accomplished by generating and inscribing cultural programme and scripts. It is important to emphasise that cultural analyses of digitisation must be able to empirically and analytically account for the different modes of inscribing scripts. To make this possible, and to thus render the proposed methodology sufficiently differentiated, I will introduce biological knowledge at this point into the theoretical framework.

As stated above, to make the theoretical considerations count as fullblown methodology, it must be possible to put them into empirical operation. Hence, the equipment specified so far will be put to the test in the section *How to make a programme run*.

Digitisation as a cultural technique

When investigating the anthropo-logic of digitisation, it is a prerequisite to specify the object of research: what shall be put to scrutiny under the rubric of 'digitisation'? The term 'digital' usually points to particular *modus operandi*, namely to processes that feature binary-digital logic. Precisely speaking, 'digital' means that a given system only knows discrete and unambiguous states (Schreiner 2009: 209); 'binary', moreover, indicates that the system only knows two such states, conventionally presented as 1/0 (ibid). Expressions, such as 'binary-digital' or 'digital', thus refer to systems of human (mathematical wizards) or non-human origin (computers) that operate on the basis of binary-digital calculations, so as to accomplish a great diversity of tasks. Having said this, 'digitisation' refers to two kinds of processes: *first*, to the integration of elements featuring binary-digital logic into (thus far) analogous processual trajectories. For instance, if someone replaces their mechanical alarm clock by one that operates digitally, this amounts to integrating a material artefact measuring time and raising alarm on the basis of binary-digital calculations into the processual trajectory 'sleeping-waking up-getting up'. In this case, 'digitisation' means replacing an analogous with a digital entity. However, the term also concerns, *secondly*, the invention of binary-digital operations from scratch that have no analogous precursors. Such is the case, for example, when it comes to search engines on the web – it is rather difficult to find an analogous archetype for those operations. Thus, when I use the terms 'digital', 'digitisation', and so forth from now on, I refer to processes featuring binary-digital fractions in the sense just specified: processes that feature, amongst others, binary-digital *modus operandi*.

Where does this *modus operandi* descend from? Generally speaking, digitisation processes take advantage of the binary-digital processing principle of the *Universal Machine* as specified by Alan Turing (1936). The theoretical specification of this machine still serves as the prototype of present-day stored-program computers and is consequently considered the foundation of today's computer technology.[3] Now, examining the *Universal Machine's modus operandi* in terms of cultural analysis, what attracts attention is the *abstract* and *formal* specification of the principle. Turing was apparently concerned with the development of a '*theory*' of computable numbers' that initially was applied to the theoretical *Entscheidungsproblem* in mathematics. The resulting processing principle turns out to be abstract insofar as it is abstracted from any actual situation in space-time. This is precisely what allows for claiming the universal validity of the principle: its abstractness results in its potential for universal re-localisation. Whereas the process of specifying and spelling out the principle in a mathematically sound way was still a culturally specific one, this process was situated within a specific socioculturally coined space-time; the goal of the exercise was to gain universal validity right from the outset. In this sense, as is always the case when it comes to producing scientific knowledge, the generation of the principle concerns 'the ways in which the local and the heterogeneous are combined to create knowledge with status of timeless and universal truth' (Akrich 1992: 205). Abstract specification in this sense may be considered as a precondition to the universal validity of some principle; nevertheless, such validity has to be produced and cannot be presupposed from the outset.[4] A *modus operandi*, such as the one of the *Universal Machine*, is therefore not about discovering universals that are existent *a priori*, but about systematically increasing the scope of local principles. Hence, abstractness is a *sine qua non*, but still not a sufficient condition for universal validity thus understood.[5]

Digitisation, for the time being, designates the formal application of the abstract *modus operandi* of the *Universal Machine*. Still, though, classifying the principle as abstract and formal does not exhaust the listing of its defining traits. S. Krämer and H. Bredekamp argue in a media theoretical account of the *Universal Machine* that the latter's conceptualisation is indeed similar to a range of mathematically equivalent approaches that were known at the time. However, the concept sets itself apart above all by its potential to bridge the gap between the symbolic and the technical, or physical world, and thus, between software and hardware, 'by showing that there are universal Turing machines capable of imitating every special Turing machine because the codes of the latter can be inscribed – that is, programmed – onto the strip of the universal machine' (Krämer & Bredekamp 2013: 23).

Hence, the abstract principle becomes operational, i.e., gains the potential to be put into actual, concrete, material operations that respond to its logic. It may possibly be translated via programming into more specific applications, and thereby inscribed into material apparatuses. In turn, those specific applications feature less abstractness, precisely because of their very specificity. So, while any software programme that controls the operations of some computer follows the same binary-digital logic specified as the *Universal Machine*, actual applications of this *modus operandi* are specifically tailored to particular situations that can be located in space-time.[6] It is therefore the very abstractness, formality, and operational character of the principle that enable actual computer operations to occur in the most diverse forms. In other words, the *Universal Machine* is potentially universal as its universal simulative potential, via programming, can be put to specific uses; whereas the 'universality' of the binary-digital principle is due to its culturally non-specific character.[7]

These remarks lead to a fourth trait of this *modus operandi*. If the processing principle as such proves to be culturally non-specific, while at the same time being empirically existent only in culturally specific ways, it gains a virtual status. As a potentiality, it is not in opposition to the real but to what is empirically actualised, just like scripture: it is downright impossible to come across scripture-as-such, for empirically one may only encounter scripting operations, or utilisations of scripture.[8] However, as it is possible to deduce from those actual operations, the defining traits of the (unobservable) virtual *modus operandi* of scripture, the latter still can be analysed: the ontological status of scripture is revealed in the uses to which scripture is put (Krämer 2005: 52).

Hence, similar to scripture, the binary-digital *modus operandi* of the *Universal Machine* may be conceived of as cultural technique: the cultural technique of digitisation. In this section, four defining traits of this technique were identified:

- *abstract* (abstracts from any concrete situation in space-time),
- *formal* (or more precisely: potentially to be formalised; the explicit formulation in Turing 1936),
- *virtual* (it empirically occurs only in specific, actualised form), and
- *operational* (i.e., potentially to be put in actual operation; thus the principle not only describes some possible *modus operandi*, such as, say, Big Bang theory, but is instead translatable into concrete operations here and now).

According to these considerations, virtual cultural techniques occur empirically only as actualisations of specific processual trajectories. For instance, the cultural technique of digitisation may possibly be performed by humans calculating mentally, on paper, or with the aid of a calculating machine. Of course, digitisation operations can also be delegated to material artefacts to a much more comprehensive extent. In extreme cases, digitally processed algorithms even may perform decision making (Preda 2000).

In this sense, the actualisation of the cultural technique of digitisation is always about concrete operations in specific spatio-temporal situations in the physical-material world (carried out by thinking note-making bodies, or by material apparatuses operating with the presence/absence of particular voltages and so forth). In other words: the actualisation of the cultural technique of digitisation is empirically always about *practices*.[9] This line of reasoning raises two different types of research question, each of which is to be dealt with in a particular methodical way:

- On the one hand, it is the task of analyses focusing on cultural techniques *in abstracto* to analyse the ways in which specific cultural techniques emerge as the condition of possibility for certain practices, what potentialities they create, and so forth. This type of analysis quite obviously tends to deal with historic research questions.
- On the other hand, there are plenty of ethnographic research questions raised by the virtual status of the cultural technique of digitisation; for, if this cultural technique empirically only occurs as practice, we may ask next, how this shift from a 'virtual' to an 'actual' mode of existence exactly takes place.

It is not quite easy to answer this latter question, and, in fact, requires the making of some theoretical arrangements. Below, readers are provided with a framework indicating how to deal with research questions of the second type. Prior to this, however, it is mandatory to conceptualise the notion of 'digital practice' in a coherent fashion. To do so, this perspective must be reversed. Let us next move from the abstract cultural technique to the concrete practices of digitisation.

Actualising cultural techniques: operational chains/practices

What is the result of the actualisation of the cultural technique of digitisation? To answer this question, it is helpful to turn to the work of French palaeontologist A. Leroi-Gourhan. Based on the central concept of the operational chain, Leroi-Gourhan combines biological, symbolic, technical, and social considerations so as to create a historical account of human evolutionary transformation. In so doing, he succeeds in telling a history of cultural evolution up to 'this point that, from being governed by biological rhythms, human cultural development began to be dominated by social phenomena' (Leroi-Gourhan 1993: 141). It is the transdisciplinary precision and humbleness that renders his general perspective and some of his central concepts so promising for a contemporary anthropology of digitisation.[10]

The point of departure for all his reasoning is the heuristic, historic, and practical priority of operational chains. The latter come first, so to speak, as they are analytically privileged in the face of those entities that are involved in their production, such as artefacts, persons, signs, technical objects, practices, or knowledge (Schüttpelz 2006: 91–92). Operational chains are composed of fractional single operations carried out by humans and/or things. In this context, the development of material technologies may be understood as a mode of delegating operations to artefacts, so as to subsequently integrate them into the operational chains of some social formation. In the course of evolution, such transferring of operations to non-human entities is initiated by the 'exteriorisation' into non-human tools of certain activities hitherto carried out by the human hand (ibid: 238–242).

The releasing of the hand began with the development of instruments exerting physical force, such as clubs, knives, hammers, etc. While human gestures in these cases still manually guided the artefacts, a fraction of the operation was exteriorised to the material thing (beating, cutting, sticking, hammering). Subsequently, the gesture itself was delegated. A progressive movement towards the exteriorisation of ever more gestures was initiated, and the result is that eventually 'the hand is used to set off a programmed process in automatic machines that not only exteriorise tools, gestures, and mobility but whose effect also spills over into memory and mechanical behavior' (ibid: 242). Human collectives thus look back on a long history of cultural evolution, in the course of which they generated complex operational chains being composed of human as well as non-human fractional operations.

In the context of Leroi-Gourhan's non-essentialist anthropology, operational chains constitute practices (ibid: 231) that may assume one of the following three forms: first, there is biologically shaped, i.e., instinctual behaviour; second, there is acquired 'mechanical behavior' based on experience and education; and third, there is 'lucid behavior' (ibid: 230).[11] Insofar as calculating artefacts (computers) act as elements of those relatively stable operational chains that we call 'practices', the latter expression does not signify 'completely digitised' operational chains; instead, 'digital practice' refers to those chains that also feature, amongst others, binary-digital single operations. In the course of everyday Internet usage, for instance, binary-digital processing of data by artefacts oftentimes connects, via interfaces, with the operations carried out by humans. This is what enables so-called 'users' to manipulate data processing, to perceive the results of the latter through representations on the screen, and to cognitively process the information thus generated (this is not to say, of course, that binary-digital operations in such a situation are generally at the complete disposal of human actors). In other words, not all the fractional operations of the operational chain follow binary-digital logic. That said, the concept of 'digital practice' opens up a perspective that allows one to analyse the entanglement of binary-digital and analogous operations as overall operational chain. So, even if 'digital practices', strictly speaking, generally must be considered digital-analogous ones indeed, I use the expression for the sake of convenience whenever operational chains feature fractional operations following the *modus operandi* of the cultural technique of digitisation.

Now, according to Leroi-Gourhan, practices of whatever kind are based on socially constituted, collective 'operational memory', which 'serves as the medium for action sequences' (1993: 413). Operational memory occurs in distributed form, including extra-somatic resources 'of an "artificial" memory in its most recent form that – without referring to either instinct or thought – ensures the reproduction of sequences of mechanical actions' (ibid: 413). Individual experience plays an important role that must not be disregarded when it comes to the development of operational chains; however, via education, those chains are collectively made durable and stable (ibid: 230). To recapitulate, operational chains are stabilised in collective operational memory that is distributed among human and non-human substrates, and that affects all the three types of behaviour being mentioned above: biologically 'hardwired' behaviour is getting culturally reshaped (ibid: 227–228);[12] conscious (reflexive) behaviour can be 'repaired' and novel sequences and contents of memory experimentally designed with the aid of language (ibid: 230); finally, 'mechanical operational sequences' significantly depend on operational memory, for those are

> sequences acquired through experience and education, recorded in both gestural behavior and language but taking place in a state of dimmed consciousness which, however, does not amount to automatism because any accidental interruption of the sequence will set off a process of comparison involving language symbols.
>
> (Leroi-Gourhan 1993: 230)

Leroi-Gourhan's ideas at this point are congruent to the greatest possible extent with central tenets of rather influential practice theories, such as the one developed by A. Giddens (1984). For example, the latter likewise emphasises the routine character of executing everyday practices, and he also points to the shared, tacit character of practical knowledge, which enables actors

> to 'go on' within the routines of social life. The line between discursive and practical consciousness is fluctuating and permeable, both in the experience of the individual agent and as regards the comparisons between actors in different contexts of social activity. There is no bar between these, however, as there is between the unconscious and discursive consciousness.
>
> (Giddens 1984: 4)

The parallels between these remarks and Leroi-Gourhan's ideas are quite obvious, yet there is a crucial difference: whereas the latter locates operational memory also in non-human substrates, Giddens focuses exclusively on human 'memory traces' (ibid: 25).

This has profound conceptual implications, of course; for, insofar as binary-digital operations are to be considered fractions of practices, that is, of more comprehensive digital-analogous operational chains, we are bound to presume that those binary-digital operations are generally entangled in, or, in case of invented sequences, encounter diverse other modes of operation. There is a meshwork constituted by routines, habits, customs, (expected) expectations, conventions, assumptions, ways of doing things, shaping any given practice; quite obviously, binary-digital operations are embedded into, and thus part of, this meshwork. It is by no means unusual to call this meshwork 'culture', and it is okay with me to keep to this convention. According to this approach, binary-digital operations have to be coordinated with, connected to, and integrated into other, perhaps analogous, and culturally stabilised operational chains; they have to be culturally 'domesticated' (Silverstone & Haddon 1996). In this sense, actualising the cultural technique of digitisation is tantamount to the production of culture. The distribution of operational memory amongst human as well as non-human substrates results in the fruitfulness of analysing human and computer agency on a level playing field (which is not to say that there was no way to analytically distinguish between different substrates of memory and different contents).

The confrontation of the abstract cultural technique of digitisation with actual digital practices thus allows for identifying a conceptual desideratum; for what the comparison leaves unexplained in terms of theory is the precise way in which the cultural technique at hand is translated into the practices that we are interested in. From the anthropological approach that is put forth here, it follows that an answer to this question generally needs an empirical base, since any theoretical claim, however plausible it may seem, must be tied down again to 'primary experience' (Dewey 1995: 22–25). But then, what is it that shall be focused upon empirically in the first place? The answer that may be given to this question is able to join the abstract dimension of cultural techniques with the actual one of operational sequences or practices, and thus yields the potential to act as a hinge. Cultural techniques are generally performed as operational chains by human and non-human elements; in this sense, they count as empirically observable practices. The latter are shaped by the contents of operational memory, which again are distributed amongst humans and non-humans. This raises the question: how to account for those contents in a theoretically differentiated manner?

Programming practices

How do we conceptualise the translation of cultural techniques into operational chains/practices? Let us approach this question by setting out from a rather simple case: the actualisation of the cultural technique called 'cooking', understood as a practical repertoire to prepare food. In practice, we do not seem to have any problem whatsoever with distinguishing between those practices and meals that belong to, say, Punjabi Cuisine (e.g., *Biryani*) and those belonging to Hessian Cuisine (e.g., *Frankfurt green sauce*). It seems just as easy to identify both practice repertoires as culturally specific instantiations of the same cultural technique: cooking. The latter observation relates to the fact that, to a certain degree, it is possible to apply the same description to both types of practice (dealing with, manipulating, and mixing together substances human bodies are able to digest, thus inducing biochemical reactions with the aid of specific sets of tools, and so forth). At the same time, however, both types of practice feature diverging, culturally specific patterns (*what* substances are manipulated, *in which way* in each case, and *what kinds of instruments* are utilised, and so forth).

According to Leroi-Gourhan (1993: 230–235), it is programme sequences stored in operational memory that shape operational chains, and thus assure the pattern-creating stability of practices: '...in humans, the mobility of tools and language has determined the exteriorisation of operational programs' (ibid: 237). Thus, on the basis of a 'shared space of common psychological ground that enables everything from collaborative activities with shared goals to human-style cooperative communication' (Tomasello & Carpenter 2007: 121–122), operational programmes are transmitted, i.e., exteriorised into other human fellows via language and body movements (gestures): chefs passing down recipes to their apprentices, parents who teach their children how to cook, and so forth. As is obvious, however, some programme sequences are also inscribed into and performed by material entities. Fitted kitchens, for instance, may be considered as material environment featuring the inscription of programmes of action that can be set into operation with help of a human. Also, at this point, we may encounter the cultural specificity characterising the translation of cultural techniques into practices by way of exteriorising programmes: to translate the cultural technique of cooking into the practices of the Punjabi Cuisine, it is indispensable to add a *Tandur* (a wood stove used for bread baking) to the material arrangement of the kitchen, otherwise certain parts of the programme cannot set into operation (no *chapati* can be produced).

The notion of programme was initially developed by Leroi-Gourhan, and, as is well known, adopted by Actor-Network-Theory (ANT) later on (see Akrich 1992, Latour 1992). Due to the widespread attention that ANT attracted the general approach may be familiar. Yet, ANT's embracing of the concept was mainly driven by questions arising in the *sociology* of technology; *anthropological* investigation into practices, in contrast, has a much stronger interest in analytically distinguishing different types of programme elements and transmission modes.[13] For this reason, I will next attempt to theorise the notion of programme for the purpose of anthropological practice research, before demonstrating its empirical usefulness in the succeeding section.

So what are the constituent parts of programmes? How do they emerge and how are they maintained and transmitted? When dealing with these questions, Leroi-Gourhan lays particular stress on motor-cognitive programme dimensions. By asserting that individuals' programmes are 'recorded in their motor memory' (Leroi-Gourhan 1993: 233), he locates them in the somatic memory stored in the neural networks of the vegetative nervous system and of the cerebellum. Such a focus on behaviour, understood as motion sequences, is not wrong, but it is certainly too narrow. To begin with, we may draw on arguments from practice theory, as well as from neurobiology to support the claim that programmes allow for purposeful bodily behaviour. A. Reckwitz (2003: 289), for instance, explains that the knowledge underlying those forms of routine behaviour practice theory is concerned with are incorporated into the bodies of acting subjects. Neurobiologist and psychiatrist J. Bauer (2006: 18) in turn holds that, biologically speaking, such incorporation happens on the basis of neurons storing programmes for purposeful actions and thus allowing for steering behaviour. Programmes, that is, are not only located in motor memory, but also in the premotor cortex, i.e., in those somatic areas that may be discretionarily accessible. Still, formulations such as these evoke 'algorithmic' images of programmes, as though the latter were tantamount in practice to explicit propositional instructions or the like, thus steering behaviour in an unambiguous, intentional way, as some theories of action might have it. Bauer is well aware that such a position is untenable, when he states that programmes, while indeed steering motor behaviour, also extend to affective and emotional processes (ibid: 31); the long-standing and ramified debate concerning the nature and status of social rules within practice theory likewise supports the more widespread applicability of the programme notion. Preda, for instance, subscribes to the view that theorising behavioural rules plays a crucial role for any theoretically sound conception of human activity, which

> means to account, in a conceptually consistent fashion, for how actors follow and reproduce rules over various contexts. Of key importance here are rules belonging to 'background understanding' (...): that is, rules which are shared by actors, orienting their actions, but not entirely reducible to explicit formulations.
>
> (Preda 2000: 271)

In line with this, Wittgenstein (1958: §23), who penned a central reference of practice theory, holds 'that the *speaking* of language is part of an activity, or of a form of life'. That is to say, the speaking of a language, which here is representative for any type of practice, is only analysable when taking into account its embedding in a form or way of life, i.e., in its cultural embedding.

The discussion so far paves the way to the reformulation of the programme notion, beyond Leroi-Gourhan's conception. Equating operational chains with practice, we may presume for the reasons just specified that these chains are stabilised by programmes, understood as knowledge incorporated in the body; such knowledge exceeds the shaping of motion sequences and does not occur necessarily in the form of propositions – but then, in what form does it?[14] A superficial glance on the discussion of rules within social theory is already sufficient to realise the danger of getting lost in endless terminological quibble and sophistry when attempting to deliver an unambiguous, exact definition of rule-following *in abstracto*. The best way to approach the issue, or so I claim, is to analytically reconstruct the *genesis* of programmes. To engage in this undertaking, it proves fruitful to take up biological knowledge. For the time being, we may focus on the infamous class of neurons that neurobiologists have christened 'mirror neurons' (Bauer 2006). These are cell structures that exhibit the same activity pattern regardless of whether the cell owner actively performs or passively observes an action; in each case, the cells are firing. Just like any other scientific theory, the theory of mirror neurons is – albeit fairly well supported by empirical data – still controversial. Nevertheless, the idea is prevalent that it is this class of neurons that biologically hard-wires in humans (and other primates) a drive below the threshold of perception to imitate behaviour. This tendency to imitate others, scholars hold, starts with the very beginning of ontogeny; moreover, according to Bauer, observation and imitation procedures carried out by children result in the creation of scripts that are stored within the brain's neural networks (ibid: 69). For cultural analysis just as for social theory, such an idea sounds familiar and rather plausible, insofar as whole sociologies haven been based upon *The Laws of Imitation* (Tarde 1903).

Thus, if we accept the premise of a general imitation drive, we get hold of an *explanans* for the emergence of relatively stable operational patterns, without having to resort to explicit or propositional social rules or the like: *programmes* may be generated and transmitted in the course of constantly ongoing imitations. 'Transmission', in this context, does not refer to the identical replication of some gene-like mental representational unit or pattern, but to variegating, fuzzy reproduction of behaviour, which is quite sufficient to provide for the relative similarity and stability that becomes empirically observable as behavioural pattern.[15]

These ideas are readily connectable to social and cultural theory – for, if our imitations may sediment to relatively stable arrangements even in the absence of explicitly formulated behavioural rules, we may assume that the resulting social order has a tacit character, at least on this quasi-foundational level. It is precisely this tacit nature of the social that Garfinkel (1984) so eloquently and virtuously has documented. When doing so, he especially points to the 'normality expectation' that social actors tend to hold in everyday practice, as they expect 'normal courses of action – familiar scenes of everyday life affairs, the world of daily life (...) taken for granted' (ibid: 35). The normality of social order is thus generally of normative quality, since actors 'refer to this world as the "natural facts of life" which, for members, are through and through moral facts of life' (ibid: 35). Garfinkel calls such tacit expectations at times 'the socially standardized and standardizing "seen but unnoticed", expected background features of everyday scenes' (ibid: 36), and we may assume that they are established with a view to the regularities of the past, thereby congealing as normative regularities. It is in this sense that they may count as tacit rules: familiar operational chains ('normal course of action') are perceived as part of the normal order of things, and violations of such normality are, at least in principle, threatened by sanctions.[16] Still, the rule-like character of these normative expectations becomes visible only in case of transgression and, in addition to their tacit nature, their negativity is also a sign of the fuzziness of such rules. It is at this point completely irrelevant, whether or not, and if so, to what extent such rules are capable of being rendered explicit; either way, they are effective. Thus, when I speak in what follows of tacit rules, I refer in the sense just specified to implicit, normatively flavoured, rule-like expectations. Programmes consequently consist, amongst others, of tacit rules that are built and reproduced via imitation. This programme dimension is the anthropological foundation in respect to all other types of programmes.

However, rules usually tend to occur also in explicit, discursive, and propositional form. A sign carrying the inscription 'Keep off the grass!', for example, utilises discursive means, so as to render a behavioural rule-piece explicit, which can be formulated also as the proposition: 'The grass must not be trespassed'. But it is precisely at this point where the highly branched discussion on the notion of rule sets in. We may approach the problem by turning to Giddens (1984: 21), who advises us to 'regard the rules of social life, then, as techniques or generalisable procedures applied in the enactment/ reproduction of social practices'. To illustrate the character of such rules, though, Giddens refers to mathematical formulas (ibid: 20). Human actors, he claims, ceaselessly apply generalisable procedures such as this, 'typified schemes (formulae)' (ibid: 2), the majority of which is tacit in nature and operates via practical (in contradistinction to discursive) consciousness (ibid: 7).

What is problematic about these ideas is Giddens's reference to mathematical formulae; for, if social behavioural rules typically operate tacitly, it is precisely the explicitness, formality, and propositional character of the mathematical algorithm that they are lacking (see Schatzki 1997: 293–300). Accordingly, Wittgenstein goes to great lengths in the *Philosophical Investigation* so as to demonstrate the impossibility of rendering explicit, or formalise as propositions, linguistic rules (which here represent all kinds of social conventions; see Wittgenstein 1958: §355). As he famously stated, 'the meaning of a word is its use in the language' (ibid: §43). In the same sense, rule-following cannot be formalised in abstraction of the context, since to 'obey a rule, to make a report, to give an order, to play a game of chess, are *customs* (uses, institutions)' (ibid: §199). The same goes for the sign with the inscription 'Keep off the grass!', because, from a practice point of view, we can see that my rule-like formalisation of the inscription as 'The grass must not be trespassed' is in fact insufficient. Strictly speaking, the sign says: 'For a certain class of actors applies the following: the grass must not be trespassed'. We could next go on and determine this class of actors as all those who do not own this property, however, this only would necessitate further specifications in turn, and so on *ad infinitum*. An exhaustive rendering explicit of the rule is neither possible nor is it required in practice, since explicit rules are underpinned with tacit expectations, and thus are perceived as sufficiently unambiguous: in most cases, the sign works in a way that does not violate most people's expectations. In this sense, operational chains and practices of all kinds always exhibit a kind of 'empirical excess': 'And hence also "obeying a rule" is a practice' (ibid: §202).

Against the background of this context dependency, any attempt to provide a trans-situational, context-free, and general definition of some semantic unit or social rule, is doomed to create an infinite regress (ibid: §29). Garfinkel, in his breaching experiments, has impressively shown where the ongoing explication of statements leads in practice: to the breakdown of practice (e.g., Garfinkel 1984: 42). As a matter of principle, exhaustive explication is impossible, which is why we tend to abstain from it in practice; instead, our language games, i.e., practices, always rest on tacit presuppositions (ibid: 179), on implicit rules – Garfinkel's 'background expectancies'.

Consequently, we should be aware of not mistaking the sociocultural rules that constitute programmes for unambiguously determinable, rule-like mental representations of some discursive consciousness that can be rendered explicit without difficulties. Explicit rules ('Keep off the grass!') should not be misconceived as implicit rules rendered explicit; rather, different types of rules refer to each other (in the Keep-off-the-Grass case rules concerning property regimes, legal rules, linguistic rules, and so forth), which is why they cannot be analysed in isolation. Any implicit or explicit rule can only fruitfully be considered as forming part of a more comprehensive cultural meshwork, within which it mutually interacts with other implicit or explicit expectations, regularities, routines, and rules. I call this meshwork a cultural programme, so as to denote its diffuse, fuzzy, culturally determined, and multiple character, which shapes operational chains and sequences. Thus, as far as the programme notion is concerned, we are to conclude that the latter may be composed of tacit expectations, regularities, routines, and rules as well as of explicit discursive rules.

However, our account of the programme notion is still not exhaustive. For, insofar as these programmes stabilise operational chains that are constituted by human and non-human contributions, we are bound to also include those programme fractions that are inscribed into material substrates (Preda 2000: 271). Without a doubt, social formations do not rely exclusively on the 'memory traces' (Giddens 1984: 25) of fallible human beings when providing their operational chains with stability. In fact, they harness a manifold of extra-somatic resources, so as to make practical order durable (Strum & Latour 1987).[17] In so doing, programme sequences are not only transmitted verbally, as is the case in instruction, training and demonstration, but those sequences are also inscribed into documents, for instance, in operating manuals, technical drawings, etc. (Latour 1992: 255). *Programs of action*, that is, are inscribed into humans and documents, yet there is another material substrate featuring programme inscription, namely technical objects (Callon 1991: 135). Again, Leroi-Gourhan (1993: 245–251) pioneered this idea by locating the operations of cultural programmes in mills, clocks, steam engines, and looms. These artefacts' operations are integrated into the operational chains of social formations in a systematic way. To achieve this, human actors exteriorise and inscribe programme sequences into non-human things; as a result, those things feature (more or less) stable operational patterns. So, inscription by way of exteriorisation takes place by manipulating and stabilising artificial modes of operation. As is well known, ANT's concept to account for stability-providing programme sequences that allow for coordinating operational chains are called 'scripts' (Akrich 1992, Latour 1992).[18] Thus, inscribing scripts into humans, documents, and material artefacts makes coordination possible, as it allows for stabilising artefacts' operations and connects these with programme sequences steering the activities of the involved human actors. Even if it is possible to distinguish human from non-human operations in analytical terms, it may be rather difficult to do so empirically – and it may be of only secondary relevance, anyway (Latour 1992: 243).

Connecting these considerations back to those types of operational chains that interest us here in the first place, namely digital operational chains, the fruitfulness of the established perspective quickly becomes clear. Not only does the notion of script seem to be heavily influenced by the widespread emergence of software, but operations shaped via software code may, in a sense, even be considered the prototype of inscription:

> The program of action is the set of written instructions that can be substituted by the analyst to any artifact. Now that computers exist, we are able to conceive of a text (a programming language) that is at once words and actions. How to do things with words and then turn words into things is now clear to any programmer.
>
> (Latour 1992: 255)

It follows that we may add inscriptions in documents and material apparatuses, i.e., scripts, to the catalogue that specifies the elements composing cultural programmes.

At this point, our catalogue is complete. To sum up, we shall determine the elements composing cultural programmes as follows: there are tacit expectations and rules, as well as explicit social rules, and furthermore, there are scripts being inscribed into documents as well as material things (technical artefacts, apparatuses, hardware, and so forth). Accordingly, transmission and maintenance of these components within operational memory may occur via implicit imitation, explicit teaching, and material inscription. All the implicit and explicit programme components that were mentioned, whether executed by human or non-human entities, interact in a complex way. For this reason, they all play a role in the composition of digital operational chains, which is why they all have to be taken into consideration analytically, even if there cannot be an exhaustive analysis of all these elements due to limitations of research capacities.

The theoretical answer to the question of how cultural techniques are translated into operational chains consequently goes as follows: by generating programmes that are composed of the elements that were specified in this section. As a result, we gain a robust hinge, which allows for taking up a research perspective from which the more abstract dimension of cultural techniques is joined with the one of actualised practices. As was argued previously, investigations into either cultural techniques or practices potentially also yield crucial findings; still, I would like to highlight here the capacity of analyses centring on the programming of practices. As it should have become at least theoretically clear, the strength of such analyses consists in their ability to thoroughly grasp the translation of one into the other. It is precisely in this way that more global or encompassing, e.g., historical analyses of cultural techniques, can be related to micro-ethnographies of digital practices. By so doing, it becomes feasible to determine, on the one hand, if digital practices feature somehow stable spatio-temporal patterns, while it still remains possible, on the other hand, to identify local specificities of those practices (see Ochs 2013).

In the introduction, I listed three quality criteria the proposed methodology is bound to satisfy: first, it shall be able to reconcile 'epochal' with 'micrological' views. As was just explained, the suggested framework is able to do this. Secondly, the methodology shall be capable of picking up and fruitfully integrating technological and biological knowledge. I attempted to demonstrate that the proposed methodology also fulfils this criterion (see the analytical inclusion of the *Universal Machine* concept and the theory of mirror neurons). However, the framework still has not submitted evidence for its capacity to be set into empirical operation. The next section aims to close this gap.

How to make a programme run

To demonstrate the fruitfulness of the methodology developed in the previous section, I refer to a case study that was researched extensively by ethnographic means (see Ochs 2012, 2013). In the course of doing ethnographic research, the terminological equipment here presented was utilised, at least, in principle. While the original case study pursued empirical interests, this section deals with methodological questions instead focusing on the development of a theoretical framework. In turn, the empirical account presented below only serves demonstration purposes and is thus less in-depth, as compared to the original case study.

The object of research was a 'technical development project' that ran for six years (with the main project phase being completed in 2010). The project's goal was to render digital Information and Communication Technologies (ICTs) usable, predominantly computers and Internet, for non-English speakers in ten Southeast Asian (so-called) 'developing countries'. To achieve this, project participants aimed at localising those ICTs; 'localisation', in the project, referred to any measure that had to be taken, allegedly, so as to generate digital operational chains on the part of the targeted 'users'. This included tasks, such as translating software and interfaces into the languages of the participating countries (e.g., Urdu/Pakistan; Khmer/Cambodia; Lao/Laos), adapting menus and navigation (e.g., adjusting box sizes to diverse linguistic demands, adapting interfaces to right-to-left script direction), in some cases adapting iconographies (colours, symbols), building translation glossaries, standardising of computer terminology, etc. Whereas such tasks were the main activities of phase I of the project, in the second phase, teams aimed to identify non-English speaking, 'digitally excluded' target groups to develop training programmes, to train them accordingly and to evaluate the whole training process.

To create digital operational chains on the part of the targeted actors, a novel cultural programme (now including digital operations/scripts) was created as an all-encompassing package; project participants strived to set this programme into operation in the course of the training. The project was organised as a large, transnational network, being funded by a North American 'development organisation'. Regional and executive project lead was located at a research centre for Urdu (Pakistan's lingua franca) computing in Pakistan. I call this institution by the pseudonym *Research Centre for Computational Linguistics* (RCCL). RCCL's staff consisted of linguists, scientists, engineers, and researchers of Pakistani origin. It was they who not only lead and organised the project in practice, but who also realised its Pakistan country component. The aim of my fieldwork was to reconstruct the strategy pursued by RCCL to design the novel cultural programme and to make it run, so as to generate digital operational chains, i.e., practices. I set out from the premise that studying such a strategic generation of digital operational chains from scratch was appropriate to reveal the general anthropo-logic underlying digitisation.[19] So *how was the novel programme designed strategically and for what reasons? What was RCCL's strategy to make it run?* To answer these questions, I did fieldwork in RCCL's computer labs.

RCCL targeted 14-year-old students (male and female) at eight public schools in the rural areas of Pakistan. By creating digital operational chains, the team aimed at connecting students to 'global society', as one team member stated. To achieve this, students' operational chains were to be connected via Internet to those constituting 'global society'. The selection of students as a target group was by no means random, as RCCL presumed that these actors feature a sufficient degree of cognitive openness, i.e., that they were able to easily pick up novel scripts, rules, regularities, and routines to thus shape their operational chains in a novel way. A survey was conducted in all eight schools, asking students about their daily activities: 'What are their likes, dislikes, their interests?' (D1).[20] In this sense, it was RCCL's intention to learn about students' previous cultural programme in order to integrate novel elements.

I would like to highlight at this point the sociocultural bipolarity of practices that become visible when analysing the latter as operational chains: whereas those chains are given *shape* by cultural programmes (rules, scripts, and so forth), it is the *interconnection* of students' operational chains with those of other actors that creates social relations.[21] The distinction helps to clarify what was meant when RCCL staff expressed their wish to connect students' operational chains to 'global society'. However, what were the programme components devised to shape students' chains? To begin with, we may state in this regard that RCCL's team pursued a 'heterogeneous engineering' strategy (Law 1987: 111), meaning that the strategy included a whole range of different engineering modes to produce programme components.

The instrumental engineering-mode: constructing a benefit

The first thing that RCCL did was to ask for the benefit that ICTs were able to bring about for the targeted group. Since the latter was never before in contact with ICTs, they could not indicate themselves how to benefit from ICT usage. For this reason, the project team attempted to construct a benefit by linking up with what was thought of as the students' previous cultural programme: 'They don't know what the Internet is, but we know. (...) That's the reason why the survey is conducted. (...) So that we can develop the content, the training, and the technology accordingly, according to the need' (D2). Drawing on the survey, ICTs were given the role of a learning tool and knowledge provider. By so doing, an element of the previous cultural programme (learning tool and knowledge provider) that was perceived as beneficial was reproduced in the reinvention of the cultural programme.

The physical-material engineering-mode: hardware, connectivity

According to RCCL, the establishing of novel operational chains also required physical-material engineering. First of all, the team furnished computer labs at the participating schools; also, antennas providing for wireless Internet connection were installed, as the telephone network does not cover the whole of the rural areas of Pakistan. The *Vfone* of the *Pakistan Telecommunication Company Ltd* (PTCL) was set into operation, a device especially tailored to these areas. As the context specificity of the hardware suggests, this engineering mode also aimed at linking up with the previous cultural programme. The necessity to do so extended to the (only seemingly) most trivial aspects:

> Schools were provided with replacements if there were any equipment problems during the course of the training program. The item most replaced was the power cable, perhaps due to the nationwide electricity supply problems (power cuts, fluctuations in voltage etc.) at the time of the training program. More robust power cables, custom-made for local context, could have provided a better alternative.[22]

We may interpret this statement as the necessity to inscribe also into material things – such as power cables – specific scripts ('robustness'), so as to make them capable of becoming part of locally specific cultural programmes.

The linguistic engineering-mode: software localisation

Linguistic engineering measures formed the bulk of RCCL's activities. The team translated Open Source Software (e.g., *SeaMonkey*) from English to Urdu to enable digital communication (email, chatting), access to the web, and the production of web content. Computer terminology ('save', 'download'), menus, and boxes were to be translated, reinvented, and adjusted. These tasks posed a manifold of challenges for the team, due to the non-discreteness (letters' shapes change depending on foregoing/succeeding letters) of the Arabic alphabet, which Urdu writing utilises. Another problem occurred when it came to reverse interfaces, as Arabic is written right-to-left. RCCL solved this problem eventually by inscribing the following code passage into the *SeaMonkey* package:

```
/*make UI RTL */
window,dialog,wizard,page { direction: rtl; }
menu { direction: rtl; }
outliner { direction: rtl; }
/* XML header shown when there's no style */
#header { direction: rtl; }
/*
 * make sure search from address bar remains in RTL
 */
#urlbar .autocomplete-search-engine
{
direction: rtl !important;
}
/*
 * keep Composer <HTML> Source tab LTR
 */
#content-source,
```

The comment /*make UI RTL */ stands for 'Make User Interface Right-to-Left' and thus describes the function of the succeeding code. A range of elements, such as window and dialogue boxes, the menu, etc., as well as the header and URL bar, are arranged from right to left by the instruction: { direction: rtl; }. The passage is part of a *Cascading Style Sheet* (CSS) file of the *Urdu Language Pack* for *SeaMonkey*. The passage is an inscription into the technical apparatus of the explicit practice rule to read and write from right to left. Human developers generated this script that then was processed by the technical apparatus so as to enable human readers to make sense from what was perceived. We can see quite plainly at this point the entangled human–non-human agency of programmes and operational chains. The tendency to reproduce elements of the previous cultural programme in a new guise when reinventing it for the creation of novel operational chains also clearly extends to the inscription of scripts into non-human components.

The cognitive engineering mode: computer literacy

RCCL was quite aware of the requirement to also enable human actors to execute fractions of the digital operational chains. Consequently, the team intended to incorporate certain skills as part of the overall cultural programme into the bodily operational memory of the students:

> A teacher should know the mental level of a student. A teacher should know: 'how can I attract them to this thing which I'm going to teach them?' And a teacher should also know: 'how can they be fascinated by this?' (...) At first I will draw their attention to me by some – you can say – tricks, even some jokes about it. And when they attend then I will try to put all those things softly in their mind (D3).

To attract the students' attention and fascination, it was necessary to link up with their previous cultural repertoire.

The semantic engineering mode: web content

However, all these measures were still not sufficient to make the novel programme run:

> At the end of the day it's the amount of content which makes the Internet meaningful, right? Because 'ICTs' have two portions: the information portion and the communication portion. (...) communication technology – you can come on board right away, right? You can start e-mailing day one. But information technology is about content and usability of that content, right? (D4).

At this point, the team faced the problem of lacking Urdu web content. To improve this situation, multiple strategies were pursued: the development of machine translation systems (English to Urdu) and of optical character recognition systems, in order to be able to put Urdu documents online quickly and easily. Moreover, students, in the course of the training, received support to produce web content. Of central relevance in our context is the observation of the project leader that students were predominantly interested in local-specific semantics.[23] In this sense, it was the local semantic index that rendered digital operational chains interesting for the students in the first place. Thus, it seems mandatory to inscribe digital ICTs with the semantics of the previous cultural programme; if ICTs cannot reproduce such semantics, stabilisation of the operational chains in question becomes improbable.

The semiotic engineering mode: making meaning

In a similar sense, it was also necessary to integrate ICTs in local meaning structures. The functioning of ICTs, for instance, was to be explained in the students' everyday language. In the printed training brochures, RCCL pointed to analogies such as equating hardware with the body and software with the soul. Another aspect of meaning production was the embedding of the origin of binary-digital technologies into an historical account of humanity that spanned from the Chinese abacus via Charles Babbage's *Difference Engine* to the first *Apple Macintosh*. Through this embedding, the novelty was in a way 'traditionalised', as it became a part of the history of all humanity and its ongoing 'tradition of change'. The local specificity of some of the meanings became most obvious in the training brochures when the functioning of the operating system, understood as a particular type of system software, was explained: 'It is the System Software due to which a user can interact with the computer. System Software is just like head of the family who not only keeps the family members together but also makes them interact with the outer world'.[24] By explaining the workings of computers via the notion of the family head, the latter usually being a male family member, the implicit rule set of the patriarchal gender regime was raised, and thus its meaning structures partly reproduced willy-nilly.

The normative engineering mode: rules of behaviour

The intended connection of the students' operational chains to those of 'global society' carried the potential to open up channels for the influx of semantics that were not in line with local norms. In this sense, Internet access threatened to reduce the social control of semantics in a technical way. The team feared that this might create conflict in regard to previous local cultural programmes: 'In our social structure people dislike free access to knowledge, to open up access to everything. [There are] some moral restrictions (...) So we are also managing this' (D3). To attenuate the problem, RCCL aimed at 'ethically programming' students' operational chains. Each and every training brochure contained an 'ethics section' that specified the normatively correct behaviour when accessing the web. What is noteworthy, though, is that it was perceived as unnecessary to formulate explicitly behavioural rules in detail; instead, students were called upon to reproduce rules which were so far also valid offline in a more general way. For instance, the ethics section of the brochure explaining the chat tool read: 'Under ethics, the same rules and regulations are applied to online chatting which are considered while talking with a group of people sitting at one place'.[25] This results in the public appeal to reproduce locally valid, implicit rules in the course of reinventing the novel programme – without considering it necessary at all to render those rules explicit.[26]

Conclusion

The goal of this contribution is to develop an anthropological methodology that allows for fruitful study of the digitisation processes. In the introduction, I indicated three 'quality criteria' such a methodology at best fulfils. I would like to take up these three criteria here again in order to specify whether the framework submits to these conditions:

- It should have become clear in the previous section that the terminological equipment can indeed be set into empirical operation (criterion 3). It is possible to draw on the proposed vocabulary to analytically describe processes of digitisation; the resulting cultural analysis poses no difficulties in including the seemingly 'most technical aspects' of the analysis.
- As I attempted to show, the methodology allows for taking up and fruitfully integrating knowledge of apparently far-away disciplinary origins that nevertheless may contribute relevant arguments to the analysis of digitisation (criterion 2). I hasten to add that including such knowledge does not lead to its privileging; cultural analysis still takes the lead, so to speak, and manages to submit technical or biological knowledge to its own anthropological interests.
- Finally, the methodology also allows one to reconcile epochal with 'micro-perspectives' (criterion 1). Considering the case in point I draw upon in the previous section as a snapshot of the massive extension of the cultural technique of digitisation in time and space, we can still see the heavy local index of this process. The globalisation of the cultural technique of digitisation takes place via local differentiation.

This results in contradictions and paradoxes that became visible when the case study was finalised: insofar as the operational chains produced in this project aimed at global connectivity precisely *by way of* locally shaping them, the resulting digital practices amounted to the generation of local culture and global sociality at one stroke. I call this somewhat contradictory double movement the paradox of digital glocalisation (Ochs 2013).

Clearly, a lesson is to be learned that may also be of relevance when it comes to addressing the question that motivated this chapter: is the ubiquity of digitisation processes to be interpreted as marking an 'epochal rupture', or shall we rather presume that we are dealing with a manifold of diverse, perhaps even contradictory digitisations that lack a common character? The precise answer to this question is: both are the case. If we look from a historical and multisited perspectives at the global similarity of the transformations on the level of the cultural technique of digitisation, we may find evidence for global, epochal rupture, similar to the analyses of the transformations induced by script or by the printing press. However, if we ethnographically stick to the diverse empirical modes of actualising the cultural technique of digitisation as multiple operational chains, we may find evidence for technocultural multiplicity and digitisations in the plural. Thus, as so many times before, our answer to the question must be: *'it depends'* – but at least we do know now *on what exactly* this answer depends.

Notes

1 I am indebted to Gertraud Koch for providing me throughout the writing process with thorough and constructive critique in a very friendly way. I believe her feedback has significantly improved this contribution. Of course, it's nevertheless me who is responsible for deficiencies.
2 The pluralisation of, say, the term 'modernity', as in the expression 'Multiple Modernities' (Eisenstadt 2002), may serve as a reference here. A similar shift occurred in globalisation research. At first, there was a rather narrow focus on economic aspects, before further dimensions (technology, culture, society) were taken into account. Moreover, criticism was raised with regard to top-down approaches, thus supporting the emergence of bottom-up perspectives. In addition, even the term 'globalisation' was questioned, modified ('glocalisation'), or replaced entirely ('transnationalisation').
3 An exclusive historical focus on Turing would, however, fall short of the manifold contingent processes that resulted in the emergence or invention of the computer (see Lévy 1994).
4 The cultural technique of scripture may serve as an analogy here: all humans are quite obviously capable of applying the abstract principles underlying this cultural technique (and cultural technology research may succeed in formally describing these principles). Yet, empirically, there are also oral cultures. In this sense, the principle has the capacity to become 'universal'. Moreover, the case of scripture clarifies that cultural techniques oftentimes are invented as actual practice without any formal description. Regarding scripture, for instance, the techniques at hand may result from time measurement or religious (and not from scientific) practices.
5 From this point of view, any principle formulated in an abstract way can be considered an attempt to exclude cultural specificity so as to gain universal validity. When setting abstract principles into actual operation, however, those

principles are 'enriched' by the (natural sociocultural) wealth of the respective space-time situation.
6 No one has better experience of this than software developers who, in the course of the design process, are faced with the task of specifying the relevant characteristics of the relevant space-time situations in order to adapt specific applications accordingly. Time and again, such specification fails, since even agreeing on what is and what is not 'relevant' is not as straightforward as it may seem.
7 Bolter (2005: 467) calls such cultural nonspecificity 'cultural ambivalence'.
8 The term 'actual' is meant to point out that the principle as such has no existence independent of its instantiation, and thus only occurs as empirically realised (actualised) potential. Nevertheless, it may become effective, which is why it is not appropriate to designate it as unreal in the same sense as the possible is unreal.
9 The view that cultural techniques are only empirically observable as practical, culturally specific actualisation is not foreign to the theorisation of those techniques. In fact, a manifold of theoretical approaches distinguishes, albeit at times in an undertheorised way, between processing principles, on the one hand, and their application, on the other, as in the case of scripture (Grube & Kogge 2005: 17), digital media (Bolter 2005: 466), and their interrelation (Krämer 2005: 46). As was said before, cultural techniques may emerge as practice without their abstract principles becoming visible at the outset (Macho 2003: 180). There is a long history of speaking, writing, and calculating in a culturally specific way, before the virtual, abstract principles underlying such practices became visible.
10 Still, there are also some obsolete aspects in Leroi-Gourhan's oeuvre (see Latour 2005: 74, Schüttpelz 2006: 93; however, note that both authors leave no doubt about their indebtedness to Leroi-Gourhan's work).
11 Leroi-Gourhan's anti-essentialism is expressed by his abstaining from assigning essential characteristics to humans. Instead, the essentials of being human are to be found in what is being *exteriorised*. For example, when stating that 'the human hand is human because of what it makes, not of what it is' (Leroi-Gourhan 1993: 240), he replaces once and for all fixed characteristics of being human by the anthropo-techniques of human becoming.
12 An extreme case of cultural reshaping is apnea diving. The latter involves culturally incorporated skills that enable one to regulate the (biologically hard-wired) drive to breathe in a way that allows for very long and deep dives without breathing gear.
13 A symptom for the slight but relevant shift in perspective is that, in ANT, there is the claim that the production of the social can be reconstructed via an analysis of the ways in which 'programmes of action' are inscribed into humans, technical objects, and documents – all three counting as 'intermediaries': *'the social can be read in the inscriptions that mark the intermediaries'* (Callon 1991: 140; italics in original). However, if we assume that programmes of action give at once shape to practices and sociality, we cannot distinguish anymore between the two at the programme level. In contrast to this, to gain a more differentiated picture, I distinguish operational chains/practices, culture as a programme *shaping* these operational chains, and, finally, sociality as an arrangement of relationships resulting from the *interconnection* of operational chains.
14 Consequently, for Wittgenstein (1958: §339), even thinking is a bodily process. At this point of the argument, it becomes once more obvious that I pursue the rather humble goal of tailoring existing anthropological knowledge to the study of digitisation. It is thus not surprising that the proposed methodology resembles heavily classical approaches. The notion of cultural programme, for example,

has strong parallels (but still differs from) Bourdieu's ideas about the incorporation of collective schemes (Bourdieu 1979: 170) and the notion of 'Hexis', which is the ensemble of bodily motions, gestures, techniques, and tools (ibid: 189–190).
15 Also, non-human primates may possess mirror neurons, but human imitation techniques still seem to be unique insofar as only humans have the capacity of shared attention (joint attention to some object and at the same time reflexive mutual knowledge of this very joint attention), and thus shared intentionality. This capacity makes it possible to not only mutually mirroring behaviour but also to include the information that mirroring takes place in the process of communication: '…some forms of social learning are mainly individual – in the sense that learners just gather information unilaterally (exploitively) from unsuspecting others. When chimpanzees learn from others how objects work, they are most often engaging in this individualistic type of social learning … . Human infants, in contrast, imitate more readily the actions of others, and they sometimes do this with the apparent motivation not just to solve a task, but rather to demonstrate to the adult that they are "in tune" about the current situation' (Tomasello & Carpenter 2007: 123).
16 'When 2-year-old children observe an adult engage in some new activity, saying something like "Now I'm going to dax", they not only imitatively learn to perform that activity, they also seem to see that activity in normative terms as how "we" do daxing. For example, Rakoczy, Warneken and Tomasello … demonstrated such a new activity for 2- and 3-year-old children, and then had a puppet enter and do it "wrong". Many of the children objected in very explicit terms, telling the puppet what it "should" be doing, and almost all protested to some degree. They saw the puppet's actions as somehow not conforming to the social norm of how we do daxing, and they enforced the norm' (Tomasello & Carpenter 2007: 124).
17 If human social formations were based exclusively on tacit expectations, they would assume the shape of baboon societies (Strum & Latour 1987).
18 As Preda (2000: 279) explains, while there is no talk of social rules in ANT, since the latter attempts to abstain from 'classic' terminology, still '[in] this new terminology, "script" is the equivalent of the notion of social rule'.
19 Obviously, I pursued the well-known ANT strategy to render visible and analyse the scripts: 'follow the device as it moves into countries that are culturally or historically distant from its place of origin' (Akrich 1992: 211).
20 When quoting from interviews with RCCL staff, I will indicate interviewees regardless of their position in the hierarchy and their actual function. I simply call them 'D', for 'developer', and a number is added in the order of appearance.
21 Latour prefers to call relations 'associations' to emphasise that they can be created by humans as well as by non-humans: '[T]here are also relations among things, and social relations at that' (Latour 1992: 257).
22 The quote is taken from RCCL's official evaluation document of the training (28).
23 The empirical source of this statement is a radio interview given by the project leader to a broadcasting station, which was based in the development organisation's home country. I took minutes while listening to the interview.
24 The quote is taken from the brochure 'Basic Computing' (54–55).
25 The quote is taken from the brochure 'Instant Messaging' (105).
26 For the sake of completeness, I would like to mention that the strategy also included an engineering mode that I call 'discursive'. Insofar as the team expected resistance to digitisation by parents, elders, or others, it was prepared to start a counter-discourse. However, as the latter concerned 'emergencies', I omit a detailed analysis of this aspect.

References

Akrich, M., 1992. The description of technical objects. In W.E. Bijker and J. Law (Eds.), *Shaping technology/building society: studies in sociotechnical change*. Cambridge: MIT Press, pp. 205–224.

Baecker, D., 2007. *Studien zur nächsten Gesellschaft*. Frankfurt am Main: Suhrkamp.

Bauer, J., 2006. *Warum ich fühle, was du fühlst. Intuitive Kommunikation und das Geheimnis der Spiegelneurone*. Hamburg: Hoffmann und Campe.

Beck, S., 2008. Natur | Kultur: Überlegungen zu einer relationalen Anthropologie. *Zeitschrift für Volkskunde*, II, pp. 161–199.

Bolter, J.D., 2005. Digitale schrift. In G. Grube, W. Kogge, and S. Krämer (Eds.), *Schrift. Kulturtechnik zwischen Auge, Hand und Maschine*. München: Fink, pp. 453–468.

Bourdieu, P., 1979. *Entwurf einer Theorie der Praxis auf der ethnologischen Grundlage der kabylischen Gesellschaft*. Frankfurt am Main: Suhrkamp.

Callon, M., 1991. Techno-economic networks and irreversibility. In J. Law (Ed.), *A sociology of monsters: essays on power, technology and domination*. London: Routledge, pp. 132–164.

Castells, M., 2001. *The internet galaxy: reflections on the Internet, business, and society*. Oxford: Oxford University Press.

Dewey, J., 1995. *Erfahrung und Natur*. Frankfurt am Main: Suhrkamp.

Eisenstadt, S. (Ed.), 2002. *Multiple modernities*. New Brunswick: Transaction.

Faßler, M., 2009. *Nach der Gesellschaft. Infogene Welten, anthropologische Zukünfte*. München: Fink.

Faßler, M., 2014. *Das Soziale. Entstehung und Zukunft menschlicher Selbstorganisation*. München: Fink.

Galloway, A., 2004. *Protocol: how control exists after decentralization*. Cambridge: MIT Press.

Garfinkel, H., 1984. *Studies in ethnomethodology*. Los Angeles: Polity Press.

Giddens, A., 1984. *The constitution of society: outline of a theory of structuration*. Cambridge: Polity Press.

Grube, G. and Kogge, W., 2005. Zur Einleitung: Was ist Schrift? In G. Grube, W. Kogge, and S. Krämer (Eds.), *Schrift. Kulturtechnik zwischen Auge, Hand und Maschine*. München: Fink, pp. 9–22.

Krämer, S., 2005. Operationsraum Schrift: Über einen Perspektivenwechsel in der Betrachtung der Schrift. In G. Grube, W. Kogge, and S. Krämer (Eds.), *Schrift. Kulturtechnik zwischen Auge, Hand und Maschine*. München: Fink, pp. 23–61.

Krämer, S. and Bredekamp, H., 2013. Culture, technology, cultural techniques: moving beyond text. *Theory, Culture & Society*, 30 (6), pp. 20–29.

Lash, S., 2002. *Critique of information*. London: Sage.

Latour, B., 1992. Where are the missing masses? The sociology of a few mundane artifacts. In W.E. Bijker and J. Law (Eds.), *Shaping technology/building society: studies in sociotechnical change*. Cambridge: MIT Press, pp. 225–259.

Latour, B., 2005. *Reassembling the social: an introduction to actor-network-theory*. Oxford: Oxford University Press.

Law, J., 1987. Technology and heterogeneous engineering: the case of Portuguese expansion. In W.E. Bijker, T.P. Hughes, and T.J. Pinch (Eds.), *The social construction of technological systems*. Cambridge: MIT Press, pp. 111–134.

Leroi-Gourhan, A., 1993. *Gesture and speech*. Cambridge: MIT Press.

Lévy, P., 1994. Die Erfindung des Computers. In M. Serres (Ed.), *Elemente einer Geschichte der Wissenschaften*. Frankfurt am Main: Suhrkamp, pp. 905–945.

Macho, T., 2003. Zeit und Zahl. Kalender und Zeitmessung als Kulturtechniken. In S. Krämer and H. Bredekamp (Eds.), *Bild, Schrift, Zahl*. München: Fink, pp. 179–192.

Ochs, C., 2012. Jenseits von technikzentrierter und anthropozentrischer Medienkultur-Beschreibung: Eine ethnographische Erläuterung der Logik medialer Transformationsprozesse. *Zeitschrift für Medienwissenschaft*, 6 (1/2012), pp. 66–84.

Ochs, C., 2013. *Digitale Glokalisierung. Das Paradox von weltweiter Sozialität und lokaler Kultur*. Frankfurt am Main: Campus.

Preda, A., 2000. Order with things? Humans, artifacts, and the sociological problem of rule-following. *Journal for the Theory of Social Behaviour*, 30 (3), pp. 269–298.

Reckwitz, A., 2003. Grundelemente einer Theorie sozialer Praktiken. Eine sozialtheoretische Perspektive. *Zeitschrift für Soziologie*, 32 (4), pp. 282–301.

Ruppert, E., Savage, M., and Law, J., 2013. Reassembling social science methods: the challenge of digital devices. *Theory, Culture & Society*, 30 (4), pp. 22–46.

Savage, M., Ruppert, E., and Law, J., 2010. Digital devices: nine theses. *CRESC Working Paper Series*, Working Paper No. 86. Availabe at: www.cresc.ac.uk/medialibrary/workingpapers/wp86.pdf (accessed 9 May 2016).

Schatzki, T., 1997. Practices and actions: a wittgensteinian critique of Bourdieu and Giddens. *Philosophy of the Social Sciences*, 27 (3), pp. 283–308.

Schreiner, R., 2009. *Computernetzwerke: Von den Grundlagen zur Funktion und Anwendung*. München: Hanser.

Schüttpelz, E., 2006. Die medienanthropologische Kehre der Kulturtechniken. *Archiv für Mediengeschichte*, 6 (2006), pp. 87–110.

Silverstone, R. and Haddon, L., 1996. Design and the domestication of information and communication technologies: technical change and everyday life. In R. Silverstone and R. Mansell (Eds.), *Communication by design: the politics of information and communication technologies*. Oxford: Oxford University Press, pp. 44–74.

Strum, S. and Latour, B., 1987. Redefining the social link: from baboons to humans. *Social Science Information*, 26, pp. 783–802.

Tarde, G., 1903. *The laws of imitation*. New York: Henry Holt and Company.

Tomasello, M. and Carpenter, M., 2007. Shared intentionality. *Developmental Science*, 10 (1), pp. 121–125.

Turing, A., 1936. On computable numbers, with an application to the *Entscheidungsproblem. Proceedings of the London Mathematical Society*, 42 (2), pp. 230–265.

Wittgenstein, L., 1958. *Philosophical investigations*. Oxford: Basil Blackwell.

2 Archive
Isto Huvila

Introduction

In colloquial discourse, almost everything can be an archive. Most blogs have 'archives' of older posts, an archive of a website is the space where all obsolete files and pages are moved to, and a shoebox of old photographs or a folder of old emails or sound files can be an 'archive'. The Internet itself has been dubbed as the archive of the archives (Allen-Robertson 2013). The conceptual intricacies are not eased by the fact that the notion of archive has captured the imagination of a large number of theorists using 'archive' as a metaphor of memory, keeping, longevity, and permanence.

Even if the gamut of the colloquial and metaphoric senses of archives might suggest that the term is too general to be useful as an analytical concept, tracing back the evolution of the idea and concept of archives and the history of the contemporary practices of archiving is helpful in framing and understanding the aspirations for organising and keeping digitised things and how these ambitions are consummated in practice. Archive is a cultural technique and archiving a practice of collecting (or accumulating), preserving, and making materials retrievable. As Derrida (1995) notes, it kills and replaces human memory. From the perspective of a critical scrutiny of digitisation, *archive* is a concept that captures in practice and in theory something very fundamental to the underpinnings of the aspirations to digitise and keep digital artefacts, whether the archives would be small, large, public, private, fleeting, or long-lived. Taking a closer look at this particular cultural technique helps us to better understand how certain things are valued and organised in contemporary society, what, how, and why things are collected and preserved, how different temporalities of things and human beings become intertwined, and how the present is related to the past and the future to the present. On an even more fundamental level, pushing forward the remark of Gitelman and Jackson (2013) on the links of archives (in Foucauldian sense) and historical epistemology, the manner in which the archives are defined and demarcated determines what is knowable of the past, both at the present and in the future of our present as the future past.

Archives in archival literature

The etymology of 'archive' and related terms in different European languages is in classical Greek. The Greek word *arkheion* (from arkhe, meaning government or rule) and its Latin translation *archivum* could refer to a place for public administration, government building, an official, a room for records keeping or to original (archival) records (Blouin Jr. & Rosenberg 2011, Lidman 2012). The purpose of archives has changed from being administrative instruments of premodern regimes to serve the historical fascination of the period of romanticism in the beginning of the 19th century and the primarily European nationalist and imperial endeavours during the following 100 years (Duchein 1992, Cox 2000). The second half of the 20th century was a period of pluralisation of archival thought in archival domain (Cook 1997, Ribeiro 2001) and a period when 'archive' turned to a cultural keyword used to denote a broad variety of repositories from databases and information systems to seed banks, libraries, museums, and archaeological excavations (Buchanan 2010). In the archival field, Körmendy (2007) sees the profusion primarily as a result of an external, societal pressure. Archives and archiving have expanded both in volume and in extent. In addition to great men and governmental history, archives are created to document public movements, local history, and marginal communities. Simultaneously, the idea of the archive allowed for a more pluralistic understanding of the audiences and purposes of keeping archives. Theory and practice of archives have shifted from the earlier positivism to functionalism (Delmas 1992) and theorising characterised by critique of earlier assumptions of the neutrality of archives. The recent theorising has acknowledged the subjectivity of archives and the influence of the choices made by their creators, custodians, and users on what archives contain and what an archive is (Lane & Hill 2010, Cook 2011, Yakel 2011). According to Cook (2013), archives have transformed in the process from passive keepers to active assessors to societal mediators to community facilitators. From the 1990s onwards, the evolution of archival thought has been influenced by new theoretical perspectives, for instance, borrowed from Giddensian sociology (McKemmish 2001), postmodernism (Cook 2001), and critical theory (Dunbar 2006). The contemporary theory has challenged the stability and persistence of archives and appropriated the ideas of processualism, life cycles (Borglund & Öberg 2006), and, increasingly, the one of continuum (Upward 1997) and participatory negotiation (Robinson 2007, Huvila 2008, Shilton & Srinivasan 2008) of what is an archive and what it contains.

Even if the contemporary discussion has extended the life span and contexts of relevance of archives and their holdings, the premise of keeping formal archives is still very much based on the *provenance* of the records and the organisational context of their office of origin (Bazerman 2012), a fundamental tenet of archival work, which dates back to the late 19th century and beyond. This emphasis marks out formal archives from informal archives, or in the archival studies parole, (proper) archives from other types of repositories including collections, libraries, and miracle chambers of the late Renaissance and Baroque. Even if the requirement of original order might not always be as compelling in the context of informal archives, theorists like Taylor (2003) and Derrida (1995) with rather different takes on archives than the one held by archival studies scholarship (Bazerman 2012) refer to stability and originality as a characteristic trait of archives. The emphasis of Blouin (Blouin Jr. & Rosenberg 2011) that archive is defined by the organic relationship of the records to their generators also characterises informal archives. Similarly, the conceptualisations of records as information, documents (Yeo 2007), evidence (Brothman 2002), transactions (Cox 2001), or speech acts (Henttonen 2007, Yeo 2010) are rooted in the actual and imagined origins of archival records. The link between records and their worth, both in terms of corporate surplus value (an important driver of corporate archives and records management, e.g., Bailey 2007, Ataman 2009, Bailey 2011) or their less tangible role as a source of societal accountability (a central aspect of the discussions of the need to strive for more inclusive and representative archives, e.g., Shilton & Srinivasan 2008, McKemmish et al. 2012), are also dependent on their provenance. In spite of its fundamental nature, provenance is a controversial and complex concept (Douglas 2010). Its apparent simplicity conceals the difficulty of determining what is original, and consequently, as Cook underlines, shifts archives far from being 'unproblematic storehouses of records awaiting historians' (Cook 2011: 631). The same problem applies to the authoritative, authentic, essential, or vital nature of the records kept in an archive (Blouin Jr. & Rosenberg 2011). These assumptions and expectations are easy to agree with, though difficult – if not impossible – to operationalise in practice and therefore often criticised in cultural analysis of archives and archival work outside of the professionally oriented discipline of archival studies (Synenko 2013). The difficulties arise from the complexity of the process on how archival records emerge in time and space, often with a plethora of individuals and institutions involved in the process. Also, the kind of record – being, for instance, a paper document, a photograph, or a seed of a plant – affects how the provenance can be conceptualised. Even if the critics make an important point in denouncing positivist ideals of provenance, the practical impossibility to determine 'true' origins of an archival record does not mean that the concept could not function as a useful guiding principle of archives and archiving.

The different perceptions of provenance and the nature of the record as, for instance, evidence, information, and persistent representations (Yeo 2007), are kin to the several competing perspectives of the nature of archives in the archival literature. They are anchored in different historical trajectories that conceptualise the premise of an archive to be information (e.g., Buckland 1991, Gilliland-Swetland 2000) or cultural heritage (Manžuch 2009), or that an archive is distinct from other types of repositories because archival records are authentic evidence (e.g., Duranti 1999) rather than information, a position which has been criticised in postmodern archives-related literature (e.g., Taylor 2003). The mission of archival institutions has been described in terms of preserving and providing access to culture and heritage (e.g., Barry 2010), memory (e.g., Cook 1997, Gilliland-Swetland 2000), and knowledge, supporting learning, promoting identity and understanding (Gilliland-Swetland 2000), and, for instance, serving (e.g., Sundqvist 2007) and empowering their users (e.g., Usherwood et al. 2005). Archives are considered to have a civic role as societal and cultural institutions (e.g., Hickerson 2001, Jimerson 2004, Johnson & Williams 2011) and access to archival records is perceived as a new civic right (Dempsey 2000) independent of the citizens' cultural background. The role of archivists has been characterised in comparable terms in the literature. The descriptions of the role of the 'new archivists' tend to emphasise the significance of such factors as outreach (Theimer 2011), technology skills (e.g., Stevenson 2008), pedagogy in formal and informal education (e.g., Zipsane 2009, Krause 2010), engagement (e.g., Prelinger 2010), and collaboration with records creators (e.g., Keough & Wolfe 2012).

In addition to broadly theoretical and societal rearticulations of archives, the rapid advance of digital technologies has raised questions on how digitality and social media will affect (formal) archives (e.g., Bailey 2008, Theimer 2011, Zhang 2012) and Derridean archives (e.g., Treanor 2009) alike in the future. The technology influences societal change and its impact on archives has been described both as an unavoidable premise (e.g., Bailey 2008, Treanor 2009) and an opportunity (Stevenson 2010). In the field of archival studies, there is a relatively broad consensus of the continuing value of the fundamental principles of archival work in the digital context (e.g., Gilliland-Swetland 2000, Duranti 2010), but as Bailey has urged, there is a need to 'fundamentally rethink the way in which we [records managers] strive to achieve them' (Bailey 2008: xv) in the contemporary context with radically divergent ideas of what an archive is and could be (e.g., Huvila et al. 2008, Theimer 2011, Zeitlyn 2012).

Archives beyond archival studies

At the present, an archive can be many things beyond the 'archives proper' discussed so far. The International Council of Archives (ICA) defines archives from the perspective of the archival profession as 'the documentary by-product of human activity maintained for their long-term value' (International Council of Archives 2009: para. 2). This definition carries repercussions for the division of (historical) archives and (current) records maintained in Anglo-American and German archival discourse, a dichotomy which does not exist, for instance, in the Netherlands, France, Italy, Spain, or the Nordic countries (Ketelaar 2000, Orrman 2007). The perspective endorsed by ICA and the somewhat similar general definition of archives 'as a collection of records accumulated by persons, corporate bodies and families in order to support their memories' in the introductory text of Thomassen (2001: 374) are indicative of, even if not entirely forthright about, the focus of interest of archival science scholarship and archival profession in the (archives as) professionally curated outcomes of processes that produce (documentary) records related to the activities of individual and collective bodies (e.g., Thomassen 2001, Craven 2008).

Even if these definitions stem from the institutional field of archivistics, they do also encompass a broader popular understanding of the archives that encompass a broad variety of collections, which either explicitly or implicitly perform an *archival function* (defining how something is an archive rather than why it is an archive). Therefore, it is possible to make a distinction between archives proper (in a strict, archival, scientific sense) and other types of repositories as two different types of archives, which both perform archival function to various degrees. By referring to the archival function, the interdisciplinary use of the term 'archive' to refer to different types of digital information systems and repositories (Breakell 2010), seed banks, databases, and collections of things (Buchanan 2010) becomes more compatible with the understanding of 'archives proper' in the archival science literature.

Besides the theoretically sometimes rather vague popular references to archives, the term has also captured the attention of many widely cited philosophers and cultural theorists (e.g., Derrida 1995, Foucault 2002, Ebeling & Günzel 2009). Parallel to the subjectivist emphases of the contemporary archival theory, the humanities scholarship has, since the 1990s, referred to the *archival turn*, a move from perceiving archives as a source to considering them as a subject (Hutchinson & Weller 2011). In spite of this general turn, the old ideas of 'archive' and 'archiving' have not disappeared and they have a certain tendency to surface as emblematic references to that which archives are supposed to be (e.g., Brockmeier 2010). What has also happened is that a critique of archival principles and remnants of archival positivism from outside has occasionally raised to heights that, as Buchanan (2010) notes, might seem hostile to archives.

Considering the extent of archives-related literature, the attempts to produce classifications of different types of archives are conspicuously few. In an attempt to elucidate the premises of different types or ideas of archives, Bowker (2010) makes a distinction between *formal archives* and *trace archives*. Bowker's formal archives are peremptory and sequential, whereas trace archives are 'about habits and customs and place rather than coordinate time and space' (Bowker 2010: 213). In contrast to a formal archive, a trace archive is inscribed in the lived environment rather than collected and curated. The idea has similarities with that of Hartley (2010), who makes a distinction between modernistic (formal, institutional) archives based on deterministic (or essence) theory and postmodernist *probability archives* (internet 'archives' like YouTube or the Internet as an archive) based on probability theory.

Bowker's idea of trace archives has certain premisory similarities with the Giddensian-inspired records continuum model (Upward 1997, McKemmish 2001). In contrast to the life-cycle approaches, records continuum emphasises that records reside in a space-time continuum and have parallel uses and roles throughout their existence that begins long before they end up in an archival repository (Borglund & Öberg 2006). Moreover, the model suggests that the process of archiving records (from records creation, to their capture in the archival domain, organisation, and pluralisation) parallels with the phases described in Giddens's theory of structuration (Giddens 1984, Upward 1997).

Even if the theoretical (including Bowker's) and often metaphorical conceptions of archives and archival work tend to differ from the practical realities of archival institutions (Ebeling & Günzel 2009), they are indicative of the cultural and societal underpinnings and implications of archives and archiving (Ernst 2008). As Synenko (2013) argues, archival metaphors – like calling the World Wide Web a library, database, or an archive – should not be considered less legitimate than empirical accounts or experiences of archival practices. Similarly, he continues, any firm distinction between a 'literal' and 'metaphorical' archive is deeply problematic and cannot be justified. 'Metaphorical' and speculative writing on archives refer to real archives as the experiences held by archival professionals and historians. The professional understanding of archive, its functions, and functioning in the contemporary and past societies is merely different from how archive is conceptualised as a metaphor by others. In spite of the apparent dissonances between practitioners and theorists and different theorists akin, the speculative literature captures the confluence and dissonance between scholars, archivists, and other stakeholders of archival records and institutions that, as Manoff (2004) notes, indeed revolve around a shared preoccupation with the function and fate of the 'record'. Theorists, including Derrida (1995) and Foucault (2002), have discussed from different perspectives the implications of the paradigmatic continuity and change of 'archives'. They can serve as monuments of an obsession to preserve, as *loci* of social and historical authority and of as much constructed as recorded (Derrida 1995) rather than unearthed – and consequently as political memory (Foucault 2002). Ernst has explicated the complex material and technological relation of archives and what they archive (Ernst 2008). Richards (1993) has scrutinised the dual role of archives as totalities of knowledge and actively constructed collections. Similarly, Synenko and Taylor problematise the concept of archive by discussing the clash of the views of professionals and cultural theorists (Synenko 2013) and the limits of 'archive' versus a non-archive 'repertoire' (Taylor 2003). Even if these observations are not primarily empirical, they capture many relevant premises of how archives are conceptualised in the literature: the situatedness of archives in the nexus of the creators and keepers of the records, the significance and perplexity of the conceptualisations of the records and their use, and the practical constraints of acting as a keeper and user of archival records.

Analytical uses of archive

Theoretical scholarship includes many examples of how archive can be used as an analytical concept to discuss prevailing and marginal memory and preservation practices in the society, how the cultural record, whether digital or non-digital, is an outcome of political negotiation and constructed rather than captured, how the use of technologies influences the practical outcomes of what is being preserved, and how the notion of archive epitomises the efforts of the societies to keep rather than forget. These theoretical insights can be used to frame the implicit and outspoken premises of digitisation and its consequences.

To highlight perhaps somewhat less obvious uses of the notion of archive in the context of cultural analysis of digitisation, archive (as related to archival institutions and work) can be a similarly powerful instrument for explicating the practices of digitisation and its different premises. Instead of (and in addition to) calling digitised repositories archives, a more careful consideration of the relation of archive defined as an outcome of certain archival practices and tracing of these changing practices and their relation to existing and emerging repositories can help to understand *how* the outcomes of digitisation could form an archive and to what extent they are something else. The debate on how archival institutions should respond to the emergence of the social web – including an abundance of services appearing and/or claiming to perform certain archival functions like YouTube, Flickr, Instagram, and Wikipedia, phenomena like Web 2.0, and the culture of participation (Jenkins 2014) – exemplifies the power of the notion of archive as an analytical tool for divesting the complexities of defining ownership, influence, and expertise in an open digital environment. Huvila's (Huvila 2008) framing of participatory archive builds on a radical redistribution of responsibilities between professional custodians and the users of (digital) archival collections and the possibilities offered by digital platforms to turn archive into an open-ended platform for curating records and their related information throughout their entire lifetime. Other authors have conceptualised digital repositories as archives in different terms. In case of the pioneering Polar Bear Expedition Digital Collections (Krause & Yakel 2007), the archive in the digital collections was conceptualised in another sense as a repository open for use, commenting and complementing but closed for direct alterations by the general public. Here, the considerations of the concept of archive and its multiple possible definitions have helped to engage in a discussion on the relations between the stakeholders of digital repositories: who should get an opportunity to have a say on what is being presented and made available, how should the holdings be organised, and which functions of the repository should be prioritised. Furthermore, a critical reflection of the archiveness of the outcomes of digitisation opens for discussion the provenance (origins and biography), multilayered continuum of the uses and multiple relevance and roles of these repositories in relation to how they came into being as a result of digitisation. Here, a researcher can find useful existing categorisations of how archival literature discusses participation (Huvila 2015b) and how the worth of digital repositories differs depending on whether their longevity or various uses are prioritised (Huvila 2015a). Similarly helpful can be the paradigmatic propositions of redefining 'archive' from the perspective of memory (Cook 1997), access (Menne-Haritz 2001), and knowledge, supporting learning, promoting identity, and understanding (Gilliland-Swetland 2000), or counter-propositions like the one of Zielinski and Winthrop-Young (2015) to define an 'AnArchive' as a locus of performative provocations, plurality, variants, lack of external purpose, and leadership, everything that a traditional archive is not. Finally, an example of cultural analytic use of the concept of archive can be found in the work of Lucas (2012), who discusses archaeological record (i.e., what is left, kept, and documented in the course of archaeological work) as an archive constituted by archaeologists to highlight the constructed and curated rather than unprocessed nature of archaeological evidence. The concept of archive could be used similarly in other contexts of cultural analysis from medical, scientific, and literary to juridical domain and beyond to draw attention to the managed nature of things.

Archive in cultural analysis

The fact that the digital discourse has eagerly embraced the term archive makes it problematic. It has become not only contested (Buchanan 2010) but also analytically meaningless if the particular sense of the term is not carefully described when it is used as an analytical concept. Different conceptualisations can be useful, but their usefulness differs from each other. Researchers who refer to the notion of archive from their different theoretical and disciplinary perspectives end up in a similar dilemma that Boltanski (2014) describes between the sociological versus ordinary (non-sociological) use of everyday life categories. They have different meanings in different analytical and theoretical contexts. The power of the metaphorical use of the term lies in the fact that archive is one of the central concepts of the contemporary Western imagination as, for instance, Derrida (1995) and Ernst (2008) persuasively demonstrate. Archive evokes impressions and the term means something for everyone. In contrast, archive in the context of the evolution of archival institutions and formal archives turns attention to how archives have been practiced in different times and how their function has changed and is changing. At the moment, it seems likely that digitisation is having a deep impact on the practices of making and keeping archives, even if the idea of archives is transmuting slower and if there is something that resists change. Similarly, to the practical value of archives, the conceptual value of the notion as an instrument of cultural analysis remains if it is not taken too lightly and without being specific of what type of an archive and what specific archival functions it is used to refer to.

Further resources

- Adema, Janneke: Open Reflections (blog) https://openreflections. wordpress.com
- Archives Library Information Center (ALIC) www.archives.gov/ research/alic/
- Archives bibliography from 1998 on http://archivschule.de/DE/service/ bibliographien/archives-bibliography-from-1998-on.html
- A Glossary of Archival and Records Terminology www.archivists.org/ glossary/index.asp
- Internet Archive Blog https://blog.archive.org
- Milligan, Ian: Digital History, Web Archives, and the History of 20th Century Canada on http://ianmilligan.ca
- Parikka, Jussi: Machinology (blog) http://jussiparikka.net
- Spellbound blog www.spellboundblog.com
- Theimer, Kate: Archives Next (blog) www.archivesnext.com/

References

Allen-Robertson, J., 2013. *Digital culture industry a history of digital distribution.* New York: Palgrave Macmillan.
Ataman, B., 2009. Archives mean money: how to make the most of archives for public relations purposes – the Yapi Kredi Bank example. *American Archivist,* 72 (1), pp. 197–213.
Bailey, S., 2007. Taking the road less travelled by: the future of the archive and records management profession in the digital age. *Journal of the Society of Archivists,* 28 (2), pp. 117–124.
Bailey, S., 2008. *Managing the crowd: rethinking records management for the Web 2.0 world.* London: Facet.
Bailey, S., 2011. Measuring the impact of records management: data and discussion from the UK higher education sector. *Records Management Journal,* 21 (1), pp. 46–68.
Barry, R., 2010. Opinion piece – electronic records: now and then. *Records Management Journal,* 20 (1), pp. 157–171.
Bazerman, C., 2012. The orders of documents, the orders of activity, and the orders of information. *Archival Science,* 12 (4), pp. 377–388.
Blouin Jr., F.X. and Rosenberg, W.G., 2011. *Processing the past: contesting authorities in history and the archives.* Oxford: Oxford University Press.
Boltanski, L., 2014. *Mysteries and conspiracies: detective stories, spy novels and the making of modern societies.* Oxford: Polity.
Borglund, E. and Öberg, L.-M., 2006. Operational use of records. In IRIS, *29th Information Systems Rresearch Seminar in Scandinavia: Paradigms Politics*

Paradoxes. Helsingør, Denmark, 12–15 August 2006. Copenhagen: IT University of Copenhagen. Available at: www.itu.dk/elisberg/Includes/Papers/6/6-3.pdf.

Bowker, G.C., 2010. The archive. *Communication and Critical/Cultural Studies*, 7 (2), pp. 212–214.

Breakell, S., 2010. Encounters with the self: archives and research. In J. Hill (Ed.), *The future of archives and recordkeeping: a reader*. London: Facet, pp. 23–36.

Brockmeier, J., 2010. After the archive: remapping memory. *Culture & Psychology*, 16 (1), pp. 5–35.

Brothman, B., 2002. Afterglow: conceptions of record and evidence in archival discourse. *Archival Science*, 2 (3), pp. 311–342.

Buchanan, A., 2010. Strangely unfamiliar: ideas of the archive from outside the discipline. In J. Hill (Ed.), *The future of archives and recordkeeping: a reader*. London: Facet, pp. 37–62.

Buckland, M.K., 1991. Information as thing. *JASIS*, 42 (5), pp. 351–360.

Cook, T., 1997. What is past is prologue: a history of archival ideas since 1898, and the future paradigm shift. *Archivaria*, 43 (1), pp. 17–63.

Cook, T., 2001. Archival science and postmodernism: new formulations for old concepts. *Archival Science*, 1 (1), pp. 3–24.

Cook, T., 2011. The archive(s) is a foreign country: historians, archivists, and the changing archival landscape. *American Archivist*, 74 (2), pp. 600–632.

Cook, T., 2013. Evidence, memory, identity, and community: four shifting archival paradigms. *Archival Science*, 13(2–3), pp. 95–120.

Cox, R.J., 2000. *Closing an era: historical perspectives on modern archives and records management*. Westport: Greenwood Press.

Cox, R.J., 2001. *Managing records as evidence and information*. Westport: Quorum Books.

Craven, L. (Ed.), 2008. *What are archives? Cultural and theoretical perspectives: a reader*. Aldershot: Ashgate.

Delmas, B., 1992. Bilan et perspectives de l'archivistique francaise au seuil du troisieme millennaire. In O. Bucci (Ed.), *Archival science on the threshold of the year 2000*. Ancona: University of Macerata Press, pp. 81–109.

Dempsey, L., 2000. Scientific, industrial, and cultural heritage: a shared approach: a research framework for digital libraries, museums and archives. *Ariadne* 22. Available at: www.ariadne.ac.uk/issue22/dempsey.

Derrida, J., 1995. *Mal d'archive: une impression freudienne*. Paris: Galilee.

Douglas, J., 2010. Origins: evolving ideas about the principle of provenance. In T. Eastwood and H. MacNeil (Eds.), *Currents in archival thinking*. Santa Barbara: Libraries Unlimited, pp. 23–43.

Duchein, M., 1992. The history of European archives and the development of the archival profession in Europe. *The American Archivist*, 55, pp. 14–25.

Dunbar, A., 2006. Introducing critical race theory to archival discourse: getting the conversation started. *Archival Science*, 6 (1), pp. 109–129.

Duranti, L., 1999. Concepts and principles for the management of electronic records, or records management theory is archival diplomatics. *Records Management Journal*, 9 (3), pp. 149–171.

Duranti, L., 2010. Concepts and principles for the management of electronic records, or records management theory is archival diplomatics. *Records Management Journal*, 20 (1), pp. 78–95.

Ebeling, K. and Günzel, S. (Eds.), 2009. *Archivologie: Theorien des Archivs in Philosophie, Medien und Künsten*. Berlin: Kadmos.

Ernst, W., 2008. *Sorlet från arkiven. Ordning ur oordning.* Göteborg: Glänta.
Foucault, M., 2002. *The archeology of knowledge.* London: Routledge. L'Archeologie du savoir first published 1969 by Editions Gallimard.
Giddens, A., 1984. *The constitution of society: outline of the theory of structuration.* Cambridge: Polity.
Gilliland-Swetland, A.J., 2000. *Enduring paradigm, new opportunities: the value of the archival perspective in the digital environment.* Report 89. Washington, DC: CLIR.
Gitelman, L. and Jackson, V., 2013. Introduction. In L. Gitelman (Ed.), *"Raw data" is an oxymoron.* Cambridge: MIT Press, pp. 1–14.
Hartley, J., 2010. The probability archive: from essence to uncertainty in the growth of knowledge. In *The Internet Turning 40: The Never-Ending Novelty of New Media Research.* Chinese University of Hong Kong, Hong Kong, 17–19 June 2010.
Henttonen, P., 2007. *Records, rules and speech acts: archival principles and preservation of speech acts.* PhD Thesis, University of Tampere.
Hickerson, H., 2001. Ten challenges for the archival profession. *American Archivist*, 64 (1), pp. 6–16.
Hutchinson, B. and Weller, S., 2011. Archive time (Guest editors' introduction). *Comparative Critical Studies*, 8 (2–3), pp. 133–153.
Huvila, I., 2008. Participatory archive: towards decentralised curation, radical user orientation and broader contextualisation of records management. *Archival Science*, 8 (1), pp. 15–36.
Huvila, I., 2015a. Another wood between the worlds? Regimes of worth and the making of meanings in the work of archivists. *The Information Society*, 31 (2), pp. 121–138.
Huvila, I., 2015b. The unbearable lightness of participating? Revisiting the discourses of 'participation' in archival literature. *Journal of Documentation*, 71 (2), pp. 358–386.
Huvila, I., Uotila, K., Paalassalo, J-P., Huurre, J., and Veräjänkorva S., 2008. Passages to medieval archipelago: from mobile information technology to mobile archaeological information. In *Information and Communication Technologies in Tourism 2008: Proceedings of the International Conference.* Innsbruck, Austria, 2008. Berlin: Springer, pp. 336–347.
International Council of Archives, 2009. *Discover ICA. About records, archives and the profession*, Paris. Available at: www.ica.org/125/about-records-archives-and-the-profession/discover-archives-and-our-profession.html.
Jenkins, H., 2014. Rethinking 'rethinking convergence/culture'. *Cultural Studies*, 28 (2), pp. 267–297.
Jimerson, R.C., 2004. The future of archives and manuscripts. *OCLC Systems & Services*, 20 (1065–075X), pp. 11–14.
Johnson, V. and Williams, C., 2011. Using archives to inform contemporary policy debates: history into policy? *Journal of the Society of Archivists*, 32 (2), pp. 287–303.
Keough, B. and Wolfe, M., 2012. Moving the archivist closer to the creator: implementing integrated archival policies for born digital photography at colleges and universities. *Journal of Archival Organization*, 10 (1), pp. 69–83.
Ketelaar, E., 2000. Archivistics research saving the profession. *The American Archivist*, 63, pp. 322–340.
Körmendy, L., 2007. Changes in archives' philosophy and functions at the turn of the 20th/21st centuries. *Archival Science*, 7 (2), pp. 167–177.

Krause, M., 2010. It makes history alive for them: the role of archivists and special collections librarians in instructing undergraduates. *The Journal of Academic Librarianship*, 36 (5), pp. 401–411.

Krause, M. and Yakel, E., 2007. Interaction in virtual archives: the polar bear expedition digital collections next generation finding aid. *American Archivist*, 70 (2), pp. 282–314.

Lane, V. and Hill, J., 2010. Where do we come from? What are we? Where are we going? Situating the archive and archivists. In J. Hill (Ed.), *The future of archives and recordkeeping: a reader*. London: Facet, pp. 3–22.

Lidman, T., 2012. *Libraries and archives: a comparative study*. Oxford: Chandos Publishing.

Lucas, G., 2012. *Understanding the archaeological record*. Cambridge: Cambridge University Press.

Manoff, M., 2004. Theories of the archive from across the disciplines. *Portal: Libraries and the Academy*, 4 (1), pp. 9–25.

Manžuch, Z., 2009. Archives, libraries, and museums as communicators of memory in the European Union projects. *Information Research*, 14(2). Available at: www.informationr.net/ir/14-2/paper400.html.

McKemmish, S., 2001. Placing records continuum theory and practice. *Archival Science*, 1 (4), pp. 333–359.

McKemmish, S., Iacovino, L., Russell, L., and Castan, M., 2012. Editors' introduction to keeping cultures alive: archives and indigenous human rights. *Archival Science*, 12, pp. 93–111.

Menne-Haritz, A., 2001. Access – the reformulation of an archival paradigm. *Archival Science*, 1 (1), pp. 57–82.

Orrman, E., 2007. The archives and the archival profession in the Nordic countries. In M. Aubry, I. Chave, and V. Doom (Eds.), Archives, archivistes, archivistique dans l'Europe du Nord-Ouest du Moyen Âge á nos jours'. Villeneuve d'Ascq: IRHIS, pp. 231–238.

Prelinger, R., 2010. We are the new archivists: artisans, activists, cinephiles, citizens. In *Reimagining the Archive: Remapping and Remixing Traditional Models in the Digital Era*. UCLA, Los Angeles, United States of America, 11–14 November 2010.

Ribeiro, F., 2001. Archival science and changes in the paradigm. *Archival Science*, 1 (3), pp. 295–310.

Richards, T., 1993. *The imperial archive: knowledge and the fantasy of empire*. New York: Verso.

Robinson, L., 2007. Abdication or empowerment? User involvement in library, archives and records services. *Australian Library Journal*, 56 (1), pp. 30–35.

Shilton, K. and Srinivasan, R., 2008. Participatory appraisal and arrangement for multicultural archival collections. *Archivaria*, 63, pp. 87–101.

Stevenson, J., 2008. The online archivist: a positive approach to the digital information age. In L. Craven (Ed.), *What are archives? Cultural and theoretical perspectives: a reader*. Aldershot: Ashgate, pp. 89–106.

Stevenson, S., 2010. Michel Aglietta and regulation theory. In G. J. Leckie, L. M. Given, and J. Buschman (Eds.), *Critical theory for library and information science: exploring the social from across the disciplines*. Santa Barbara: Libraries Unlimited, pp. 1–13.

Sundqvist, A., 2007. The use of records – a literature review. *Archives & Social Studies*, 1 (1), pp. 623–653.

Synenko, J., 2013. Archive cultures: technicity, trace and metaphor. In M. Parrot and J. Derry (Eds.), *The everyday: experiences, concepts, narratives*. Newcastle-upon-Tyne: Cambridge Scholars Publishing, pp. 228–244.

Taylor, D., 2003. *The archive and the repertoire: performing cultural memory in the Americas*. Durham: Duke University Press.

Theimer, K. (Ed.), 2011. *A different kind of web: new connections between archives and our users*. Chicago: Society of American Archivists.

Thomassen, T., 2001. A first introduction to archival science. *Archival Science*, 1 (4), pp. 373–385.

Treanor, B., 2009. What tradition, whose archive? Blogs, googlewashing, and the digitization of the archive. *Analecta Hermeneutica*, 1(1), pp. 289–302.

Upward, F., 1997. Structuring the records continuum, part two: structuration theory and recordkeeping. *Archives and Manuscripts*, 25 (1), pp. 10–35.

Usherwood, B., Wilson, K., and Bryson, J., 2005. Relevant repositories of public knowledge?: libraries, museums and archives in 'the information age'. *Journal of Librarianship and Information Science*, 37 (2), pp. 89–98.

Yakel, E., 2011. Balancing archival authority with encouraging authentic voices to engage with records. In K. Theimer (Ed.), *A different kind of web: new connections between archives and our users*. Chicago: Society of American Archivists, pp. 75–101.

Yeo, G., 2007. Concepts of record (1): evidence, information, and persistent representations. *American Archivist*, 70 (2), pp. 315–343.

Yeo, G., 2010. Representing the act: records and speech act theory. *Journal of the Society of Archivists*, 31 (2), pp. 95–117.

Zeitlyn, D., 2012. Anthropology in and of the archives: possible futures and contingent pasts – archives as anthropological surrogates. *Annual Review of Anthropology*, 41 (1), pp. 461–480.

Zhang, J., 2012. Archival context, digital content, and the ethics of digital archival representations. *Knowledge Organization*, 39 (5), pp. 332–339.

Zielinski, S. and Winthrop-Young, G., 2015. AnArcheology for anArchives: why do we need – especially for the arts – a complementary concept to the archive? *Journal of Contemporary Archaeology*, 2(1), pp. 116–125.

Zipsane, H., 2009. Lifelong learning through heritage and art. In P. Jarvis (Ed.), *The Routledge international handbook of lifelong learning*. London: Routledge, pp. 173–182.

3 Imperfect imaginaries
Digitisation, mundanisation, and the ungraspable

Robert Willim

The relatively short history of network-connected digital technologies has been related to a number of imaginaries, some fantastic, evocative, and utopian while others nightmarish.[1] Visions about cyberspace, digital matrixes and electronic superhighways, augmented and virtual realities, and assemblages of humans, things, and machines, as well as autonomous smart technologies, have been evoked parallel with business appeals to purchase or upgrade to the latest model of specific products or to connect to services and systems. Bold and sometimes lofty business visions have been enmeshed with words and images promoted by technological evangelists, worlds of science fiction, and popular culture. Evocative imaginaries of the digital have been flourishing parallel with very prosaic, everyday actions of users by screens, keyboards, and other interfaces.

The question I depart from is how imaginaries evoking ungraspable and enchanting dimensions of digitisation have been related to everyday practice and the mundane. I will begin by elaborating on the concept of imaginaries, to then introduce the concept *mundanisation*, in order to explain how technological systems – at times imagined as ungraspable, enchanting, and sublime – are transmuted into the commonplace while being enmeshed in people's everyday practices. I will draw upon some examples that are all related to various research projects I have been working on in the last few decades. The aim is to discuss the interplay between evocation of imaginaries and mundanisation.[2]

The presented examples should be seen as illustrations to the relationship between imaginaries, mundanisation, and digitisation. Even if the examples' points of departure are placed in different times, they should not be read as a unified chronological story of digital technology development and cultural practices. I will however start with examples from the 1990s and move along to phenomena in the second decade of the 21st century. The examples and my argumentation illustrate a variation in the understandings of imaginaries.

Several of the examples brought up deal with imaginaries evoking potential futures. As a contrast, I will therefore utilise the potential of the imperfect verb form. How does it influence how we imagine things? If we describe even present things in past tense, as something that has already happened, a tension and a sort of distance is engendered. It will position the reader in a certain relation to the narrative and to empirical accounts, giving perspective on notions of change, turnover time, and temporality. The imperfect tense also implies that something can be continuous and progressive, even provisional, incomplete, and ever shifting, which harmonises with the way I outline and envision imaginaries.

Imaginaries

Any attempt to fully define a broad concept like imaginaries is preordained to disappoint or even fail in an epic way. My intention here is to briefly outline the concept in a way useful for cultural analysis and empirical studies of digitisation. I will intentionally advocate keeping the very concept of imaginaries somewhat fuzzy, open-ended, and not too distinct and delineated. The point of departure is how imaginaries have been approached from disciplines like social and cultural anthropology and ethnology. This said, my own background in the theoretically and empirically eclectic practices of European ethnology in Sweden is based on an open approach to disciplinary boundaries, which makes me reluctant to rigidly frame the discussion within a specific academic discipline (Ehn & Löfgren 2010: 217ff). Imaginaries, as I use the concept, will therefore be tinted by the ways it has been utilised in a number of contexts, but I will not inscribe it in social or cultural anthropology, science and technology studies (STS), sociology, or any other major discipline. Instead, I will maintain an in-betweenness and provisionality that I argue also characterises the way imaginaries should be understood.

When it comes to imaginaries, a particular fuzziness is its ambiguous locus as either an individual or social feature. They can range from being extraordinarily inconspicuous to wide-ranging. Imaginaries can emerge through activities of daydreaming (Ehn & Löfgren 2010: 123ff). They can also be coupled to extensive constructs like religion and belief systems. Imaginaries and imagination can be seen as a universal human trait or capacity, which all human beings deal with. Mark Harris and Nigel Rapport use this broad characterisation in the preface to the volume *Reflections on Imagination: Human Capacity and Ethnographic Method* (2015). According to them, imagination '…is a common practice, something to which human beings attend whenever they make sense of their environments and situate their life-projects in these environments: a human facility' (ibid: xiii). This sense-making quality of imaginaries has also been advocated in a number of writings on social imaginaries. A groundbreaking work in this vein was *Imagined Communities* (1983) by Benedict Anderson, who wrote about the way technologies like the printing press, new businesses, standardisation of time, as well as systems of dissemination and mediation of information were crucial for the way nationalism became possible. According to Anderson, nations can be understood as entities held together partly by social imaginaries. In the mid-1990s, Arjun Appadurai, inspired by Anderson, extended the concept in his book *Modernity at Large* (1997) by writing about imagined worlds and imaginaries of a postcolonial, deterritorialised world of globalisation and global flows (1997). Another acclaimed scholar who has been inspired by the thoughts of Anderson is Charles Taylor, who introduced *Modern Social Imaginaries* (2003) as capacities that *make sense* for people, for example, by instilling moral order, thereby enabling the practices and organisation of a society.

In all of these examples, imaginaries are not about escapism and make-believe. Instead, they are intimately coupled to intentions, tangible actions, expressions, and events in the world. By focusing on these accounts, imaginaries can be seen as something that primarily homogenises, holds together, and forms congruity in social worlds and people's lives. This predilection is also characterising the concepts *technoscientific* and *sociotechnical imaginaries*, which can both be associated with various strands of STS and other endeavours to interweave the social with technological development. Sociotechnical imaginaries have been defined by Sheila Jasanoff as: 'collectively held, institutionally stabilised, and publicly performed visions of desirable futures, animated by shared understandings of forms of social life and social order attainable through, and supportive of, advances in science and technology' (Jasanoff 2015: 4). The normative and aspirational dimensions of sociotechnical imaginaries makes them comparable with vision statements in business management. Organisations and corporations can describe their visions and strategic intents as:

> A clearly understood statement of the direction in which a firm intends to develop. It should be both understood and interpreted by each employee in relation to their work and is a crucial element in the strategic management of a firm.
>
> (strategic intent 2009)

Vision statements by organisations are more specifically defined and more precise than the wider sociotechnical imaginaries. The concepts have different genealogies, implications, and usages, but are both about the aspirational, 'collectively held', 'institutionally stabilised', and normative.[3]

The focus on shared understandings and the normative, as well as its description as collectively held visions of desirable futures, makes sociotechnical imaginaries into something that homogenises and unifies groups of people. This homogenising dimension of imaginaries has been criticised for being more or less a substitution for the concept of *culture*. When culture has become less used as a fruitful analytical concept, imaginaries has been said to replace the word as something holding societies, communities, and groups of people together. Claudia Strauss writes: 'to a certain extent *the imaginary* is just *culture or cultural knowledge* in new clothes' (Strauss 2006: 322, italics in original). Sneath, Holbraad, and Pedersen have explained this feature in the anthropological usage of imaginaries as a specific difficulty (2009). A holistic notion of imaginaries has been part of a move through which 'a fixed totality of explicit meanings ("culture") has been substituted with a fluid totality of implicit ones ("the social imaginary")' (ibid: 8). They also present two other difficulties. One is the way that imaginaries are defined as something primarily purposeful, as something that can '...fulfil a certain purpose, whether in terms of social function or existential potential' (ibid: 9). Instead, they advocate that imaginaries should be seen as something more indeterminate.

The third difficulty or shortcoming they outline is a romantic tendency to approach imaginaries as something primarily positive, an autonomous force of creativity or even wonder. They stress that imaginaries could also be understood as something more disturbing, as part of '...dystopian potentials of imaginative engagement' (ibid: 19). The question is why this would even have to be stated. Good or bad morals, paranoia, hope, dreams, or nightmares, the eerie as well as the reasonable, the seditious, and the calming could all be encompassed by imaginaries, depending on how we value and understand the human condition in various contexts. There is no scheme, diagram, or flowchart of imaginaries. They might not be autonomous and totally out of control, but neither are they easily framed, logical, nor programmable. Imaginaries are instead provisional and imperfect.

A notion of imaginaries that has to do with a certain closure, and which diverges a bit from the approaches just mentioned, is the particular *building* of imaginary worlds. In his extensive book *Building Imaginary Worlds: The Theory and History of Subcreation* (2012), Mark J.P. Wolf deals with the ways imaginary worlds, like the ones induced by J.R.R. Tolkien and George Lucas, have been generated through different media. He starts the book by framing the scope:

> Since the advent of daydreaming, imaginary worlds have drawn us away vicariously to fantastic realms culled from endless possibilities. The allure of such wayward speculation, conjuring new wonders, strange terrors, and the unexplored byways of beckoning vistas, has grown stronger over time along with our ability to render them into concrete forms, albeit mediated ones. Books, drawings, photographs, film, radio, television, video games, websites, and other media have opened portals through which these worlds grow in clarity and detail, inviting us to enter and tempting us to stay, as alive in our thoughts as our own memories of lived experience.
>
> (Wolf 2012: 1f.)

Wolf's book delves into the ways possible worlds are made through processes of subcreation (a concept he takes from Tolkien's elaborations on imaginary worlds). He writes about primary and secondary worlds, where the first is the lived world and the second the imaginary constructed world (e.g., the universe of Star Wars). Built imagined worlds can house stories, beings, places, and events that can be experienced by people through various media, ranging from books to computer games (see also Neumann & Zierold 2010). This kind of world-building requires specific choices and a certain degree of detail. In the worlds of a computer game or a film, landscapes and characters need definite form and texture. Some kind of mapping is required. Things will have to unfold according to certain laws and patterns. But not even these kinds of constructs are totally completed. They mutate as long as people engage in them, and moreover, in a built imaginary world there's always a horizon with an unspecified beyond.[4]

The imaginary worlds of fiction are enmeshed with the ways several other imaginaries unfold. Imaginaries as I use them here, however, stretch beyond built constructs. I use imaginaries as open-ended, indeterminate, as capacities that are impossible to complete. In the empirical contexts I will present, they are mutable capacities that mediate between the ungraspable and the mundane. I will partly draw inspiration from the ways that Vincent Crapanzano has elaborated on *Imaginative Horizons* and how imaginaries can be seen as frontiers, as elusive boundaries that never can be transgressed or reached (2004). However, these imaginaries have a constant influence on the way reality is perceived, approached, and handled, and on the way practices are spawned. According to Crapanzano, frontiers, unlike borders and boundaries, cannot be crossed or transgressed. They make a change in ontological register. 'They postulate a beyond that is, by its very nature, unreachable in fact and in representation' (ibid: 14). He stresses what lies beyond the horizon and the possibilities it suggests, 'the licit and illicit desires it triggers, the plays of power it suggests, the dread it can cause – the uncertainty, the sense of contingency, of chance – the exaltation, the thrill of the unknown it can provoke' (ibid: 14). The elusiveness, the ephemerality of imaginaries, is crucial to the way I use them in relation to my examples of digital phenomena and the ways digitisation can be approached. There is something *beyond* the use of any fairly sophisticated technology that energises or sometimes disturbs practices, but which remains teasingly out of hand, challenging any attempt to final resolution.

Mundanisation

All attempts to catch, describe, utilise, or reify imaginaries (or the beyond) lead to a shift or change in them. 'Our images, dream projections, calculations, and prophecies may give form and substance to the beyond, but as they do, they destroy it; for, as they construct it, they assure its displacement' (Crapanzano 2004: 14). The empirical examples I will present in this chapter all relate to notions of imaginaries and the beyond. I will start describing how imaginaries about elusive digital frontiers and the spatial construct cyberspace emerged during the end of the last century, and how they were enmeshed in the practices of Internet-related business. These imaginaries somewhat resemble sociotechnical imaginaries including normativity and an aspirational direction. I will then go on to relate imaginaries to economic dimensions and practices of using networked technologies. Example by example, the indeterminateness of imaginaries will be increased. I will relate imaginative frontiers to suggestive visualisations and stories about remote and intricate infrastructures, about disturbing as well as fascinating intangible frontiers that have characterised processes of digitisation. I will now, however, bring up an important counterpoint to how the ungraspable is imagined.

Since the technologies and systems described in this chapter all characterise complex assemblages of humans, technology, spaces, materialities, objects, and flows, it is not surprising that they can be coupled to evocative imaginaries. It is maybe even more intriguing that those enormously complex assemblages, when used by people in everyday settings, become part of the mundane and the commonplace.

Geographer Nigel Thrift has pointed out that: 'All human activity depends upon an imputed background whose content is rarely questioned: it is there because it is there. It is the surface on which life floats' (Thrift 2007: 91). Utterly complex systems and technologies are enmeshed in people's lives, and when technology and infrastructure works, it is experienced as a backdrop to life, as uncomplicated ingredients of daily practices. Interconnected digital technologies housing embedded software became entangled with life in several societies during the decades around the year 2000. These entanglements might be coupled to extensive imaginaries, but they might also be part of mostly unremarked and overlooked patterns, movements, and structures of everyday life (cf. Ingold 2013a, Hodder 2016).

Since the 1990s, the word 'domestication' has been used to describe how technology is incorporated in users' everyday life (see Berker et al. 2006). Introduced by Roger Silverstone in 1992, domestication theory captures how technology is adopted, how negotiations take place, and even how users may influence future strategies of producers. But the question is, to what extent is technology use about anything wild becoming domesticated or tamed? What are the ends of a technology? We still need a word for the processes through which unfathomable complexity, sometimes mediated through an artefact, is turned into the ordinary, a word saying something about how a protective layer between overwhelming, even ominous, complexity and commonplace everyday life is engendered when it comes to technology use. I suggest the word *mundanisation*.[5] The way I use it is to capture how unfathomable, complex arrangements of technologies and human organisation are transmuted into the ordinary, the mundane, the commonplace in people's everyday lives. Mundanisation might even be seen as a basis for a sane life in a world of all-encompassing, dizzying, complex systems. It is about conceptually creating a thin shielding membrane, hiding the incomprehensible, engendering what metaphorically could be described as the everyday empire of *Mundania*.⁶ Mundanisation is the process through which the ungraspable is ignored and how imaginary frontiers are (at least for some while) overlooked.

Ephemeral digital frontiers

At the end of the 1990s, I started studying the emerging business of Internet consultancies, particularly in Sweden. These fast-growing companies were soon labelled dotcoms, and they were imagined to be part of a new economy, which was seen as a novel way to organise and conduct business in a digitally networked world. Proponents of these businesses referred to older industry as 'dinosaurs' or 'respirators'. They also promoted societal investments in new infrastructure, and the need to act fast in order to not be sidestepped in the race towards the future. Jonas Birgersson, the CEO of the Swedish Internet consultancy Framfab that I studied between 1998–2001, was called 'Broadband Jesus' because of his visions and rhetoric about the coming of a new world of the Internet (see Willim 2002, 2003b). Birgersson promoted certain imaginaries, and his efforts can be compared to those of other technological evangelists, who pitched and endorsed real or imagined products, systems, and recommendations for action. For some years, a certain imaginary had been characterising the rhetoric and understanding of businesses like Framfab. This imaginary was based on a new digital spatiality and about ephemeral digital or electronic frontiers.

In countries like Sweden, between the 1950s and 1980s, the digital had often been associated with engineering, large-scale systems, and the possibilities of programming and computability (Lundin 2012). During the coming decades, computers were transformed. The personal computer, graphical user interfaces, and later Internet-based services made computing and the digital part of the lives of larger parts of the public. The complexity of the systems had increased immensely. It became harder than ever to grasp how software-based interconnected technologies really worked.

At this time, imaginaries about an ephemeral digital space grew stronger. Ideas like cyberspace and virtual reality had been either derived from or strengthened by the imaginations of popular culture, art, and science fiction. Science fiction writer William Gibson's use of the concept cyberspace in his novel *Neuromancer* was one strong reference (1984). Several other popular cultural products were invoked as inspiring visions for upcoming IT businesses and promoters of new technologies (Flichy 2007).

The networked future was envisioned as virgin land that was up for grabs. A colonial pioneer rhetoric, characterised by frontier myths resembling discourses from 'The Wild West' of the North American 19th and early 20th centuries, was used as a reference and metaphor now applied in the world of digital technology. It stemmed from contexts mainly in California and the Bay Area around San Francisco. The magazine *Wired* and the social environment around it was a nexus of cyberspatial imaginaries. In his account of *Wired* and its founder Louis Rossetto, Gary Wolf compared the dynamics around the ideological (and ontological) keystones of the magazine as a romance, a kind of evocative constellation of affects and ideas that emerged at this time:

> Original ideas often appear unexpectedly in several places at once; like ghosts, they are in the air. Once such ideas begin to spread, they can have an uncanny, almost demonic effect, causing otherwise rational people to act strangely. The story [of *Wired*] that follows is a romance in exactly this sense: it traces the effect of a fantastic idea – the idea that computers will make every existing authority obsolete – as it worked through and upon the man who conjured it up.
>
> (Wolf 2003: xi)

A number of actors and organisations that influenced global imaginaries about the potential of a growing Internet were, in one way or another, connected to *Wired* and the people around the magazine. So was John Perry Barlow, a former rancher and lyricist and also one of the founders of EFF (Electronic Frontier Foundation). In 1996, he wrote his famous *A Declaration of the Independence of Cyberspace*. The manifesto starts by Barlow positioning himself in opposition to 'the old world':

> Governments of the Industrial World, you weary giants of flesh and steel, I come from Cyberspace, the new home of Mind. On behalf of the future, I ask you of the past to leave us alone. You are not welcome among us. You have no sovereignty where we gather. We have no elected government, nor are we likely to have one, so I address you with no greater authority than that with which liberty itself always speaks. I declare the global social space we are building to be naturally independent of the tyrannies you seek to impose on us. You have no moral right to rule us nor do you possess any methods of enforcement we have true reason to fear.
>
> (Barlow 1996: para. 1)

The declaration is clearly techno-liberal and particularly tinted by North American frontier ideas (cf. Gouge 2007). It is about freedom from earlier structures, institutions, and authorities. In some paragraphs, the ephemerality of cyberspace is stressed: 'Cyberspace consists of transactions, relationships, and thought itself, arrayed like a standing wave in the web of our communications. Ours is a world that is both everywhere and nowhere, but it is not where bodies live' (Barlow 1996: para. 6). This imaginary, captured in words by Barlow, characterised the framing of a number of practices, from dotcoms around the turn of the millennium to hacker movements and initiatives like The Pirate Bay some decade later (cf. Larsson 2013).

The character of this techno-liberal imaginary could be readily related to the way Jasanoff and others outline sociotechnical imaginaries as aspirational and normative constructs (Jasanoff & Kim 2015). However, I will now present how imaginative frontiers could be understood as becoming gradually blurry, when they intersect with processes of mundanisation and when the imaginaries become ever more indeterminate. By the end of the chapter, the directions pointed out by the likes of Barlow will be harder to recognise.

Escape velocity and mundane practice[7]

During the years up until 2000, imaginaries about the rise of a cyberspatial networked reality were merged with ideas of a new economic logic, especially in Europe and North America. Several commentators advocated that societies were entering what was called 'the new economy' (Kelly 1999, Tapscott 1999). The Industrial Society had been declared obsolete already in the 1970s by several commentators, scholars, and from different political perspectives (Toffler 1970, Bell 1973, Touraine 1974). Heavy industries were of course still prevailing at the end of the century, but they were seldomly brought into visions about future possibilities. Instead, the Western world was supposed to fully enter the Information or Network Society, the era of the Internet and networked digital infrastructures (Castells 2000).

When I did fieldwork at the Internet consultancy Framfab from 1998 to 2001, the ephemeral digital was used as a kind of prerequisite for building up a certain kind of business rhetoric based on need for speed. The companies that were supposed to thrive in the rising world of digital networks and a new economy would have to be speedier and bolder than their predecessors. Like pioneers staking out land, while moving towards the imaginary frontier, they should move ahead, onward and upwards, boldly towards the unknown where no one had stepped before. They should be able to leave earlier structures, industries, and stakeholders behind through a kind of escape velocity towards the new digital frontier. Framfab explicitly referred to high speed by calling themselves a fast company and by incorporating the fast forward symbol in their logotype. They stated that there were no models for the new operations pursued by the venture, so the company had to set its own standards and take the role of pioneers (Willim 2003a).

While these imaginaries of a networked ephemeral frontier circulated and influenced the way the entire endeavour of Framfab was understood, what took place in the workplaces was strikingly mundane. Mostly young people worked by computers in an office of IKEA furniture, playful props, and some paraphernalia evoking 'geeky' creativity. Cyberspace and the virtual digital frontier were imagined to take place beyond the glowing screens with HTML code and colourful graphics. The speed so often alluded to was hard to sense and experience. A relative tranquillity prevailed as people were normally absorbed by the gentle actions of fingers on keyboards and eyes locked at occurrences on screens. The digital frontier was not just beyond reach in the everyday practices of the office, it was hidden by the commonplace mundanity of digital labour. What had taken place even in these so-called fast companies was a process of mundanisation.

The imagined next technology

There were occasions when the relative tranquillity of the office was stirred up. When the network connections at some occasion stopped working, people had to rise from their workplaces and the movement in the office space increased. People started to talk louder and a nervous energy spread. While troubleshooting as well as speculation about what could be wrong was taking place, the protective layer of 'Mundania' was temporarily dissolved. It became apparent how the computer workstations were dependent on invisible infrastructures, stretching far beyond the office.

Most of the time, infrastructure is located in the backdrop of everyday life. It can be defined as 'the basic physical and organisational structures and facilities (e.g., buildings, roads, power supplies) needed for the operation of a society or enterprise' (Oxford Dictionaries 2015). Nicole Starosielski, who has studied negotiations between visibility and invisibility of critical infrastructures, proposes that: 'Whether infrastructure is materiality hidden or simply ignored, invisibility has been naturalised as its dominant mode of visuality' (2012: 39f.). Through its integration in routinised behaviour, infrastructure is seldom visible, and to a large extent, everyday life for most people takes place where infrastructure is designed to be outside the centre of attention. But there is definitely no clear limit between visibility and invisibility, especially not when infrastructures or technologies are brought to attention and related to various imaginaries. At Framfab, there were a number of occasions when imaginaries were evoked among the web programmers and designers at the office in Lund, the city that I mostly studied. It could be when things stopped working and when the infrastructural dimensions of the digital had to be approached. When something peculiar or strange occurred in the daily computer practices, some users had to enter areas that were beyond their own competencies. And when their own knowledge and experience were not enough, external expertise had to be summoned. Troubleshooting challenged the invisibility and the whereabouts of infrastructure. System logs with cryptic messages had to be scrutinised and interpreted in order to comprehend what was taking place in certain situations. Opaque software assemblages of daemons and algorithms doing invisible work were brought to attention and interpreted. Sometimes, problems were related to hardware, wirings, actions, and connections far beyond the office space. The workers had to imagine in order to decide what and who to trust. The membrane that shielded the routinised everyday practices of Mundania was challenged and imaginaries about 'the beyond' poured in.

Imaginaries could also trickle into Mundania during the most prosaic of events. While the digital was associated with speed and with a coming future, everyday computer practices were often characterised by waiting. At the end of the 1990s, it took several minutes to start a computer. Progress bars and other ways to visualise for the user that they had to wait while the system was processing some task were common ingredients of computer practice at Framfab, like in other environments where computers were used (Willim 2003b). While this everyday waiting took place, future models of technology, products, and inventions were promoted as much faster than the ones available. Product pre-announcements and the early constitution of consumer expectations have been characterising IT business for decades.

Evocative imaginaries about upcoming fascinating products that would be speedier than ever transformed the very practices of waiting in front of the screen. The technology at hand seemed even slower and the waiting felt longer when related to imagined technologies beyond grasp (Willim 2003b: 133f.). The boundaries of Mundania were constantly challenged by the work of imagination, summoning a possible future where users could co-mingle and interact with new technologies and products.

Software industry has been characterised by product cycles involving frequent upgrades. First, 1.0 is followed by 1.1, and after some time a larger upgrade will result in 2.0 and so on. The upgrading logic, and its supposed progress, engenders recurring expectations through which imaginary upcoming upgrades puts the existing technology in a different light. In situations of involuntary waiting, or when even smaller glitches and peculiarities in functionality emerge, an imagined future version might be longed for. The longing and desire for imagined futures are fuelled by pre-announcements of products by different producers. The imaginary comes jostling for attention, entwining mundane action with imagined possible worlds.

The upgrading logic evokes a potential one-way track to a possible future frontier. The future could be speedier than the present – upgraded, a version 2.0. But the single direction was often challenged. There were imaginaries that blurred the possible frontier. Some pre-announced future technologies happened to be more elusive and challenging than others. These were called vapourware, since they were virtually discernible but almost hidden in a hazy possible future (Willim 2003b: 132, Atkinson 2013). A number of IT companies have announced upcoming products that have never reached users and consumers. Often, they remained by some imaginary frontier that could be located in very different futures. However, in their pre-announced imaginary form, they also transformed the experience users had of existing media and technology.

Before Framfab was deconstructed during the years after the great stock market plunge in 2000, they also produced some vapourware. At a special event in Stockholm in February 2000, just some months before the rapid decline, the CEO entered the stage of a movie theatre in a spectacular way. He sported a hockey jersey in Sweden's national colours, and while talking about the bold future of Framfab and how Sweden should beat the Americans in Internet business and digital broadband-based services, he dropped a brick tile through a pane of glass. The message was 'Brikks breaks Windows'. He was alluding to an upcoming product of Framfab called Brikks. It was a coming infrastructure for networked services that the company would launch. It would possibly render operating systems like Windows obsolete.

Brikks was launched later on, but not in the revolutionary form announced. The disruptive game changer of IT business remained by the imaginary frontier, in a hazy, never realised world. Fifteen years after the presentation in Stockholm, it did however still exist. It did not break Windows, but it was the core of the product portfolio of the company Labs[2] and was called a 'complete business support system for broadband services' (What is Brikks? 2015).[8] However, it was a product in the periphery of global Internet business, especially compared to products launched by other major players who emerged after the fall of the dotcoms. In the first decade of the second millennium, these companies presented products and services that came to challenge the way users understood digital environments, operating systems, digital spaces, the elusive ends of a computer, and the imaginary frontiers of digitisation.

The clouds

The year 2000 was the climax of visionary cyber-imaginaries and rhetoric about a new economy connected to dotcoms. In the spring of that year, the speculative economic bubble burst. Stock values plunged and a huge restructuring of IT business took place. While companies like Framfab disappeared from the centre stage of Internet business, a new colourful stakeholder appeared from the rubble of the dotcom period.

Californian company Google had avoided the worst economic disaster of 2000, probably because it was still too small and not part of the volatile stock market at that time. It was maybe saved by not being too fast. During the coming years, Google became a giant in the search engine business, and by acquiring a number of smaller companies, it grew into other parts of digital business. Google also became one of the major creators of emerging networked services referred to as located in 'The Cloud'.

The cloud was provided through a converged infrastructure based on networked technology connected through scalable and dynamic structures. Services from companies like Google, Facebook, and Amazon, to name some of the major providers, were based on a so-called cloud architecture and were introduced and in growth during the first decade of the millennium. Also, companies like Microsoft and Apple, as well as a large number of business-to-business cloud providers, built up vast infrastructures to keep up their own clouds. These digital entanglements were not called cyberspace in the rhetoric anymore, but the clouds could still be imagined as an ungraspable void, only reachable by mediated interaction through customised user interfaces of computers, smartphones, and other kinds of devices.

It's easy to think of the Internet as an ethereal, otherworldly place. Even the way we talk about it speaks to its inherent everywhereness and

nothingness – it's all "wireless" this, "in the cloud" that. We used to reach the Web by sitting down at a large, slow-moving machine and waiting while screeching noises let us know our phone lines were being put to work. Now we carry the Internet around in our pockets, we have Wi-Fi on planes, and we can Instagram photos from the middle of nowhere.
(Berman 2012: para. 1)

These were the first sentences of a *Washington Post* review of the book *Tubes – A Journey to the Center of the Internet* by Andrew Blum (2012). The book details a trip to a number of places showing the materialities of the Internet. Data centres, network hubs, Internet exchanges, landing points of transoceanic cables, and a number of other locations that are critical points in the infrastructure of the Internet. This was part of another imaginary world of digitality emerging beneath or beyond the cloud. It was, of course, very real in its materiality, but for the users of services, it became part of an immense, evocative, and ungraspable complex system that was connected to the mundanity of everyday uses of digital technology. A small caress on a touchscreen could generate processes in networks and infrastructures in ways almost incomprehensible.

Evoking trust through distant tangibility

During the 1990s (characterised by cyberspatial frontier imaginaries), companies aiming their businesses at broader markets and the public did not often evoke the tangibility of the Internet. Major companies like Google had concentrated their marketing and branding on what people experienced mainly on screens and through other user interfaces. When production was exhibited it was mostly profiled as an innovative practice, taking place in cool and tastefully designed offices by clever and creative (mostly young) people. The backbone of the Internet was held out of attention in campaigns and promotions of products. In 2012, however, Google decided to launch a visualisation and information campaign, through which the infrastructure behind their products was presented. This was the first time that Google's production environments beyond their office complex in Mountain View were highlighted and pointed out as part of the company's public image. Now the 'nuts and bolts' of the Google verse should be displayed. On a website, onlookers were 'welcomed inside' (Google 2015). During the last years, the website was coupled with certain campaigns: 2010 was 'Renewable energy', 2011 'External certification', 2012 was 'Transparency'.

> Showing what we're made of – inside and out. For the first time, we're giving everyone a glimpse into where we run our products. Now you can take a virtual tour of one of our data centers in Street View. Or browse a photo gallery of the technology, people, and places that keep Google's products running.
>
> (ibid: para. 1)

Figure 3.1 Servers at Google's Douglas County, Georgia, data centre.[9]
Source: Google 2016. Google and the Google logo are registered trademarks of Google Inc., used with permission.

Evocative images from their facilities were presented and coupled to stories and information that should make people trust the business of the corporation (Figure 3.1). The facilities appear tangible, well organised, yet very distant and unreachable. In the book *Tubes*, Blum takes the reader to some of the places making cloud-based services possible. One of the places he travelled to was Google's data centre in Dalles, Oregon. For Blum, the place was mysterious, generating associations to spirituality and strange powers.

> I had come to Dalles because it is home to one of the Internet's most important repositories of data, as well as being the de facto capitol of a whole region devoted to storing our online selves. The place struck me as a kind of Kathmandu of data centers, a foggy town at the base of a mountain that happened to be the perfect jumping-off point for an exploration of the massive buildings where our data is stored. Even better, the Dalles was evocative and mysterious enough – a natural nexus – to highlight these buildings' strange powers. A data center does not merely contain the hard drives that contain our data. Our data has become the mirrors of our identity, the physical embodiment of our most personal facts and feelings. A data center is the storehouse of the digital soul. I liked the idea of data centers tucked away up in the mountains like wizards – or perhaps warheads. And Kathmandu felt right in another way: I was looking for enlightenment: for a new sense of my digital life.
> (Blum 2012: 229)

Imperfect imaginaries

Blum notes that data centres seemed to be everywhere, as part of infrastructures holding representations of our most intimate elements. Looking at Google's map of some of its facilities, they seemed to be widely dispersed in countries like Germany, Belgium, and Chile. Likely, all the other cloud service providers' data centres must have been located in numerous places, ranging from the warehouse around the corner to structures hidden faraway in secret rock shelters. They were secluded, due to security reasons, housed in inconspicuous buildings as well as in spectacular brandscapes (Klingmann 2007). But in the everyday use of networked technologies, they have also been hidden behind the invisible walls of Mundania. Like industries producing toilet paper, door knobs, or wall paint, they have seldom been reflected upon in people's everyday life. But sometimes, facilities, infrastructures, and 'the whereabouts of data' might be brought to attention and incorporated in mind-boggling imaginaries. Blum's account is one example, but there are others.

Imagine industries

Google was not the only company displaying its facilities, even if its campaign stood out as especially conspicuous. With this campaign, the visibility of Google was also inscribed in a longer visual tradition, spanning back in time to early industrialism, a tradition evoking certain imaginaries. Several of the pictures from data centres that Google presented in 2012 resemble how early industry and technological structures gained cultural significance and became part of movements of aestheticisation and enchantment over 100 years ago.

In *American Technological Sublime*, David E. Nye shows how large technological endeavours like the railway, electrically lightened cityscapes, or early industrial facilities became the focus of fascinated attention. Awe-inspiring and fascinating, slightly frightening encounters with places like the Niagara Falls or the Grand Canyon had been described as sublime. '…the sublime is a mental state caused by our inability to fathom the power, vastness, magnitude, and magnificence of an object witnessed' (Peeples 2011: 379). According to Nye, a new kind of sublime grew in strength with the rise of the industrial society. This sublime was fuelled by the awe and amazement over the scale and power of human achievements reached through the rationality of engineering and technological advances. What were brought up, as parts of the enchantment, were the vastness, the symmetry, and the dynamic rationalism that could be witnessed and associated with factories and mills:

> Most visitors to early textile mills testified to the powerful impression created by many rows of spinning machinery, each containing uniform lines of spindles that became a visual metaphor for the promised cornucopia of industrial production. (In later years photographers often exploited such repetition in depicting the mills.) In an age when one seldom

saw more than one or two machines together at one time, the view of a large factory room humming with incessant activity created astonishment at the ingenuity and apparent perfection of the arrangements.

(Nye 1996: 114f.)

Mostly, onlookers and visitors not working in the very places experienced the enchantment of factories. For workers, a textile or steel mill of the early 20th century was part of a routinised (probably arduous) everyday life, in which the possible awe-inspiring and fascinating qualities of technologies and arrangements were mundanised. For them, the factory was a place of labour more than a landscape that could be associated with the sublime. 'Outsiders might perceive it as a form embodying certain abstract ideas, but to the laborer the factory was a place of action' (ibid: 116).

Visiting a factory and experiencing it as a place of sublime experience required some distance to the facility. This distance was made even broader through the photographs and stereoscopic images of industry that became popular in the early 20th century. The ways factories were pictured in these has a number of striking resemblances with Google's visualisations of their facilities 100 years later. Long lines of machinery or equipment stretch out towards the horizon, creating an impression of visual depth and the repetition of elements, and evoke how man-made structures reach out towards some distant imagined point.[10]

The stereoscopic images from the steel works in Pittsburgh show long steel bars, machinery, and repeated architectonic elements, creating the impression of an almost infinite factory hall (Figure 3.2). The image showing cooling pipes from one of Google's data centres has a similar composition (Figure 3.3). Here, the winding, colourful pipes also give the

Figure 3.2 Steel works, Pittsburgh, PA.[11]
Source: Wikimedia commons.

Figure 3.3 Pipelines at Google's Douglas County, Georgia, data centre.
Source: Google 2016. Google and the Google logo are registered trademarks of Google Inc., used with permission.

impression of vastness and a rationalised complexity. Google also brought in the bright colours from their logo into the space, creating a playful and imaginative contrast to the robustness and rigor of industry. In order to communicate ingenious rationality paired with simplistic logic, the company described how the blue pipes are for cool water on its way to cool down the servers, while the red pipes are for heated water coming back from the server hall.

Despite the difference in colour scale, it is obvious that there is a resemblance between early visualisations of industrial grandeur and Google's way to make facilities visible a century later. In one of Google's data centres pictured on their website, the colours and constructions of the newer systems collide with the patinated structures of earlier industrial structures. This data centre was located in a former paper mill in Hamina, Finland. In the images, some of the structures and signs from the early heavy processing industry of the facility are juxtaposed with rows of servers tinted by glowing lights (Figure 3.4).

The data centre located in the former paper mill also appeared in a *Wired* article in 2012. Steven Levy, an acclaimed journalist based in the United States, wrote about the facilities under the heading *Where Servers Meet Saunas*. He described how Google started large-scale server operations in unexpected or interesting places. The article is an example of an inclination

Figure 3.4 Servers at Google's Hamina, Finland, data centre.
Source: Google 2016. Google and the Google logo are registered trademarks of Google Inc., used with permission.

among influential people at the time to describe cloud industries of companies like Google as part of an exotic geography that opened up imaginary vistas about 'other places'. In one passage, Levy was seduced by the possible exoticness of the Nordic country: 'There were geographic anomalies in building a data centre in wintry Finland. At one point, the team working on the fibre connection to Helsinki encountered a polar bear' (Levy 2012: para. 9). There are no polar bears between Hamina and Helsinki in southern Finland. The landscape characterised by woodland is quite far from any polar regions. I do not know about the origin of the anecdote, but Levy, as well as editors and staff of *Wired*, must have been fascinated by the imaginary wild frontiers evoked by the supposed testimony from a faraway country. The anecdote also hints at the circumstance that influential accounts, fuelling certain imaginaries, are often evoked and dispersed from quite specific cultural contexts characterised by biases and certain preconceptions.

Imagined industries engender a plethora of imaginaries, where any frontier might be harder to pinpoint. Internet industries become related to earlier accounts of the industrial sublime and to specific geographic places that are either hard to reach or depicted as fascinating in some sense. The ephemeral digital spaces from earlier Internet industries are still somewhat looming in the background. The question is how to now understand the whereabouts and status of imaginary frontiers of digitisation.

Beyond the digital

The visualisation efforts by Google did not randomly occur in 2012. Part of the reason might be a broader interest in the physical and tangible matters behind digital services and the Internet. Google's campaign began the same year as Blum's book on Internet infrastructures was published and, at about the same period of time, academic interest to research the social and cultural aspects of materiality of the digital started to grow (see van den Boomen et al. 2009, Dourish & Mazmanian 2011, Starosielski 2011, 2012, Reichert & Richterich 2015, Pink et al. 2016). In 2012, the book *The Art of the Data Center: A Look Inside the World's Most Innovative and Compelling Computing Environments* by Douglas Alger was released. The Internet had also become dispersed beyond computers and designated devices like smartphones. What was called the Internet of Things opened up a new scope for imaginaries, and bodies were linked to digital systems through trackers and sensors of various kinds.

In 2015, *Alphabet* was created as Google's parent company. It should be the start of a conglomerate that could invest and innovate in a number of areas beyond Internet operations that once had characterised Google – health and life sciences, artificial intelligence, the development of drones, autonomous cars, automated homes, and humanoid robots. All this and more could be contained under the Alphabet umbrella. While developing and experimenting in an extremely broad manner, Google continued to offer services for free, if they could just get hold of the data. In 2015, users were offered the opportunity to upload all their photos and videos into Google's network of services. There would be free, unlimited storage in the cloud – free, at least, to some extent. Monetary cost was substituted with other potential, more indeterminate costs. Alluring offers to upload as much (personal data) as possible into the networks of service providers like Google kept recurring.

Alphabet was merely one corporate player in a field of evolving innovation, through which the digital became almost all encompassing. Almost everything could be possibly interconnected, part of a digitised industrial infrastructure. A once-imagined ephemeral cyberspace had not just become infrastructural and connected to structures like data centres, satellites, and transoceanic cables, it had spread into human bodies, homes, and lifeworlds in ungraspable ways. The once-ethereal digital was enmeshed with the very core of human life. The most intimate and private could become dispersed in ungraspable networks. How to make sense of, comprehend, and analyse this scenario of potential risks, benefits, and trajectories into the future? Imaginaries of the digital could become imaginaries about almost everything.

These entanglements also reflected the way scholarly and scientific accounts played out at the time. Media, including processes of digitisation, were brought beyond notions of deep time and analysed as part of geology and large-scale global processes, most notably by Jussi Parikka in *A Geology of Media* (2015). Worlds of particles and dust were related to dispersed processes of networked digital systems and the geopolitical. 'Now a geopolitical turn is happening that takes into account that data have a material and legal territory and that we can speak of geophysics of information' (ibid: 25).

Media was also explained as something more than message-bearing institutions like radio, TV, or the Internet. In *The Marvelous Clouds: Toward a Philosophy of Elemental Media* (2015), John Durham Peters argued that we should understand the digital in relation to water, earth, fire, and air. According to him, media 'are vessels and environments, containers of possibility that anchor our existence and make what we are doing possible' (Peters 2015: 2). Part tongue in cheek, he introduced the term *infrastructuralism* as a kind of subsequent hybrid of structuralism and post-structuralism (ibid: 33). Hereby, the elements, geology, and the infrastructural became enmeshed with imaginaries about media, technology, and digitisation.[12]

Along the way, with scientific accounts and various suggestive and haunting imaginaries being evoked, people had to pursue everyday lives characterised by digitisation. They could not constantly worry or try to continuously imagine the consequences and conditions of evolving possible worlds. Increasingly complex technological assemblages were mundanised and housed in people's everyday lives within the often messy routines and habits of Mundania. Like the workers in the factories David E. Nye wrote about, or the workers at Framfab for that matter, they had to overlook the potentially sublime and ungraspable and approach a digitised everyday life as a place of action (Nye 1996: 116).

An imperfect finale

What I've learned of my work with imaginaries of digitisation is to not too easily define the boundaries and directions of certain imaginaries. At certain times, a kind of conceptual congruity might occur, which seems to outline the contours of an imaginary with a specific direction and transformative power (Willim 2003b). This appeared during the years of the dotcom bubble. Then, speed and the move into the ephemeral digital seemed to be the whole lot. The electronic cyberspatial frontier was the goal towards which businesses were to strive. Rapid growth and an upgrading ethos based on the longings for a possible coming escape velocity that would leave history behind, as well as a fetishisation of the new and the potential of specific technologies, prevailed for some years. It soon became more complex as the digital was absorbed into lifeworlds and infrastructures in new ways. Imaginary frontiers remained out of grasp.

Another lesson is that the recurring force of mundanisation is especially powerful. Even the most spectacular imaginative scenarios and constructs might be hidden behind the protective membrane of Mundania. Mundanisation seems to protect against the imaginative onslaught of dizzying complex systems and the possible anxiety that attempts to understand how things *really* work might generate. The ungraspable is generally ignored in everyday life and imaginary frontiers are often overlooked.

At last, to fruitfully deal with imaginaries in cultural analysis, we have to tolerate ambiguity and even blur as epistemological dimensions (see Ingold 2013b). Meanwhile, it might be rewarding to study how attempts to transmute complex indeterminate imaginaries and unreachable frontiers develop (Willim 2014). How is the indeterminate given form? Which practices deal with the imaginary? It is often when the imaginary gets a shape or direction that it is possible to get a hint even of the ungraspable. It can be in the form of the rendered worlds of science fiction, through stories, or in specific vision statements and strategies of organisations. How are imaginaries transmuted into renditions and actions, into schemata, plans, figures, and procedures, and when are they obscured by mundanisation? Imaginaries definitely offer potential for cultural analysis, even if they are ever imperfect, provisional, and mutable.

Notes

1 Some parts of this chapter are based on the paper *Floating Points of Reference*, which was presented at 'Beyond the frame: the future of the visual in an age of digital diversity', Nordic Network for Digital Visuality (NNDV) conference in Stockholm, April 2014. It was financed by The Pufendorf Institute at Lund University. Further work with the text has been financed by SCACA (Swedish Centre for Applied Cultural Analysis) at Halmstad University, Sweden
2 For the concept of mundanisation, see endnote 5.
3 In an anthropological elaboration on how designers approach development of so-called 'smart technologies', Débora Lanzeni separates the role of imaginaries and visions of the future in design processes. She suggests that 'imaginaries are connected with a broader global future and have less significance in the design, whereas the visions of future are informed by the local (quotidian) experience of the designers and have a crucial role in the materialisation of design projects' (2016: 456). Lanzeni defines the social formation of imaginaries as vital for conceptualising designs (ibid: 59). The way I use imaginaries in this chapter is not limited to conceptions of possible futures or designs, instead they have a more imprecise, and even imperfect character and temporality. They might be *blurry* projections of possible futures, but also affectively conditioned ideas about the present world beyond what is cognitively graspable.
4 Worth mentioning here is the World Building Institute (WBI). It was formed in 2008, and furthermore became part of the USC School of Cinematic Arts, University of California, Los Angeles. It was described by its director Alex McDowell as a: 'non-profit Organized Research Unit dedicated to the dissemination, education, and appreciation of the future of narrative media through World Building' (World Building Institute 2016: para 1). The generation of fictive worlds described by Wolf, as well as the ones associated with WBI in Los Angeles, can be related to wider discussions on world-making (Nünning et al. 2010), as well as to discussions on speculative design, how (probable, plausible, or possible future) worlds, artefacts, and constructions might be imagined, and the role between design and science fiction (Sterling 2005, Dourish & Bell 2013, Dunne & Raby 2013, Galloway 2013).

5 The word *mundanisation* has been used in various contexts before. Edmund Husserl, for example, used it to describe how humans constitute themselves as entities or subjects in the world (Moran & Cohen 2012: 215). It has also appeared in discussions on disenchantment or in relation to the ways knowledge work has become bland when moved between contexts (Bell 2007).
6 I use the notion of Mundania in a partly speculative way, to see how far it can be used as part of methodological experimentation and as a cultural analytic probe (cf. Willim 2013; Lury & Wakeford 2014; Vannini 2015).
7 Escape velocity is a concept derived from physics, defining the minimum velocity that is required from a body (like a spacecraft) to escape the gravitation from another massive body (like a planet). It was also used in the title of a book by Mark Dery with the subtitle *Cyberculture at the End of the Century* (1997).
8 Labs[2] was a company that sprung from the former Framfab enterprise. When Framfab was diminished and acquired by other stakeholders after the dotcom-burst, its former CEO Jonas Birgersson, together with some of the founders, moved to what was in 2015 the company Labs[2].
9 Figures 3.1, 3.3, and 3.4: Google, 2016. The tech. Data centers. Available at: www.google.com/about/datacenters/gallery/#/tech.
10 These ways to evoke the industrial sublime can also be seen in other visual examples, most notably in the photos of Edward Burtynsky. He has used similar compositions as in the early examples of the industrial sublime to picture large-scale production environments, but also beautiful yet frightening motifs from exploited environments of the extraction and mining industry. A good example is his book *China* (2005), in which he presents images of Chinese manufacturing facilities and industrial environments; see also *Manufactured Landscapes* (Burtynsky & Pauli 2003). The photos by Burtynsky are thought-provoking acquaintances to the visualisations of Google infrastructures and the imaginaries they might be related to.
11 Licensed under the Wikimedia commons 'public domain' license. Wikimedia Commons. Steel works, Pittsburgh, PA, beam of hot iron in rolling mill, drawn out (00?) feet long. Available at: https://commons.wikimedia.org/wiki/File:Steel_works,_Pittsburgh,_Pa.,_beam_of_hot_iron_in_rolling_mill,_drawn_out_(00%3F)_feet_long,_from_Robert_N._Dennis_collection_of_stereoscopic_views.png.
12 Peters took a wide and thought-provoking grip on media, combining theories and empirical examples beyond disciplinary boundaries. Peters argued that 'the notion of elemental media advanced in this book is more than an interdisciplinary gesture; it is also a bid in a long philosophical, religious, and political debate about the nature and location of meaning' (2015: 380). Another work with similar ambitions as Peters, published the same year as *The Marvelous Clouds*, was *The Stack: On Software and Sovereignty* by Benjamin H. Bratton. *The Stack* was, according to Bratton, an unapologetically interdisciplinary work, combining 'political philosophy, and architectural theory, and software studies, and even science fiction' (Bratton 2015: xvii). By addressing clouds, minerals, climate, cities, and more, Bratton articulated a project of 'geodesign' by discussions on planetary-scale computation and geopolitical realities (ibid: xix).

References

Alger, D., 2012. *The art of the data center: a look inside the world's most innovative and compelling computing environments*. Westford: Prentice Hall

Anderson, B., 1983. *Imagined communities: reflections on the origin and spread of nationalism*. London: Verso.

Appadurai, A., 1997. *Modernity at large*. New Delhi: Oxford University Press.
Atkinson, P., 2013. *Delete: a design history of computer vapourware*. London: Bloomsbury Academic.
Barlow, J.P., 1996. A declaration of the independence of cyberspace. Electronic Frontier Foundation. Available at: www.eff.org/cyberspace-independence.
Bell, D., 1973. *The coming of post-industrial society: a venture in social forecasting*. New York: Basic Books.
Bell, D., 2007. Fade to grey: some reflections on policy and mundanity. *Environment and Planning A*, 39 (3), pp. 541–554.
Berker, T., Hartmann, M., and Punie, Y., 2006. *Domestication of media and technology*. Maidenhead: Open University Press.
Berman, M., 2012. Review: Tubes – a journey to the center of the Internet. *The Washington Post*, 2 November 2012.
Blum, A., 2012. *Tubes: a journey to the center of the Internet*. New York: Ecco.
Bratton, B.H., 2015. *The stack: on software and sovereignty*. Cambridge: MIT Press.
Burtynsky, E., 2005. *China*. Göttingen: Steidl.
Burtynsky, E. and Pauli, L., 2003. *Manufactured landscapes: the photographs of Edward Burtynsky*. New Haven: Yale University Press.
Castells, M., 2000. *The rise of the network society: the information age: economy, society and culture*. Hoboken: Wiley.
Crapanzano, V., 2004. *Imaginative horizons: an essay in literary-philosophical anthropology*. Chicago: The Chicago University Press.
Dery, M., 1997. *Escape velocity – cyberculture at the end of the century*. New York: Grove Press.
Dourish, P. and Bell, G., 2013. Resistance is futile: reading science fiction alongside ubiquitous computing. *Personal and Ubiquitous Computing*, 18 (4), pp. 769–778.
Dourish, P. and Mazmanian, M., 2011. *Media as material: information representations as material foundations for organizational practice*. Paper presented at the Third International Symposium on Process Organization Studies, Corfu, Greece.
Dunne, A. and Raby, F., 2013. *Speculative everything: design, fiction, and social dreaming*. Cambridge: MIT Press.
Ehn, B. and Löfgren, O., 2010. *The secret world of doing nothing*. Berkeley: University of California Press.
Flichy, P., 2007. *The Internet imaginaire*. Cambridge: MIT Press.
Galloway, A., 2013. Emergent media technologies, speculation, expectation, and human/nonhuman relations. *Journal of Broadcasting & Electronic Media*, 57 (1), pp. 53–65.
Gibson, W., 1984. *Neuromancer*. New York: Ace Books.
Google, 2015. Data centers. Available at: www.google.com/about/datacenters/.
Gouge, C., 2007. The American frontier: history, rhetoric, concept. *Americana: The Journal of American Popular Culture* (1900 to Present), 6 (1). Available at: http://www.americanpopularculture.com/journal/articles/spring_2007/gouge.htm.
Harris, M. and Rapport, N., 2015. *Reflections on imagination: human capacity and ethnographic method*. Farnham: Ashgate Publishing.
Hodder, I., 2016. *Studies in human-thing entanglement*. Available at: www.ianhodder.com/books/studies-human-thing-entanglement.
Ingold, T., 2013a. *Making: anthropology, archaeology, art and architecture*. Abingdon: Routledge.
Ingold, T., 2013b. Dreaming of dragons: on the imagination of real life. *Journal of the Royal Anthropological Institute*, 19 (4), pp. 734–752.

Jasanoff, S., 2015. Future imperfect: science, technology, and the imaginations of modernity. In S. Jasanoff & S.-H. Kim (Eds.), *Dreamscapes of modernity: sociotechnical imaginaries and the fabrication of power.* Chicago: The University of Chicago Press, pp. 1–33.

Jasanoff, S. and Kim, S.-H. (Eds.), 2015. *Dreamscapes of modernity: sociotechnical imaginaries and the fabrication of power.* Chicago: The University of Chicago Press.

Kelly, K., 1999. *New rules for the new economy: ten radical strategies for a connected world.* New York: Penguin.

Klingmann, A., 2007. *Brandscapes: architecture in the experience economy.* Cambridge: MIT Press.

Lanzeni, D., 2016. Smart global futures: designing affordable materialities for a better life. In S. Pink, E. Ardèvol, and D. Lanzeni (Eds.), *Digital materialities. Design and anthropology.* London: Bloomsbury, pp. 45–60.

Larsson, S., 2013. Metaphors, law and digital phenomena: the Swedish pirate bay court case. *International Journal of Law and Information Technology,* 21 (4), pp. 354–379.

Levy, S., 2012. Where servers meet saunas: a visit to Google's Finland data center. *Wired,* 24 October 2012.

Lundin, P., 2012. *Computers in Swedish society: documenting early use and trends.* London: Springer.

Lury, C. and Wakeford, N., 2014. Introduction: a perpetual inventory. In C. Lury and N. Wakeford (Eds.), *Inventive methods: the happening of the social.* Abingdon: Routledge, pp. 1–24.

Moran, D. and Cohen, J., 2012. *The Husserl dictionary.* London: Bloomsbury Academic.

Neumann, B. and Zierold, M., 2010. Media as ways of worldmaking: media-specific structures and intermedial dynamics. In V. Nünning, A. Nünning, and B. Neumann (Eds.), *Cultural ways of worldmaking.* Boston: De Gruyter, pp. 103–118.

Nye, D.E., 1996. *American technological sublime.* Cambridge: MIT Press.

Nünning, V., Nünning, A., and Neumann, B. (Eds.), 2010. *Cultural ways of worldmaking.* Berlin: De Gruyter.

Oxford Dictionaries, 2015. Infrastructure. Available at: www.oxforddictionaries.com/de/definition/englisch/infrastructure.

Parikka, J., 2015. *A geology of media.* Minneapolis: The University of Minnesota Press.

Peeples, J., 2011. Toxic sublime: imaging contaminated landscapes. *Environmental Communication: A Journal of Nature and Culture,* 5 (4), pp. 373–392.

Peters, J.D., 2015. *The marvelous clouds: toward a philosophy of elemental media.* Chicago: The University of Chicago Press.

Pink, S., Ardèvol, E., and Lanzeni, D. (Eds.), 2016. *Digital materialities: design and anthropology.* London: Bloomsbury.

Reichert, R. and Richterich, A., 2015. Introduction – Digital material/ism. *Digital Culture & Society,* 1 (1), pp. 5–17.

Sneath, D., Holbraad, M., and Pedersen, M.A., 2009. Technologies of the imagination: an introduction. *Ethnos,* 74 (1), pp. 5–30.

Starosielski, N., 2011. Beaches, fields, and other network environments. *Octopus Journal,* 5, pp. 1–7.

Starosielski, N., 2012. Warning: do not dig. Negotiating the visibility of critical infrastructures. *Journal of Visual Culture*, 11 (1), pp. 38–57.
Sterling, B., 2005. *Shaping things*. Cambridge: MIT Press.
Strategic intent, 2009. In J. Law (Ed.), *A dictionary of business and management*. Oxford University Press. Available at: http://www.oxfordreference.com/view/10.1093/acref/9780199684984.001.0001/acref-9780199684984-e-6182?rskey=Zzz4JM&result=1 (accessed 29 October 2015).
Strauss, C., 2006. The imaginary. *Anthropological Theory*, 6 (3), pp. 322–344.
Tapscott, D. (Ed.), 1999. *Creating value in the network economy*. Boston: Harvard Business Review Press.
Taylor, C., 2003. *Modern social imaginaries*. Durham: Duke University Press.
Thrift, N., 2007. *Non-representational theory: space, politics, affect*. Abingdon: Routledge.
Toffler, A., 1970. *Future shock*. New York: Random House.
Touraine, A., 1974. *The post-industrial society: tomorrow's social history: classes, conflicts and culture in the programmed society*. London: Wildwood House.
Van den Boomen, M., Lammes, S., Lehmann, A-S., Raessens, J., and Schäfer, M.T. (Eds.), 2009. *Digital material – tracing new media in everyday life and technology*. Amsterdam: Amsterdam University Press.
Vannini, P., 2015. Non-representational research methodologies: an introduction. In P. Vannini (Ed.), *Non-representational methodologies: re-envisioning research*. London: Routledge.
What is Brikks?, 2015. Available at: www.labs2.com/brikks/about-brikks/a-complete-business-support-system-for-broadband-services.
Willim, R., 2002. *Framtid.nu: Flyt och friktion i ett snabbt företag*. Stockholm/Stehag: Östlings bokförlag Symposion.
Willim, R., 2003a. Tools for the electronic frontier: artefacts and associations in IT business. *Ethnologia Scandinavica*, 33, pp. 21–29.
Willim, R., 2003b. Claiming the future: speed, business rhetoric and computer practice in a Swedish IT company. In C. Garsten and H. Wulff (Eds.), *New technologies at work: people, screens and social virtuality*. Oxford: Berg.
Willim, R., 2013. Out of hand: reflections on elsewhereness. In A. Schneider and C. Wright (Eds.), *Anthropology and art practice*. Oxford: Bloomsbury.
Willim, R., 2014. Transmutations of noise. In B. Czarniawska and O. Löfgren (Eds.), *Coping with excess: how organizations, communities and individuals manage overflows*. Cheltenham: Edward Elgar Publishers.
Wolf, G., 2003. *Wired: a romance*. New York: Random House.
Wolf, M.J.P., 2012. *Building imaginary worlds: the theory and history of subcreation*. New York: Routledge.
World Building Institute, 2016. About WBI. Available at: http://worldbuilding.institute/about.

4 Ethnography of digital infrastructures

Gertraud Koch[1]

The number of things we surround ourselves with has grown enormously today, and their complexity also continues to increase – for instance, if we think of appliances such as televisions, cars, electric toothbrushes with integrated apps, and many more with a vast array of extensions of functions. Even more, though, things these days no longer stand for themselves but are increasingly integrated into infrastructures, without which some would not function at all. Telephones, radio, and television would be useless if the transmitters, cables, SIM cards, and so forth were not there to guarantee their functionality. More and more, however, other objects are also integrated into digital infrastructures; the car, for instance, always transmits via a 'direct line' to the manufacturer and the garage; a parcel can always be localised because of the close tracking of the delivery chain. The number of such infrastructures continues to rise and constitutes a central element in the everyday lives of many countries, people, and organisations. From the point of view of cultural analysis, they initially seem to be entirely meaningless. However, it is often such seemingly insignificant things on the outskirts of the shimmering trends that allow for interesting insights about cultural processes (Löfgren & Ehn 2010). Infrastructures are just such boring, however highly productive, objects of knowledge for cultural analysis.

Infrastructures distribute people, goods, and ideas in a highly efficient manner across cities, countries, or the world. Yet, even more, they constitute a material basis for everyday action and widespread social practices. Today's notions of cleanliness, grooming, and hygiene would hardly be thinkable if electricity, water supply, and waste removal were not as enormously elaborate as they are and if, in this part of the world, highly reliable infrastructures did not guarantee the preconditions for daily showers, uncomplicated laundry, and the removal of the manifold wastes of today's consumer society in an unproblematic and relatively effortless manner.

Infrastructures, it is becoming apparent, are a precondition for social practices of hygiene. They are not necessarily digital. However, an increasing digitalisation of analogous infrastructures can be recorded, and vice versa it can be said that digitality can only be realised so potently because of the expansion of infrastructures such as the Internet. Furthermore, the idea, the research, and the theoretical elaboration of the 'Ethnography of Infrastructures' (EoI) has principally arisen in the context of software development and the implementation of information systems in different work contexts (hospitals, information systems in different academic disciplines, and cultural institutions). Since the mid-1990s, central contributions in the elaboration of cultural analytical infrastructure research have been developed, with American research and technology sociologist Susan Leigh Star and information scientist Geoffrey Bowker being central instigators of this concept and having elaborated it further in close cooperation and co-authorship in an interdisciplinary team (Star 1996, Star & Ruhleder 1996, Bowker & Star 1998, Star 1999, 2002, Star & Bowker 2006, Edwards et al. 2009, Bowker et al. 2010).[2] With their relational understanding of technologies, practices, and forms of organisation, the EoI is to be conceived of as a part of science and technology studies (STS) and, as with most STS approaches, has been strongly influenced by Actor-Network-Theory (ANT); however, it is not synonymous, but constitutes an independent research tradition that has originated from the intensive dealings with infrastructures of information, partly due also to collaboration in their development and implementation.

What are digital infrastructures?

> People commonly envision infrastructure as a system of substrates – railroad lines, pipes and plumbing, electrical power plants, and wires. It is by definition invisible, part of the background for other kinds of work. It is ready-to-hand. This image holds well enough for many purposes – turn on the faucet for a drink of water and you use a vast infrastructure of plumbing and water regulation without usually thinking about it.
> (Star 1999: 380)

This general and somewhat superficial view of infrastructures can, upon closer observation, effortlessly be complicated, if one considers the social and cultural constitution of such material substrates. At a very fundamental level, infrastructures can thus already be characterised as composed of different elements – technological, organisational, and social, and thus as relational (Bowker et al. 2010). They are characterised by the following traits: (a) embedded into social connections (embeddedness), (b) transparent with respect to their use (transparency to use), (c) have a certain reach or scope (reach or scope), (d) must be learned in the sense of a community of practice in order to use them (learned as part of membership), (e) are linked with conventions in dealing with them (links with conventions of practice), (f) set certain standards and norms (embodiment of standards), (g) build onto existing infrastructures (built on installed base), (h) only become visible once they cease to function (becomes visible upon breakdown), and (i) are not global or holistic but are constructed incrementally and in layers (is fixed in modular increments, not all at once or globally) (Star 1999: 381f., cf. also Bowker et al. 2010: 98).

In this, infrastructures reconfigure space-time relationships where spatial dimensions and scopes of infrastructures can be placed on an axis of local to global, and where the relevance of historic references in the seemingly revolutionary development of digital infrastructures is always emphasised (Jackson et al. 2007). Furthermore, their sociotechnical character needs to be considered, that is, an axis from technical up to social dimensions of infrastructures. Their use requires the users to know, on the one hand, the technical standards and incorporate them, such as acquiring the necessary knowledge and skills in order to use an elevator, an escalator, or a paternoster. Even if incorporated knowledge here constitutes a marginal element and is quickly learned, it is nevertheless indispensable. On the other hand, different conventions have originated around this use of infrastructures: standing on the right side of the escalator so that those in a rush can walk past on the left, or the polite glance downwards or sideways so that the confined space of the elevator does not become too oppressive. The sociotechnical character of infrastructures is thus to be understood on an axis with two poles, with the incorporated technical standards on the one side and the social practices with their conventions on the other (Bowker et al. 2010: 101).[3] Technical and social aspects are closely interwoven in this and operate in union; the safeguarding of email inboxes by passwords serves as a good illustration of this mutual relatedness, as the technical limitation of access requires appropriate ways of dealing with its use – which are easily countered by a sticky note with the access codes next to the keyboard and other practices of sharing.

In the context of digital infrastructures and information structures (Bowker et al. 2010), the Internet prominently springs to mind as it constitutes a central, but not the only network of relevance here. Rather, there are also other networks for distributing information, which more or less all operate digitally as well. The Science Museum in London presents in its exhibition six selected networks as technical infrastructures that, according to the subtitle of the exhibition, changed the world: The Cable, The Telephone Exchange, Broadcast, The Constellation, The Cell, and The Web (Blyth 2014). These networks represent the ever-growing desire to command over more information and to communicate ever more immediately. All these networks are operated digitally today and function in a combination of hardware and software while possessing, due to their histories, very different characteristics for the transmission of information. The change from analogous to digital functioning means that the ontology of information infrastructures with their different media formats is changing to become more mutually convergent, their content more calculable and easier to store and process.[4]

Due to its topicality, however, the Internet is at the centre of research on digital infrastructures. Here, different emphases have emerged. These focus particularly on: (a) new forms of social life, which are facilitated and characterised by the technological means, (b) the social, ethical, and political values that are inscribed in the infrastructures, and (c) how knowledge work changes with the new information technologies (Bowker et al. 2010: 105). It is becoming apparent that research on digital infrastructures, which is to be understood in the sense of interconnected partial and special areas and which makes perfect sense in the face of the complexity of digital infrastructures, will continue to differentiate. Keywords in this area are Software Studies in the sense of individual systems for supporting work processes (e.g., Fuller & Goffey 2012), Mobile Media Studies (e.g., Goggin 2011), or by now also Platform Studies, which turn to the cultural and social analysis of social media platforms (e.g., Boellstorff 2008, Miller 2011, Wilken 2014). It is noticeable that the intensity with which work is actually undertaken relationally and with which an actual, multiple perspective view of the different elements of infrastructures is incorporated into the analysis varies. This is most likely due to capacity reasons.

Cultural and social dimensions of digital infrastructures

Governmentality, exclusion, and distinction

Infrastructures are not equally distributed across the globe. Particularly for digital infrastructures, there are strong asymmetries between centres and peripheries (Sassen 2002), on a national as well as a regional scale, as well as in the global North and South (Schnitzler 2013, Donovan 2015). In cities, many infrastructures converge. Accordingly, in the context of urban studies, there is a strong interest in the digital and the non-digital. In this 'infrastructure turn' in urban research (Graham 2010), the focus of the analyses lies on the circulations of people, goods, and ideas, as well as the questions of governmentality and power relations that become apparent in the infrastructures. Reference is made to cultural and political economies of infrastructures (Rodgers & O'Neill 2012, Wilken 2014), which not only enable social practices but also partially limit them for individual segments of the population or are used as a means of distinction – for instance, if access to public spaces is regulated via 'Software Sorting', that is, digital access codes in the shape of number pin codes, QR or bar codes, or if chip or NFC cards are required in order to enter them. Such options for accessing public spaces, which are regulated via software, can frequently be found at airports and subway stations, as well as banking facilities with automated teller machines and many other urban places (Graham 2005). User-related configurations of spaces, for instance in the shape of consumer-related advertising images in shopping centres, which are calculated based on already purchased contents of shopping bags, can be classed as 'Software Sorting'. In the attempt to capture the importance of the increasing implementation of digital information infrastructures in cities, three different perspectives are suggested: from a technology centred perspective, it is the substitution and transcending of places; from a cultural studies research tradition, it is the co-evolution of space and information infrastructures, as well as in the understanding of Actor-Network-Theory a recombination of locality (Graham 1998, Graham & Marvin 2002).

The concept of 'infrastructural violence' is intended to make these effects of neoliberal infrastructure politics describable from a different viewpoint and thus make them available for critique:

> What is very promising about the idealism of infrastructural violence, however, is its implicit steer towards practical, material recommendations that have the potential to improve dramatically the lives of those who are pressed upon by the city. Our hope is that by orienting future discussions about social suffering and the just city towards the notion of 'infrastructural violence', we will generate not just more talk about social responsibility but directed action towards its realization.
>
> (Rodgers & O'Neill 2012: 406)

Infrastructures in this sense are ascribed a central function for the resilience of urban neighbourhoods against the neoliberal economisation of social spaces, as political design can occur with and via infrastructures, and open access to these is an important factor (Sage et al. 2015). The design decisions and orientations at specific qualities, respectively norms in infrastructure development, are viewed as decisive in this, while an openness of infrastructures towards the use by people with different values is also postulated (Knobel & Bowker 2011).

Appropriation, idiosyncratic use, and breaking rules

In all this, infrastructures are not set once and for all and then just function; they are a constant process. Particularly, the appropriation of infrastructures is a necessary process in which these are adapted, as far as possible, to the user and thus modified accordingly. In this, context and creativity of the appropriation process are decisive aspects (Bar et al. 2016). How are infrastructures 'moved into', 'made habitable', 'maintained', and 'renovated'? These are questions that should be asked more frequently in research to then bring the social practices into view, which emerge with and around infrastructures and which can also shed light on missing matches, resistances, and impositions of infrastructures. Politics of infrastructures are by no means set but are evaded or reinterpreted via the breaking of rules of set standards, by the creative evasion of structures, or through the bending of rules (Latham & Wood 2015). Practices of regulation are thus partly overridden and partly trigger the development of new social practices. Hacking can be seen as one such purposeful breaking of rules in order to gain access to areas to which one has none or for which one does not fulfil the criteria (Nguyen 2016).[5] The idiosyncrasy with which the appropriation of infrastructures can occur also becomes comprehensible in the so-called Darknet or Deep Web, which has constituted itself on the Internet. Here, intensive efforts are undertaken to counter or entirely override the sociotechnical qualities of the Internet – transparency, openness, and easy access to information. Hardware and software precautions and organisational arrangements have been developed specifically for the purpose of safeguarding the use of the Internet in total anonymity. In order to reach this side of the Internet, certain router programmes and guides for how to proceed are necessary while the tracking procedures via cookies and so forth, which are common in the Internet, are overridden. The Darknet itself is split again into pages that can be found and others that have been screened entirely from others by their operator. With the Darknet, an infrastructure has emerged that is oriented at anonymity and freedom but is often criticised for this because nothing there can be traced and thus it has a certain attractiveness for criminals (Gehl 2014). It is becoming apparent how central sociotechnical ontologies of infrastructure are, and the questions arise – which positings are made here that then cause specific possibilities for action, and how are these positings then taken up, overridden, or renegotiated in different areas of work such as in journalism but also in regard of the global transfer of these technologies (Srinivasan 2013, Rodgers 2014).

Infrastructuring

In this context, the active, customised design of infrastructures is also brought up for discussion, particularly when thinking of Open Source and new forms of Commons in cities, which represent the possibilities for an open urban development policy. The postulate is that these should also be represented accordingly in infrastructures. The idea behind this perspective is mainly generated through Henri Lefebvre's 'right to the city' approach, which is extended to infrastructures here.

> Rather, the right to infrastructure allows us to escape the human–nonhuman and epistemology–ontology dichotomies altogether by opening up the agential work of infrastructures as a source (an open source) of possibilities *in their own right*.
>
> (Corsín Jiménez 2014: 343)

The idea of 'infrastructuring', that is, the active design and (co-)construction of infrastructures, is one that is strongly inspired by the research activity and the involvement of the authors of the EoI in the development of information systems (Bowker & Star 1998, Star & Bowker 2006). In this, 'infrastructuring' emphasises, on the one hand, the contribution of infrastructures to the solving of existential as well as social problems (Donovan 2015), as well as, on the other hand, the importance of participatory approaches via the inclusion of users into the design of infrastructures (Karasti 2014) so that they can function robustly in different social contexts (Karasti et al. 2006, Le Dantec & DiSalvo 2013). 'Infrastructuring' is understood here as a specific practice of knowledge that materialises specific social and cultural forms and supports the social practices connected with it and thus, to a certain degree, also sets social conditions. It is the architects of cultural analytical infrastructure research themselves, Susan Leigh Star and Geoffrey Bowker, who formulate this idea of 'infrastructuring'. They emphasise the culture-producing character of infrastructures.

> Information infrastructure is a great tool for distribution of knowledge, culture, and practice. Homesteading the space it has slowly opened out over the past two centuries involves building new kinds of community, new kinds of disciplinary homes, and new understandings of ourselves.
> (Bowker et al. 2010: 114)

For digital media, they undertake the attempt to formulate the central requirements in the building of digital infrastructure for those who still have little experience in this area. Here, they stress particularly the necessity of a prior theoretical understanding of infrastructures and the sustainability of infrastructures, which are to be adaptable and flexible for new requirements (Star & Bowker 2006: 241).

Experience, embodiment, augmentation

In the manifold developments and implementations of information infrastructures, which by now have established themselves in almost all areas of work and life, and considering that absence or abstinence equates to social, cultural, economic, and political marginality, it is one aspiration that these systems merge as seamlessly as possible into the respective everyday connections and take on the invisibility that is immanent to infrastructures. The design paradigms of information science – such as 'ubiquitous computing' (Ubicomp), 'pervasive computing', or of 'embodiment' in the sense of an increasing merging of information infrastructures with their users – stand for this aspiration (Dourish 2001, Dourish & Bell 2014). It becomes apparent in the shape of the Google Glasses or other so-called 'wearables' that, on the other hand, such as in the case of the Google Glasses, again led to significant debates and finally the retraction from the market, where cultural conventions such as privacy, in this case, are questioned too much. In the face of the dynamic development of the technological basis of digital infrastructures that we are currently experiencing, and which we can expect to continue for some time, a permanent reworking of them is underway, which is – in return – also experienced in everyday life and here demands a permanent reorientation in dealing with them, or conversely entails an adaptation of design paradigms, appliances, and applications for a better compatibility with social practices.[6] The intensive pervasion of work and lifeworlds with information infrastructures and appliances of all kinds – which are also largely driven by smartphones as small portable computers, but also by many other digital technologies and appliances (Koch 2012, 2013) – also implies a reorientation of perception and experience. Because digital technologies intervene between human perception and the world, they mediate these in a specific way, e.g., by showing images of the world that would never be visible as such beyond their technological generation (Ihde 1993), or also are the essential prerequisite for enabling sensory perception by adding a layer to real spaces in the sense of an 'augmented reality' (Milgram et al. 1994) and thus generating new ways of knowing (Bowker et al. 2010), for instance, when apps blend the past and the present in augmented realities in digital city tours.[7]

Operationalisation and empirical research

Infrastructures tend to remain in the background, unnoticed, yet they are well observable. For the empirical cultural analysis of infrastructures, a relational view of all elements of the infrastructures is necessary – that is, the composition of infrastructures of social, organisational, and technical components whilst taking into account the temporalities and processes of aging of infrastructures. In this, the relationships between these elements are of particular importance – that is, how technologies and organisational aspects, respectively business models and social practices, interact in order for the infrastructures to function. For it is only in the interplay of these different elements – the technological possibilities, the operation, and the distribution of these possibilities, as well as the users and uses – that infrastructures as such constitute themselves.

> This vision requires adopting a long term rather than immediate timeframe and thinking about infrastructure not only in terms of human versus technological components but in terms of a set of interrelated social, organisational, and technical components or systems (whether the data will be shared, systems interoperable, standards proprietary, or maintenance and redesign factored in).
> (Bowker et al. 2010: 99)

For such a methodological approach, it is necessary to inspect the 'backstage' area and thus to decipher the design logic of the respective infrastructure. 'Understanding the nature of infrastructural work involves unfolding the political, ethical, and social choices that have been made throughout its development' (ibid: 99). Geoffrey Bowker terms this analysis of the positings in the development process, which have become implicit in the infrastructure, 'infrastructural inversion' (Bowker 1994). Classifications and terminologies that are taken up and depicted in the infrastructure are of central importance in this (Star 1998, Lampland & Star 2009). The manner in which quantifications, classifications, and formalisations are undertaken, which attributions and valuations are connected with this, and according to which principles these form the world thus constitute important information for cultural analytical studies (Star 2002).

A further possible approach is to focus on the work and activities that are undertaken in the area of operation in which infrastructures are implemented. Frequently, this work remains invisible and only comes into the focus of attention when it cannot be represented in the infrastructures, yet it is nevertheless essential for the functioning of the processes and the social practices. The informal communication of nurses with patients in hospitals, for example, provides additional, often crucial information on their health status (Star & Ruhleder 1996, Star & Strauss 1999). Such fractures are particularly visible in the development and implementation phase of new information infrastructures or in their revision for new versions or updates.

Furthermore, the core concepts that Bowker et al. (2010: 106ff.) name also facilitate relevant analytical approaches, which, in the sense of 'sensitizing tools', orient the research approaches as heuristics in certain areas of research and here can mean very different things and therefore must be adapted to the specific research contexts.[8] These core concepts are:

a mediation, in the sense of technological mediation that addresses the manifold new dimensions of spatial reach and the connections, interweavings, and mediations between systems, networks, people, and organisations that are facilitated by infrastructures,
b process development, in the process of which new types of roles and responsibilities in work and life contexts emerge, such as that of the information manager, the digital librarian, and so forth, which in turn let characteristic aspects of infrastructure become visible,
c the stability of the system, i.e., everything that contributes to its functioning: the technology-related resources, maintenance, repair, and also, in regard to hardware, software, and staff, as well as the connected experiences with technology, its mediated character and the interfaces as points of intersection between man and machine,
d the social and political values that are obvious for the design of such systems and here determine, for instance, the ontologies of databases and are indeed also subject to controversy, such as in the case of ethnic classifications in the framework of the US census or also content on Wikipedia, and last but not least,
e temporal relationships and sustainability are a problem of data infrastructures when these, such as in academia, are based on strong discontinuities and different temporalities of organisational aspects, such as financing periods, staff availability, institutional contexts, technological aspects such as the constant innovation of approaches in data maintenance, and social practices such as in the course of methodological reorientations.

Beyond the decided interest in infrastructures, the EoI in general offers access to highly different questions from social and cultural anthropology, but particularly to those of processes of social order, social power, and hierarchies:

> As such, infrastructure emerges as an ideal ethnographic site for theorising how broad and abstract social orderings such as the state, citizenship, criminality, ethnicity and class play out concretely at the level of everyday practice, revealing how such relationships of power and hierarchy translate into palpable forms of physical and emotional harm.
> (Rodgers & O'Neill 2012: 402)

It is therefore very different points of access and approaches that can be taken up in an EoI, depending on the conditions in the respective research field and the analytical emphasis that is to be pursued. Also, very different scales can be chosen and thus, specific spatial and temporal sections of infrastructures can be set.

Perspectives on future research

It can be stated that, currently, only a little research is undertaken on social practice and the political economies of infrastructure (Bulkeley et al. 2014); however, these have the potential to tap into a rich field for cultural analytical research. One such particular need for research in regard to mobile infrastructures is formulated by Heather Horst (2013), who would particularly like to see research on practices of regulation in corporations and government from the point of view of the users – that is, research that starts at the organisational dimension of infrastructures in which she is particularly interested is the relations of governmentality and power. Further need for research is also formulated for the radio-frequency identification (RFID) technologies and the area of the 'Internet of Things' (Frith 2014), which deal with the tracking of people and objects or the limitations into certain social spaces, such as departure halls in airports, spaces in banks or buildings in general, subway stations, and so forth in 'software sorted realms' (Graham 2005).

The methodological challenges in this are certainly manifold. A central one will be working on the variety of simultaneously present and overlapping digital infrastructures.

> [W]hen the ethnographer is confronted with so many infrastructural systems in play in a field site, each of which contributes to action and interaction in a particular way, how can they locate 'where the action is?'
> (Vertesi 2014: 268)[9]

Also, dealing with overlapping infrastructures is a new and largely unexplored question. The concept of 'seamful spaces' (Vertesi 2014) – that is, spaces in which many different infrastructures are present simultaneously – initially formulates that such a juggling with different infrastructures exists as a practice. The combination and the breakdown, as well as the limitation of different digital infrastructures, video conferences, satellite images, PowerPoint, and so forth, constitute an everyday life that continuously has to be remastered by means of improvisation and flexibility. The necessity to improvise using different infrastructural options is likely to remain an everyday experience in the digital transmission of data. It refers to a set of problems that increasingly poses itself in such situations of overlap from the point of view of ethnography: the question of what constitutes co-presence when there is traffic on many channels simultaneously, how it can be established, and with which spectrum of mediated and unmediated forms can the research process be designed (Koch 2012).

Observations of infrastructure mostly concentrate on the analysis of the political and cultural economy of infrastructures and thus leave the social practices, which are connected with them, unconsidered. Particularly, the often cited relational perspectives on infrastructures, which bring together technological, organisational, and social dimensions, have so far found little reflection in theory and little implementation in empirical contexts. It is in this area in which we can hope for exciting research to emerge in the future.

Notes

1 Translated from the German by Dr. Stefanie Everke Buchanan.
2 Sadly and unexpectedly, Susan Leigh Star passed away in 2010, aged 56.
3 In this context, see the visualisation in Bowker et al. 2010: 101.
4 Digital, contrasted with analogous, means a representation of data onto discrete values and thus a specific quality of data that is electronically stored and processed in binary codes (zero and one). This technical fact has far-reaching consequences for the processing, transmission, and use of data.
5 See Chapter 5 by Murillo and Kelty on Hacking in this volume.
6 See Chapter 9 on Interfaces by Nishant Shah in this volume.
7 Available at: www.ibtimes.co.uk/london-streetmuseum-app-free-street-scenes-then-now-photos-1438209 (accessed 01 May 2016).
8 The term 'sensitizing tools' is here used in reference to Adele E. Clarke and Susan L. Star (2008: p. 118), while the reference to Herbert Blumer takes on particularly the situative applicability and thus a somewhat broader horizon of interpretation of the terms.
9 With the question 'Where the action is?', Vertesi refers to academic contributions of the interaction researcher Erving Goffman (1971) and information scientist and interface researcher Paul Dourish (2001).

References

Bar, F., Weber, M.S., and Pisani, F., 2016. Mobile technology appropriation in a distant mirror: baroquization, creolization, and cannibalism. *New Media & Society*, 18 (4), pp. 617–636.

Blyth, T. (Ed.), 2014. *Information age: six networks that changed our world*. London: Science Museum.

Boellstorff, T., 2008. *Coming of age in second life: an anthropologist explores the virtually human.* Princeton, Oxford: Princeton University Press.

Bowker, G.C., 1994. *Science on the run: information management and industrial geophysics at Schlumberger, 1920–1940.* Cambridge: MIT Press.

Bowker, G.C., Baker, K., Millerand, F., and Ribes, D., 2010. Toward information infrastructure studies: ways of knowing in a networked environment. In J. Hunsinger, L. Klastrup, and M.M. Allen (Eds.), *International handbook of Internet research.* Dordrecht: Springer, pp. 97–117.

Bowker, G.C. and Star, S.L., 1998. Building information infrastructures for social worlds – The role of classifications and standards. In T. Ishida (Ed.), *Community computing and support systems: social interaction in networked communities.* Berlin: Springer, pp. 231–248.

Bulkeley, H., Castán Broto, V., and Maassen, A., 2014. Low-carbon transitions and the reconfiguration of urban infrastructure. *Urban Studies*, 51 (7), pp. 1471–1486.

Clarke, A.E. and Star, S.L., 2008. The social worlds framework: a theory/methods package. In E.J. Hackett, O. Amsterdamska, M.E. Lynch, and J. Wajcman (Eds.), *The handbook of science and technology studies.* Cambridge: MIT Press in cooperation with the Society for the Social Studies of Science, pp. 113–138.

Corsín Jiménez, A., 2014. The right to infrastructure: a prototype for open source urbanism. *Environment and Planning D: Society and Space*, 32 (2), pp. 342–362.

Donovan, K.P., 2015. Infrastructuring aid: materializing humanitarianism in northern Kenya. *Environment and Planning D: Society and Space*, 33 (4), pp. 732–748.

Dourish, P., 2001. *Where the action is: the foundations of embodied interaction.* Cambridge: MIT Press.

Dourish, P. and Bell, G., 2014. *Divining a digital future: mess and mythology in ubiquitous computing.* Cambridge: MIT Press.

Edwards, P.N., Bowker, G.C., Jackson, S.J., and Williams, R., 2009. Introduction: an agenda for infrastructure studies. *Journal of the Association for Information Systems*, 10 (5), pp. 364–374.

Frith, J., 2014. Communicating behind the scenes: a primer on radio frequency identification (RFID). *Mobile Media & Communication*, 3 (1), pp. 91–105.

Fuller, M. and Goffey, A., 2012. Digital infrastructures and the machinery of topological abstraction. *Theory, Culture & Society*, 29 (4–5), pp. 311–333.

Gehl, R.W., 2014. Power/freedom on the dark web: a digital ethnography of the Dark Web Social Network. *New Media & Society*, doi:10.1177/1461444814554900, pp. 1–17.

Goffman, E., 1971. *Interaktionsrituale: Über Verhalten in direkter Kommunikation.* Frankfurt am Main: Suhrkamp.

Goggin, G., 2011. *Global mobile media.* Abingdon: Routledge.

Graham, S., 1998. The end of geography or the explosion of place? Conceptualizing space, place and information technology. *Progress in Human Geography*, 22 (2), pp. 165–185.

Graham, S., 2005. Software-sorted geographies. *Progress in Human Geography*, 29 (5), pp. 562–580.

Graham, S. (Ed.), 2010. *Disrupted cities: when infrastructure fails.* New York: Routledge.

Graham, S. and Marvin, S., 2002. *Telecommunications and the city: electronic spaces, urban places.* London: Routledge.

Horst, H.A., 2013. The infrastructures of mobile media: towards a future reseach agenda. *Mobile Media & Communication*, 1 (1), pp. 147–152.

Ihde, D., 1993. *Postphenomenology*. Evanston: Northwestern University Press.

Jackson, S.J., Edwards, P.N., Bowker, G.C., and Knobel, C.P., 2007. Understanding infrastructure: history, heuristics and cyberinfrastructure policy. *First Monday*, 12 (6). Available at: http://firstmonday.org/issues/issue12_6/jackson/index.html.

Karasti, H., 2014. Infrastructuring in participatory design. In H. Winschiers-Theophilus (Ed.), *Proceedings of the 13th participatory design conference: research papers – Volume 1*. New York: ACM, pp. 141–150.

Karasti, H., Baker, K.S., and Halkola, E., 2006. Enriching the notion of data curation in e-science: data managing and information infrastructuring in the long term ecological research (LTER) network. *Computer Supported Cooperative Work (CSCW)*, 15 (4), pp. 321–358.

Knobel, C. and Bowker, G.C., 2011. Values in design. *Communications of the ACM*, 54 (7), pp. 26–28.

Koch, G., 2012. Kybernetische Imaginationen. Zur Notwendigkeit einer virtuellen Ethnografie. In K. Braun and C. Schönholz (Eds.), *Umbruchzeiten: Epistemologie und Methodologie in Reflexion*. Marburg: MAKUFEE, pp. 144–159.

Koch, G., 2013. Studying heritage in the digital era. In M-T. Albert, R. Bernecker, and B. Rudolff (Eds.), *Understanding heritage. Perspectives in heritage studies*. Berlin: De Gruyter, pp. 169–182.

Lampland, M. and Star, S.L., 2009. *Standards and their stories: how quantifying, classifying, and formalizing practices shape everyday life*. Ithaca: Cornell University Press.

Latham, A. and Wood, P.R.H., 2015. Inhabiting infrastructure: exploring the interactional spaces of urban cycling. *Environment and Planning A*, 47 (2), pp. 300–319.

Le Dantec, C.A. and DiSalvo, C., 2013. Infrastructuring and the formation of publics in participatory design. *Social Studies of Science*, 43 (2), pp. 241–264.

Löfgren, O. and Ehn, B., 2010. *The secret world of doing nothing*. Berkeley: University of California Press. Available at: http://lup.lub.lu.se/record/1775039.

Milgram, P., Takemura, H., Utsumi, A., and Kishino, A.F., 1994. Augmented reality: a class of displays on the reality-virtuality continuum. In H. Das (Ed.), *Proceedings of SPIE Volume 2351 Telemanipulator and Telepresence Technologies*. Bellingham: SPIE Press, pp. 282–292.

Miller, D., 2011. *Tales from Facebook*. Cambridge: Polity Press.

Nguyen, L.U., 2016. Infrastructural action in Vietnam: inverting the techno-politics of hacking in the global South. *New Media & Society*, doi:10.1177/1461444816629475, pp. 1–16.

Rodgers, D. and O'Neill, B., 2012. Infrastructural violence: introduction to the special issue. *Ethnography*, 13 (4), pp. 401–412.

Rodgers, S., 2014. Foreign objects? Web content management systems, journalistic cultures and the ontology of software. *Journalism*, 16 (1), pp. 10–26.

Sage, D., Fussey, P., and Dainty, A., 2015. Securing and scaling resilient futures: Neoliberalization, infrastructure, and topologies of power. *Environment and Planning D: Society and Space*, 33 (3), pp. 494–511.

Sassen, S., 2002. *Global networks, linked cities*. New York: Routledge.

Schnitzler, A., 2013. Traveling technologies: infrastructure, ethical regimes, and the materiality of politics in South Africa. *Cultural Anthropology*, 28 (4), pp. 670–693.

Srinivasan, R., 2013. Re-thinking the cultural codes of new media: the question concerning ontology. *New Media & Society*, 15 (2), pp. 203–223.
Star, S.L., 1996. *The cultures of computing*. Oxford, Cambridge: Blackwell.
Star, S.L., 1998. Grounded classification: grounded theory and faceted classification. *Library Trends*, 47 (2), pp. 218–232.
Star, S.L., 1999. The ethnography of infrastructure. *American Behavioral Scientist*, 43 (3), pp. 377–391.
Star, S.L., 2002. Infrastructure and ethnographic practice: working on the fringes. *Scandinavian Journal of Information Systems*, 14 (2), pp. 107–122.
Star, S.L. and Bowker, G.C., 2006. How to infrastructure. In L.A. Lievrouw and S. Livingstone (Eds.), *Handbook of new media: social shaping and social consequence and consequences of ICTs. Updated student edition.* London: Sage, pp. 230–245.
Star, S.L. and Ruhleder, K., 1996. Steps toward an ecology of infrastructure: design and access for large information spaces. *Information Systems Research*, 7 (1), pp. 111–134.
Star, S.L. and Strauss, A., 1999. Layers of silence, arenas of voice. The ecology of visible and invisible work. *Computer Supported Cooperative Work*, 8 (1), pp. 9–30.
Vertesi, J., 2014. Seamful spaces: heterogeneous infrastructures in interaction. *Science, Technology & Human Values*, 39 (2), pp. 264–284.
Wilken, R., 2014. Places nearby: Facebook as a location-based social media platform. *New Media & Society*, 16 (7), pp. 1087–1103.

Part II
Doing digital culture

5 Hackers and hacking

Luis Felipe R. Murillo[1] and Christopher Kelty[2]

Introduction

Hackers seem to be everywhere today.

Fifty to twenty years ago, 'hacking' was an underground practice, associated with a particular politics and defining a set of individuals usually characterised as adolescent, white males obsessed with computers. Today, literally anyone could call themselves a hacker, or any action a 'hack'. In recent decades, we have experienced the extension of the term to encompass many ordinary technical practices in various domains, such as education, healthcare, humanitarian response, farming, parenting, bodily modification, among many others. In Silicon Valley, companies like Facebook describe a 'hacker way' to reference ordinary practices of coding, engineering, or entrepreneurship. In public relation campaigns, hackers are described simply as 'doers' since 'hacking just means building something quickly or testing the boundaries of what can be done' (see Figure 5.1, Funders and Founders Notes 2016). According to this very generous definition, we can all be called 'hackers', since we have been making artefacts for, at least, 2 million years with the creation of mode-1 stone tools (Clark 1961).

We have also witnessed a proliferation and globalisation of hackfests, hackathons, hackerspaces, and gatherings around the symbol of hacking for a myriad of purposes, including but not limited to the design of open hardware devices in medical settings, the corporate-sponsored challenge of reinventing the soda fountain with support from big companies such as Coca-Cola, the collective work around public data sets to fight corruption or help with issues of public administration, or simply coming up with an inventive use of a product that is about to be released in the market. Hacker conventions – like conferences in academic settings – serve the purposes of exchanging knowledge and bringing small and fringe groups of computer aficionados together to find peers, in order to share questions and findings that emerge when playing and working with information and communication systems (Coleman 2010b). In the contemporary, many 'hacker marathons' have been organised by companies to identify programmers for hire, offering comparatively small sums as prizes for new software or hardware

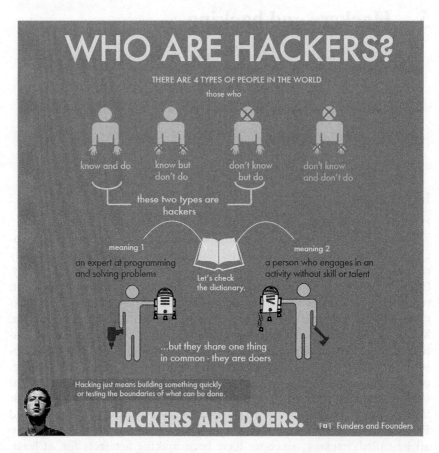

Figure 5.1 'Hackers are Doers' Funders and Founders website.
Source: Funders and Founders Notes 2016.

solutions that would necessarily cost considerably much more in research and development. A new business has been created around the work of headhunting for software engineering talent under the rubric of the 'hackathon', despite its original connotation of a community-led sprint for developing technologies.

These facts suggest that the figure of hacking and hackers has become fundamental to a contemporary technopolitical imaginary. Whatever hacking is, it is a very appealing figure and explanation for something. Hacking has become both a rubric for something very general, while at the same time designating a very specific set of practices from specific genealogies. In this chapter, we explore some of the ways hackers and hacking have been studied by academics, as well as the forms of self-narration that different hacker groups have themselves forged. We argue that the subjectivities involved in cultivating 'hacking skills' are not implicated in the range of things that can be called 'hacks' – or put differently: these days, not all hacks are perpetrated by hackers. We then ask what the relationship is between *hackers*, as a particular elaboration on what it means to be a 'technical person', and *hacking*, as a particular practice. We end by suggesting one possible way to decompose hacking into a 'stack' of practices that can be used to diagnose technical and political thresholds indicative of a mutation in the 'topology' of power in the world today. To put it differently, there is a reason why hacking seems to be spreading everywhere: because the forms and affordances of technical and political power are themselves changing, and hackers are at one forefront of experimenting with such mutations.

Stories of hackers and hacking

From the early 1950s, experimentation with communication and computing systems to the present-day hacker activist initiatives in the Global North and South, the narratives of hacking have been given different genealogies, supporting different positionings with respect to who and what counts as a legitimate expression of hacking. Most of these influential narratives have been provided by journalists and self-designated hackers, but canonical scholarly works generally focus on the 'culture' of computing – and not specifically on hacking: Sherry Turkle (1984, 1995); Diana Forsythe (2001); David Hakken and Barbara Andrews (1993); Lucy Suchman (1987); S.L. Star and K. Ruhleder (1996); Stephen Helmreich (1998); Joseph Dumit (2004); and Gary Downey (1998). At the most basic level, this literature has helped to address the so-called 'myth of autonomous technology', which presupposes a modernist ontology that separates it from human culture and society (Winner 1978, Latour 2008). Pfaffenberger (1992) pioneered the anthropological study of sociotechnical systems in this tradition. In his studies of digital technology (such as the Usenet and the personal computer), he argued for the processual analysis of technological design as invariably embedded in cultural systems. With the exception of Pfaffenberger and Turkle, little of this scholarly work is directly focused on hackers or hacking as such, but nonetheless constitutes some of the most significant ethnographic studies of computing in the English language.

The journalist Steven Levy is one of the most authoritative sources on the early history. His mid-1980s book *Hackers: Heroes of the Computer Revolution* (Levy 1984) is exemplary in offering early heroic narratives of the exploration of computer systems. It is based on life histories, tracing the origins of the hackerdom to the 'Tech Model Railroad Club' (TMRC) of the Massachusetts Institute of Technology (MIT) of the 1950s. *Hackers* has been translated into several languages and accepted among distinct hacker communities worldwide. Its narrative had the performative force of instituting a return to the figure of the 'virtuous hacker' through descriptions of the experience of early hackers at MIT and Stanford, in Northern California collectives such as 'People's Computer Company' and 'Homebrew Computer Club', and companies such as Apple Computer and Sierra Games. Levy has also popularised a positive definition of hacking as grounded in the 'hands-on imperative' and the 'hacker ethic'. This ethic includes the commitment to information freedom to facilitate technical exchange and to promote further hacking; a rebellious attitude with respect to authority, centralisation, and control of computing infrastructures; and the idea that technical work could be used to bring forth beauty and effect positive social change. Similar stories are told in the widely read books *Where Wizards Stay up Late* (Hafner & Lyon 1998) and later in *The Hacker Ethic and the Spirit of the Information Age* by the Finnish philosopher Pekka Himanen (2001). While the former offered a heroic tale of the early days of the Internet engineering research, the latter was calling attention to the subjective dimensions of a cultivation that is distinctive of the computer hacker ascesis.

A more recent account of the early origins of hacking was given by the technologist Phil Lapsley (2013) in his book *Exploding the Phone: The Untold Story of the Teenagers and Outlaws who Hacked Ma Bell*, which reconstructed the early history of 'phone phreaking', the precursor of computer hacking, which consisted in the exploration and information sharing about phone systems. Lapsley describes a genealogy that connects the direct action of Yippies of the 1960s with exploration of information and communication systems in the context of corporate control and centralisation of computing in the 1960s and 1970s. His account calls attention to a fundamental aspect of phone phreaking: a shared experience in which the telephone became the very embodiment of curiosity, and the phone network, a space for exploration, discovery, and socialisation.

A distinctive feature of hacker collectives resides in their effort of self-organisation around publications and gatherings. Akin to other independent groups, many phone phreaking and hacker groups engaged in the practice of self-documentation, with the publication of 'electronic zines', manifestos, and, in a few cases, with the enthusiastic adoption of an anthropological and historiographic mode of inquiry. Jason Scott (see further resources) has emerged as a self-appointed archivist of much of this material, maintaining extensive archives of natively produced electronic documents, series of life histories on hacking, Bulletin Board Systems (BBS), text-based adventure games, and much else. Eric Raymond, a well-known hacker and writer, is also regarded by many as a native anthropologist of hackerdom, having written on the culture and language of hackers, and the moral norms associated with Open Source and Free Software (Raymond 1999, 2004). Raymond did much to preserve and popularise the collaboratively produced document known as the 'Jargon File', which documented the rich language of early Internet, usenet, and hacking terminology, and was republished as *The New Hackers Dictionary* (Raymond 1993). Two major sociological and historiographic contributions in the literature depicting the rise and fall of the 'hacker underground' of the 1980s and 1990s were *Hacker Culture* by the communications scholar Douglas Thomas (2003) and *Hackers* by the sociologist Paul Taylor (1999). These two books are complementary in the sense that they describe the underground hacker scene of the United States and the United Kingdom in the period of popularisation of hacker techniques and criminalisation of its practice. Taylor's work is focused on the relationship between the nascent computer security industry of the 1990s and the computer underground, giving a fruitful description of the duality that is characteristic of the underground lifeworld in which hackers are ambiguously the chaser and the chased, both on the side of law enforcement and on the side of the curious hacker collectives. Douglas's work is particularly useful in describing the discursive strategies in which the figure of the hacker as an unpredictable and uncontrollable 'criminal' was instituted – a mythology created around the alleged superpowers of curious adolescents with access to a personal computer, a modem, a phone line, and certain information about flaws and holes in computer and communication security.

The work of the science fiction writer Bruce Sterling, alongside others in creating the 'cyberpunk' literary genre, led him to cross paths with hacker groups in the United States in a key moment of its history: the late 1980s and early 1990s, in which the wave of 'crackdowns' on hacker collectives had become widespread, especially in the wake of the US Computer Fraud and Abuse Act (CFAA) of 1986 and the UK Computer Misuse Act of 1990. His book *Hacker Crackdown* narrates the story of police chasing teenage hackers (Sterling 1993). Key to his depiction is the argument of how misguided the police attempts were in framing computer hacking as a serious criminal offense without understanding of the practice and its consequences. This period also saw the rise of several media darlings and misfits, the most famous being Kevin Mitnick, who became the mainstream media martyr after an incredible story of playing cat and mouse with the federal police in the United States. Following his arrest in 1995, a mobilisation of hackers around the slogan 'Free Kevin' took over the computer underground to clarify the misguidance and overreach of prosecutors in Mitnick's case. Many other hackers' stories rose to prominence in this period, such as the hacker collective Legion of Doom (LOD) and their New York-based rival offshoot Masters of Deception (MOD).

In the 2000s, scholarly attention to hackers and related communities of computer users picked up significantly, especially around Free and Open Source Software, on the one hand, and online gaming, on the other hand (the latter being too large a field to touch on here, but see Coleman [2010a]). Hackers in Free Software projects formed the subject of work by Kelty, Coleman, Auray, Broca, and others (Auray 2003, Kelty 2008, Coleman 2012, Broca 2013). This literature connected work on hackers directly to issues of intellectual property and activism around it, on the one hand, and also to questions about the liberal underpinnings of Free Software and the question of the putative liberalism and/or libertarianism of hackers themselves in the Euro-American context, on the other hand. Coleman and Golub (2008) argued most explicitly for refining our understanding of the differences *within* hackerdom by proposing several 'genres' of hacking to get at distinctions with respect to the moral and technical orders different hacker groups inhabit.

More recent publications and public debates around Free Software and hacker communities has shifted the focus to questions of gender discrimination and imbalance. Alongside the work of feminist and female hackers around computer collectives such as Systers, LinuxChix, Ada Initiative, and Geek Feminism, new publications have addressed the question of extreme disparity in Free and Open Source projects where it is estimated that less than 2% of the contributors are women and other gender minorities (Ghosh 2005). Nafus (2012) has discussed the question of gender with respect to the ways in which the organising symbol of 'openness' represents more than an alternative to the intellectual property regime and a mode of managing the collaborative efforts in software development. According to the author, the question of openness is accompanied with the insistence that gender plays no role in software development, serving in fact to disguise the mechanisms of exclusion of women and gender minorities from Free and Open Source projects.

The most recent publication on the topic of 'hacking' includes work by Michel Lallement on the topic of 'hackerspaces' and the rise of the discourse and practice of 'making' (Lallement 2015). Lallement has offered an ethnography of the 'maker movement' in the San Francisco Bay Area with a focus on the question of the transformations labour and its reorganisation with the creation of independent spaces for collaborative work with digital technologies. The book describes the community space 'Noisebridge' in San Francisco, which has been one of the most influential autonomous spaces for the recreation of hacking as a political symbol for self-organisation and wider access to expert computing knowledge around the globe.

Hackers...

One of the key insights of recent anthropology of hackers in the Euro-American context is how they represent a distinctive elaboration of liberalism – especially in the domains of free speech and a cultivation of self that is directed towards freedom, autonomy, privacy, and other liberal values (Kelty 2008, Coleman 2012, Coleman 2014). Hackers are not uniformly libertarian or simply privacy advocates, but articulate a relationship to technology with a different range of values depending on their 'genre' (Coleman & Golub 2008). Little work has been done on the way this relation to technology articulates with different political and philosophical traditions outside the Euro-American world, though a handful of people have advanced such questioning (Xiang 2007, Takhteyev 2012, Chan 2013, Murillo 2015).

Complicating this question of the productive relation to technology, however, is the question of whether 'hacking' is a specific practice, skill, or domain of knowledge. How are hacks valued and assessed, and how are they shared, learned, improvised, recorded and displayed, or circulated? Hackers are said to have a culture of perpetrating a particular kind of act that can supposedly be distinguished from the actions of other kinds of people in other professional domains (especially, those in bureaucracies, corporations, or other hidebound organisations of the past). The 'hack', however, can be mobile and reusable, and it often, but not always, takes the form of a tool or set of tools; it can be a one-off, round-about way of getting something done, but it is often something that is reusable, which serves as yet another element for generalisation. Many ordinary practices of repurposing, in different contexts, have different expressions, such as the 'gambiarra', the Brazilian term for an improvisation of and deviation from technical knowledge, the 'juugad', which expresses a similar improvisation of technical nature in India, and the 'shanzhai', which is a more specific form of repurposing mobile devices in mainland China.

In these cases, the 'hacker' persona is troubled by the recognition that 'hacks' are frequently borrowed, reused, reconfigured, and redeployed by people who only subsequently come to self-identify as hackers – or maybe do not do so at all. Conversely, depending on the context of occurrence and the participants of the exchange, you could be called a hacker regardless of your expertise, but solely on the basis of the technical feat you have performed. Hacks are contextually visible to hackers and depend on such assessment and attribution to legitimately be called hacks.

The tension in hacker personhood arises when someone claims to be a hacker, but of an unfamiliar sort. Most classic hackers hew to a definition of hacker that is open to inclusion in a particular way: through the convivial but competitive demonstration of skill in hacking. The definition of what can be included as a hack is always extensible, but only by demonstration to some set of witnesses capable of judging it. So when the word 'hack' is used to refer to something that does not seem to manifest particular skill or convivial competitiveness, then a tension emerges. For example, when participants in a hackerspace assert that they 'hack politics' or 'hack food' – but do not provide a suitably 'hackish' example of having done so – they might fail to be recognised as hackers.

Such a complicated identification of who counts as a hacker poses a dual problem for anthropological research. On the one hand, it creates a challenge for making sense of personhood in a domain where conventional idioms of what it means to be a hacker are highly contested and subjected to critique and extension. In particular, hackers are often fond of rejecting the traditional credentials of schools, employers, states, or other entities that confer expertise; they prefer the recognition of the hack itself. On the other hand, our anthropological attention can be directed to the act itself – 'hacking' – and to the question of what it is and how it might be studied anthropologically, or for instance, with the tools of science studies. When are the actions of pirates, activists, criminals, spies, and other such figures considered 'hacks'? When do people call mid-level engineers, librarians, scholars, and designers 'hackers' and when do they not? Do all of these people, with a wide range of computing expertise, share a particular set of cultivated dispositions, or do they share a milieu of technical devices and communication infrastructures at their disposal? How can we disentangle the sudden dispersal of hacking and hackers, in the sense of a wide circulation of devices, persons, and discourses, around the world?

As we briefly explored in the previous section, there have long been competing definitions of hackers. Consider two distinctive definitions: one from the Internet engineering community and the other from the 'hacker underground'. For the former, a hacker is defined as a person 'who delights in having an intimate understanding of a technological system' (see Request for Comments 1392 in further resources), whereas his or her opposite, a 'cracker', is an individual who attempts to access computer systems without authorisation. These hackers pride themselves on building complex systems out of available parts or on getting around an engineering problem in a simple and elegant way – and they deride 'crackers' as adolescents with only enough knowledge to cause trouble. As Coleman (2012) details, they are highly individualistic (in the Euro-American context), but strongly oriented towards a community of other hackers – as in the global case of Free Software hackers.

By contrast, for underground hacker collectives, hacking may signify exclusively 'systems' penetration' and exploitation of vulnerabilities as a display of technical ability and mastery or, increasingly, for monetary gain. For them, hackers are people who gain unauthorised access to computer systems, a practice which is evaluated in terms of technical aptitude and virtue, measured up against the value of contributions in software code, information, and documentation. 'Exploits' become objects of value that circulate – at one time, only for a kind of cultural capital amongst hackers, but increasingly today as part of a robust market for 'zero-day' exploits. Increasingly, these forms of hacking are also well established in the military and defence world, both defensively and offensively in geopolitical arenas.

Members of the early Internet engineering expert communities claiming hacker status would strongly disagree with the definition given by or of the hacker underground. Many of these communities were described in the origin stories of Free and Open Source development as a 'natural attitude' of pioneer computer technologists. They are strongly associated with research universities and with a Mertonian understanding of scientific practice; they point back to avatars such as the Digital Equipment Computer Users' Society (DECUS) in the sixties, or SHARE, the IBM mainframe computer users' group (Akera 2001), the MIT community around the Tech Model Railroad Club, the Artificial Intelligence Laboratory operating system development staff, and the early community around the Berkeley Software Distribution (BSD) version of Unix, among many others. The period that extends from the 1950s to the 1970s is identified in the literature with the work of early, pioneer hackers in pushing the boundaries of computing on many fronts: from hardware hacking to personal computers, from new operating systems to video games and graphical user interfaces as we discussed earlier.

By contrast, the 'hacker underground' has its own mythical histories, associated with the history of phone phreaking, bulletin board systems (BBS), zines like '2600' and 'Phrack', and movies like 'War Games' and 'Hackers'. These stories are more likely to reference the criminalisation of hacking under the US Computer Fraud and Abuse Act, and the fabled exploits of people like Phiber Optik, Kevin Mitnick, Markus Hess, Dark Dante, Erik Bloodaxe, and many others. While many Internet Engineers will point to Stephen Levy's *Hackers* as canonical, the hacker underground may point to Bruce Sterling's *Hacker Crackdown*. Over the years, these two versions of hacking have competed and collaborated: they cross paths at conferences like DEF CON, Chaos Communication Congress, and Hackers on Planet Earth (HOPE), as well as at major Free Software and Open Source events worldwide. The *cause célèbre* of intellectual property activism amongst hackers—Dmitry Sklyarov—was arrested at DEF CON 9, but became a symbol for hacking as political form: 'code as speech' (Coleman 2009).

Both of these communities reject with derision the widespread use of the term today in mainstream culture. Facebook employees who 'hack', 'brogrammers' in Silicon Valley, and the generalised use of the term 'hack' to mean 'do something' are seen as corruptions of various competing versions of a tradition. The Vitra 'Hack' desk (see Box 5.1), for instance, performs an attenuated and distorted version of hacking that emphasises Silicon Valley neoliberal start-up culture, an 'individually adaptable private sphere', and a vision of flexibility in which the desks convert into sofas as part of a longed for collapse of work, leisure, and political authenticity, primarily amongst white, upper-middle-class technology employees in Europe and North America.

But the mainstreaming of hackerdom is just as likely to expose the sexism and/or racism of past communities of hackers. An 'elite hacker', in an interview to an influential hacker zine, declared the demise of the hacker underground with the xenophobic and misogynistic observation that 'today it is claimed that the Chinese and even WOMEN are hackers. Man, am I ever glad I got a chance to experience "the scene" before it degenerated completely' (.:: Phrack Magazine ::. 2016). Symbolically violent in its own terms, this observation indexes not only the ethnocentric and gendered lifeworld of most hacker collectives, but the fact that hacking is no longer limited to the virtual play-fight among Anglo-American suburban adolescents, pointing also to questions of personhood, ethnicity, and gender with the evaluation of who gets to be considered a hacker.

What is important to emphasise in these examples is not the adoption of one definition or another, but the evidence of a series of disparate, conflicting, and generative differences for the contextualisation of hacking. This is the evidence of an oscillation in respect to the value of 'hacking' over

> **Box 5.1 Promotional language for the Vitra 'Hack' desk**
>
> 'With its raw wooden panels, Hack presents an unfinished aesthetic at first glance, like a snapshot of an experimental project under development. The system reflects the attitude of companies that similarly define themselves in terms of constant change. Each Hack unit forms an autonomous element whose adaptability allows it to satisfy various needs: companies value Hack for its flexibility, since it can be folded up into a practical, flat "box" in just a few simple steps. This makes Hack easy to dismantle and transport and enables space-saving storage. Individual users appreciate Hack's expansive work surface, as well as its provision of a private sphere that can be personalised. The height adjustment feature offers standing, sitting and lounge options and thus defines distinct niches for work, meetings and relaxation. Hack is not only functional and flexible, but is also fabricated in a manner that reflects the environmental priorities of young companies: thanks to its manually operated mechanism, the production and utilisation of Hack are ecologically sustainable. In addition, the tables are manufactured on site in an energy-efficient manner, with wooden parts that are locally produced and assembled by Vitra using prefabricated metal hardware'.

Source: From the website description for the Vitra Hack desk designed by Konstantin Grcic, https://www.vitra.com/en-us/product/hack.

time: fluctuating from positive to negative moral valences, that is, from elite, exclusive groups of technologists to online and offline communities marked by the rhetoric of openness and transparency, or, from what the communications scholar Douglas Thomas (2003) called the 'culture of secrecy' of the hacker underground in the height of the Cold War to a culture of collaboration, openness, and transparency as a shared utopian horizon in neoliberal times.

... or hacking?

What is a hack? How is it different from leaks, breaches, exploits, or other actions of information disclosure and circulation, both constructive and destructive?

Leaking, for instance, has become associated with hacking more strongly in the last decade. The furore around WikiLeaks brought the fun-loving, underground, anti-collective Anonymous to worldwide attention via its coordinated attacks on corporations and governments. Similarly, piracy: The May 15 movement in Spain was spearheaded alongside protests of

La Ley Sinde – an anti-piracy law widely opposed by musicians and consumers alike. These protests ultimately merged with anti-austerity protests of the 'Indignados' who shared key activists in the North African revolutions, who were also involved in the movements organised by Anonymous, who were often connected to actors in both May 15 and Occupy, as well as hacker activists of many groups including the group 'Telecomix'. Taken together, such actions have often been labelled 'hacktivism'.

Hacktivist movements, sometimes autonomous and sometimes part of larger movements like Occupy, borrow tactics, technologies, slogans, and ideas that both explicitly and implicitly reference the 'hacking' of Free Software, of the Pirate Party, of anti-surveillance, pro-privacy hackers (like Tor) and copyright reform movements like Creative Commons or activists fighting for Net Neutrality. All of this has occurred at the same time that US and Israeli spies were infecting Iranian nuclear power plants with the StuxNet virus (Zetter 2014), and the National Security Agency (NSA) and the Government Communications Headquarters (GCHQ) were cataloguing all of this, and perpetrating some of the greatest 'hacks' the Internet has ever seen – and this we know mostly because of the even greater intelligence leak – or 'hack' – of Edward Snowden.

Hacking, as a practice, is not confined to hackers, but increasingly a practice essential to the social fabric in surprising ways. One might turn to 'practice theory' in its various forms to explore this – are hackers part of a 'community of practice' constituted by hacks-as-practice? Is hacking a political expression of an emergent sociocultural field, in Bourdieusian terms? Are hackers heterodox actors in the context of a mainstream computing field, wildcards in various professional domains, akin to the figure of Luther Blissett or Guy Fawkes who can come in and out of a mysterious identity? Is hacking now mainstream itself?

To speak of hackers as a community of practice, or of hacking as a field, combines questions of technical and political cultivation with the identification of a distinctive practice in order to suggest that the identity of a hacker is based in the practice of hacking. Indeed, such definitions are common (we ourselves have proposed them – in the case of a 'recursive public', for instance in Kelty [2008]). But this approach cannot accommodate the fact that hacking and hacks have become more obstreperous and unruly in their circulation: hackers fight hackers (vigilante-style or as collaborators of law enforcement), trolls troll trolls, and pirates steal from pirates today. To conflate the question of hacker identity with the question of the practice – political, technical, or otherwise – would miss the mark.

Instead, there are multiple and intersecting moral and technical orders inhabited by people who self-identify or are identified by peers as hackers – from the underground hacker collectives to 'grey hat' security researchers to spam-slinging criminal actors to the hard-core free speech and privacy cryptography defenders; from the diehard Free Software activist to the business-oriented Open Source evangelist; from the über-cool Northern European design artists to the goofy-but-terrifying Anonymous hackers, and so on. As Coleman and Golub (2008) point out, there is no single liberal ideology that hackers adopt, but rather a range of 'genres' in which any given individual or group might operate, implying both the freedom and the constraints that word signals.

The notion of genre is useful for integrating hackers and hacks in a complex whole – a kind of story that integrates elements of personhood with material forms, technical practices, and political rationalities. But it is also the case that 'hacks' can be independent, modified, shared, and reappropriated across genres. Hacks imply a range of technological affordances that make practices of hacking less dependent on embodied skill than many other kinds of practice (e.g., glass blowing or cabinet making) and as a result more mobile and modifiable. The expanded field of hacks, leaks, exploits, breaches, ops, online campaigns, and so on form the very substance of political and working life today in its complex entanglements with things, protocols, perspectives, and sensibilities of digital technologists and technologies.

In the next section, therefore, we turn to the question of how one might distinguish hacks as different forms and expressions of power. Because hacks are a regular feature of work, play, and politics, it is important to look at the kinds of hacks that cross these divisions and represent mutations in the configuration of power, knowledge, and everyday practice in a Foucauldian sense. Hacking, leaking, breaching, hacktivism, and so on each imply different sets of tools, tactics, and practices, and engage overlapping genres of hackers. Foucault's approach to power has generally been used to make a claim for certain general or epochal changes as a result of such mutations in the field of tactical power: sovereignty, biopower/disciplinary power, and the governmentality of neoliberalism. But recent work on Foucault reflects his own interest in the mutual configuration of these different political rationalities (Macmillan 2011). Instead, we suggest that there are reconfigurable elements in the domains in which 'hacking' is relevant – a 'topology' in which some forms of 'hacking' respond to and constrain others. Hacks are often a recombination of the elements of power, a stretching of this topology in which action and response create tensions and thresholds of power related to particular modes of hacking. Approached this way, one can better narrate the 'recombination of elements' that Foucault (at least later in his work) recognised as a way to stretch, transform, or warp the patterns of correlation that make up a given mode or configuration of power (Collier 2009).

Viewed this way, we can better account for the fact that many 'hacker practices' do not originate with hackers, nor do they confine themselves to use by hackers. Tactics or practices are picked up by computer security scientists and researchers, security firms, police forces, political campaigns, anti-piracy outfits, analysts of all sorts, as well as spreading globally through networks of activists, hacker and maker spaces, legal firms, musicians and artists, or consumers and pirates. Hackers go to work for Google or Facebook and anti-piracy companies appropriate the tools of hackers and pirates. Pirates hack Free Software tools, while large corporations engage in technical and legal cat and mouse games with pirates. Power, in this sense, is about the recognition, appropriation, recomposition, and redistribution of tactics and practices: not ideologies or genres as much as recomposable instances of power. The technique of distributed denial-of-service (DDoS) attack, for instance, or the use of a copyright infringement cease-and-desist letter, the creation of commons-producing copyleft licenses, or the use of the BitTorrent protocol can all be called hacks, but they look different when employed in response to other hacks, leaks, or breaches.

The first three practices listed here (invention, inversion, and figuration) are quasi-direct forms of action; they are accessible to anyone and do not require large investments of money or stable organisations to mobilise. The last two (regulatory action and enforcement), however, are often more second-order or 'representative', in that they often require physical, financial, or organisational resources and a certain scale and depth of involvement to perform. But we suggest that even these two forms have a 'hackish' character in the contemporary.

Invention

Invention is the broadest possible meaning of hacking. It includes building and making things, and not only material things, but especially so-called 'immaterial' products such as software, legal licenses, or organisations. Invention might include those practices that arise out of a lack, and at least in the conventional language of research and development, it comes with extensive planning, study, and investment of time and money. But considered as a form of 'hacking', invention often implies a possibility based in the existence of multiple existing tools and components. Creating software, hacking together hardware components, or starting up an organisation are all practices of invention insofar as they are carried out to solve a problem or respond to a lack that makes certain possibilities clear.

Invention in the era of hacking has become much simpler and cheaper, as toolkits, frameworks, patterns, languages, and other easily reusable and either cheap or free to use tools create a material culture of ready-made, accessible, and often lego-like parts. Easy access to tools is by now an ethic (as in the case of the Whole Earth Catalog, 'Access to Tools' described by Turner [2006]) of mutual aid and instruction, and an appreciation for both the DIY possibilities of our world and sometimes a respect for (and desire to contribute to) the coordinated engineering necessary to bring them into being.

Furthermore, the practice of invention implies an affinity based in shared understandings of how things work. Whether that is a classic engineering culture (based in University and/or corporate practice), a DIY geek culture, a UNIX culture, a glass-blowing culture (O'Connor 2005), invention demands not only skill but the capacity to recognise others with more or less the same skills, habits of practice, and commitments to certain kinds of technological or material choices (Sennett 2008). It is this version of hacking that is most clearly identified with, for instance, the Internet Engineering communities described earlier, but also, because of its generally positive valence, the aspect that is emphasised by Silicon Valley start-ups.

Inversion

Inversion is a term meant to signal a practice that – at least under the label of 'hacking' – is often insufficiently distinguished from invention. Inversion is the kind of hacking that involves finding a way to use existing tools or technologies to achieve something they were not meant to do. Under this aspect, exploits of vulnerabilities are using systems against themselves, inverting the intended purpose of a system, or remixing something for purposes of critique, parody, ridicule, or something more practical (Galloway & Thacker 2007). In terms of hardware, it includes the practices of modding and customisation; in terms of software, it includes the practice of finding and exploiting weaknesses in software (for good or evil intent), or recombining software elements in a surprising and clever way in order to achieve a new goal.

A famous example of inversion is neither hardware nor software but a so-called 'legal hack': the General Public License (GPL), which uses existing statutory copyright law to accomplish something it was not intended to do.

Inversion generally assumes institutional structures or infrastructures with a certain transparency (one must be able to see how something works in order to exploit it). Some tactics exist for inverting the non-transparent. Leaking – such as the actions of Manning or Snowden – might be considered an inversion, as might some forms of reverse engineering of secret techniques or closed technologies. For technologists of the 'hacker underground' ilk, a whole range of tools exist for attempting to break technologies, either to gain access or to control them (the creation of Botnets, zombie servers, and so forth). Inversion also implies faith in engineering and in the rule of law: for something to be inverted is not the same as for it to be destroyed or disabled. GPL licenses work because copyright law is legitimate and enforced outside hacker circles. Remix or recombination of software is done because the challenge is often to make something work better or meet higher standards of practice or security. By extension, the faith is that things can always be and become better; only imperfect things can be hacked – inverted – in this sense.

Similarly, inversion implies patterns of regularity that can be exploited – the very thing upon which the practice of invention often depends as well. Inversion works upon certain technological or organisational preferences shared amongst a large group of people – for instance, the use of SCADA control systems amongst process engineers who design large industrial plants allows for hackers to imagine exploits with powerful effects on the material world. Similarly, entrenched social practices are often exploited in 'social engineering' hacks that rely on the regularities of organisational design and human behaviour.

Figuration

It is also clear that not all hacking is restricted to technological manipulation. We suggest *figuration* to capture the more traditional and recognisable forms of political action: rhetorical persuasion, ideological argument, political advocacy, and so forth. Classic descriptions of the functions of the public sphere tend to characterise it as an issue of sovereignty – the ability of 'the people' to speak to and force changes amongst established domains of power like the state, the church, the military, and so forth (Anderson 2006). All of the tactics involved here are about visibility and the legitimacy of a political process.

To speak of figuration as a mode or component of 'hacking' today, however, is to recognise that such classical forms of discursive and persuasive power are also used as forms of, and in response to, hacking: operations by Anonymous, for instance, can be restricted to the widespread circulation of messages, videos, and manifestos. The protests against the Stop Online Piracy Act (SOPA) and the PROTECT IP Act (PIPA) in 2012 in the United States were largely traditional responses by hackers (and others) to an attempted regulatory action.

Materially speaking, figuration depends on shared communication structures. Very often today, hacking can take the form of *inventing* or *inverting* communication practices as part of or in response to practices of figuration. In many ways, open government data advocates of the last few years are demanding that longstanding institutional structures (such as public hearings or requests for comments) be hacked in order to make interacting with the government and its agencies simpler or to make agencies more responsive. To do this, they rely on a figuration of government as slow, non-responsive, elitist, or rigidly bureaucratic.

Substantively (at least in the domains of intellectual property and information technology), practices of figuration have been heavily focused on issues such as privacy, transparency, free speech, freedom to operate (or innovate), and network neutrality. These issues subtend a long and rich discourse that includes both scholarly debates and conventional understandings of these concepts and their value to our lives. Examples of 'hackish' figuration include the Electronic Frontier Foundation. EFF was created in the context of 'hacker crackdowns' of the early 1990s to articulate a discourse on the importance of defending civil liberties online and as a fund to legally defend hackers from prosecutorial overreach. Similarly, the death of Aaron Swartz provides a figure of openness and Open Access as vital global struggles to which hackers can and should contribute.

Regulatory action

Regulatory action is not historically a form of hacking, and in many ways, it is its most important opposite. Nonetheless, a hackish attitude towards regulatory action has emerged as a possibility – the strategic attempt to regulate (either formally as part of government action or through software, or informally through other means) represents another tool in the hacker kit. The language of regulation-as-control was most clearly articulated in terms of hacking by Lessig's famous work *Code, and Other Laws of Cyberspace*, which argued that many different kinds of things regulate behaviour (law, morals, architecture) and are amenable to change to different degrees (Lessig 2000).

At the most general level, regulatory action includes any form of policy change intended to introduce, maintain, or extend control. State-based forms of regulatory action are the most obvious and familiar but large corporations also engage in regulatory action of particular kinds routinely – especially those industries who control networks or technologies in widespread use.

Regulatory action almost necessarily implies large bureaucratic organisations of a classic Weberian type – rule-based, hierarchically ordered, and subject to regimes of oversight and transparency. As a result, regulatory action is relatively rare and comparatively complicated to carry through. The tactics of invention, inversion, and figuration are often oriented towards influencing, responding to, or disrupting regulatory action of various kinds – the 2012 case of protests against SOPA/PIPA being a clear case; another case would be Operation Payback conducted by participants of Anonymous against the global credit card companies MasterCard and Visa, who had engaged in the regulatory action of systematically blocking PayPal donations to the WikiLeaks organisation.

Inversion often borders on regulatory action when it is used strategically to achieve something in the interests of a particular entity, but so too does figuration, which is the tactic most often employed to support or protest a proposed legal change (both were used in the case of SOPA/PIPA). Cases of significant interest include those where regulatory action looks more like cases of inversion – i.e., where a legal action, for instance, is used to threaten, intimidate, or censor a particular group. The injunctions filed against Megaupload and its owner, Kim Dotcom, for instance, are represented as mere enforcement of the law, but in reality represent the effective mobilisation of state power by industry organisations like the Motion Picture Association of America (MPAA) and the Recording Industry Association of America (RIAA).

Enforcement(s)

Finally, there are practices of enforcement. Often only implicitly or metaphorically included in a Foucauldian analysis (discipline is held to be a more insidious and more profound kind of force, a 'government of the soul' for instance). While classic displays of sovereign power are relatively rare (helicopter raids by antiterrorist forces on New Zealand mansions of flamboyant wannabe hackers notwithstanding, such as the case of Kim Dotcom), they remain a central tactic in the repertoire of power and they are by no means restricted to a state and its repressive apparatuses.

Many forms of enforcement are widely available today. A range of tactics and practices, of which the DDoS is only the most common, are not confined to any particular segment of society, but available to anyone who might, for instance, download and install a copy of the Low Orbit Ion Cannon (or its predecessor, FloodNet, written by the Electronic Disturbance Theater group in 1994 to flood the Mexican government website with requests in order to send a message of support for the Zapatistas). Legal tools like cease-and-desist letters are also routinely used as a tactic of force; even patent litigation can be understood this way when conducted by so-called patent trolls. Network-based forms of disruption are very often simultaneously tactics of invention or inversion – coming into existence in response to the actions of states or corporations. Piracy and cracking generally might be said to move from being a tactic of inversion when a vulnerability in a copy-protection scheme is merely demonstrated (e.g., Sklyarov demonstrating holes in the eBook Reader) to a tactic of force when that vulnerability is routinely exploited for gain, protest, or sabotage.

These five practices fit together as one possible description of how 'hacking' is part of a topology of power today. The software programmer's metaphor of a 'stack' is useful here – in common parlance, it refers to a frequently deployed but heterogeneous collection of tools, libraries, and interfaces. Such tools interface with each other and can, in some cases, come to depend on one another (nothing would happen without an operating system in place, but there are many variants available).

The image of the stack we employ here, however, is meant to provide a basic map of push and pull, of action and reaction, or of provocation and response (Figure 5.2). We intend it to be used to help diagnose certain technical or political thresholds in the recent past of hacking; and we take it as axiomatic, as anthropologists, that hackers are historically situated subjects who experience these thresholds in their own lives and practices under different conditions, despite sharing many technical devices, protocols, and infrastructures.

Conclusion

Are we experiencing the distancing between the 'hacker' as a particular manifestation of personhood and the 'hack' as a set of tactics of power? Conversely, are we witnessing a global expansion of the conditions for cultivation of hacker expertise, leading to a proliferation of differences in the context of what it means to be a hacker and to hack? How is the global difference in technical and moral cultivation of computer technologists related to the global accessibility of hacks? What are examples of the reverse process in which the actions of global hackers (say, book pirates in Russia, hacktivists in Tunisia, or Anonymous's attacks on ISIS) have an impact on mainstream practices?

To return to these open questions, we conclude with the following observations. First, we suggest that the study of hacking as a practice of ethical

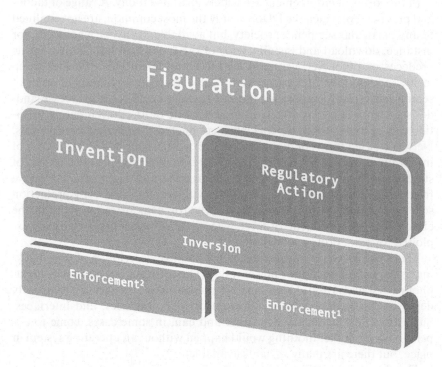

Figure 5.2 Stack of power: an analytic decomposition of hacking practices and how they might relate.
Source: Murillo & Kelty 2016.

and technical enskillment can be advanced to describe and interpret manifestations of hacking outside the Euro-American centres of technical and discursive production on digital technology. Such work is necessary if we seek to displace the imperial imperative of digital technology, which often transforms sociotechnical *difference* into resemblance (or poorly made copy of Euro-American 'sources') in other parts of the world.

Anthropological studies of the global circulation of digital technologies and expert technologists have demonstrated the naturalisation of Euro-American assumptions in digital design: from user interfaces to data models; assumptions of usefulness of particular technologies based on the prestige of their place of creation; and the imposition of particular projects from centres of production to disconnected peripheries of the global South (Xiang 2007, Takhteyev 2012, Chan 2013, Murillo 2015). We suggest that such assumptions might well be built into the global stack of power we describe above, and to look to different traditions would be to ask how hacks themselves might be hacked.

The recent past of hacking clearly overlaps with other practices – those of piracy, activism (whether labelled cyber-activism or hacktivism), trolling, leaking, and breaching. Such practices should be understood not as simple forms of hacking, but as part of a topology of power that is locally stable, but historically changing. If history has tended towards the stabilisation of forms of power, the manifest speed and ease with which the contemporary topology can be stretched and deformed suggests a perpetual oscillation or turbulence, limited only by the energy and enthusiasm that can be committed to the various practices of invention, inversion, enforcement, figuration, or regulation. Secondly, we suggest that the widespread embrace of hacking reflects changes in three traditional domains of study: work, education, and political action. The language of hacking and the question of hacker personhood provides a window on the changing meaning of work, career, and expertise. Rather than a lifelong career, grounded in training and acquisition of expertise, maintained through repeated trials of problem-solving within the basic framework of the division of labour, the embrace of 'hacking' (as in the case of the Hack desk) is clear evidence of anxieties about work that is temporary, flexible, transient, and often includes highly individualised ways of demonstrating one's creativity. They remain focused on problem-solving, but disdain a division of labour and a hierarchy of expertise, which often mirrors actual changes in work environments in the last several decades (Boltanski & Chiapello 2005). In this respect, hacker personhood may represent a vanguard of sorts for changes affecting large parts of a global labour force.

By contrast, when hacker personhood is considered in light of political action, it highlights a different set of changes, and possibly a political threshold that might be described as a new form of sabotage. It is this aspect – especially under the label of hacktivism – that raises the central question of the link between hackers and hacking, precisely because of the proliferation of hacks that can simultaneously be used for diametrically opposed political purposes. Sabotage does not always imply a form of class resistance and consciousness – it is also often a competitive tactic within capitalism and a form of warfare engaged in by state actors. But whether this form of hacking-as-sabotage is related to the production of a particular form of hacker personhood remains an open question.

Finally, there are important methodological lessons to be learned by anthropologists and other social scientists through the study of hacking. The obsession with 'the digital' in contemporary scholarship is a kind of Stockholm Syndrome: we have been kidnapped by shiny new technologies and the discourse of innovation, and as a result we have come to love our captors. The digital (as in 'digital humanities') is no panacea for the problems of collaborative research in anthropology and the human sciences at large; even if digital technologies must be redesigned to further promote collaborative engagements between ourselves and with the communities with which we conduct research. Digitisation of field research carries potential benefits with respect to the ease of archiving and data sharing, allowing for longitudinal and extensive comparative studies. But it also carries serious issues of data privacy and anonymity as most researchers (in both the sciences and humanities) are not well informed and trained to handle problems of information security. Hacking, in this regard, is an important source of practices and workarounds – in the sense of invention and inversion – to deal with questions of information security, sharing, and remote collaboration across transnational lines of exchange, but it also represents a danger that transcends the immediate ethical relations that anthropologists have traditionally been concerned with. In the future, we may have more to learn from hacking than about it.

Further resources

- Original MIT Jargon File, kept by Paul Dourish: www.dourish.com/goodies/jargon.html
- Extended Jargon File at Eric Raymond's site: www.catb.org/jargon/html/
- Jason Scott's textfiles: http://textfiles.com/
- Hackerspaces wiki: http://hackerspaces.org
- Request for Comments 1392: www.rfc-base.org/txt/rfc-1392.txt

Notes

1 Berkman Center for Internet and Society, Harvard University
2 Institute for Society and Genetics, Department of Anthropology, and Department of Information Studies, UCLA

References

Akera, A., 2001. Voluntarism and the fruits of collaboration: the IBM user group SHARE. *Technology and Culture*, 42 (4), pp. 710–736.
Anderson, B., 2006. *Imagined communities: reflections on the origin and spread of nationalism*. London: Verso.
Auray, N., 2003. Communautés epistemiques d'innovation – La regulation de la connaissance: arbitrage sur la taille et gestion aux frontières dans la communauté Debian. *Revue d'économie politique*, 113, p. 161.
Boltanski, L. and Chiapello, E., 2005. *The new spirit of capitalism*. London: Verso.
Broca, S., 2013. *Utopie du logiciel libre: du bricolage informatique à la réinvention sociale*. Neuvy-en-Champagne: Éd. le Passager clandestin.
Chan, A., 2013. *Networking peripheries: technological futures and the myth of digital universalism*. Cambridge: MIT Press.
Clark, G., 1961. *World prehistory: an outline*. Cambridge: Cambridge University Press.
Coleman, G., 2009. Code is speech: legal tinkering, expertise, and protest among free and open source software developers. *Cultural Anthropology*, 24 (3), pp. 420–454.
Coleman, G., 2010a. Ethnographic approaches to digital media. *Annual Review of Anthropology*, 39 (1), pp. 487–505.

Coleman, G., 2010b. The hacker conference: a ritual condensation and celebration of a lifeworld. *Anthropological Quarterly*, 83 (1), pp. 47–72.

Coleman, G., 2012. *Coding freedom: the ethics and aesthetics of hacking*. Princeton: Princeton University Press.

Coleman, G., 2014. *Hacker, hoaxer, whistleblower, spy: the many faces of Anonymous*. London: Verso.

Coleman, G.E. and Golub A., 2008. Hacker practice: moral genres and the cultural articulation of liberalism. *Anthropological Theory*, 8 (3), pp. 255–277.

Collier, S.J., 2009. Topologies of power: Foucault's analysis of political government beyond 'governmentality'. *Theory, Culture & Society*, 26 (6), pp. 78–108.

Downey, G.L., 1998. *The machine in me: an anthropologist sits among computer engineers*. New York: Routledge.

Dumit, J., 2004. *Picturing personhood: brain scans and biomedical identity*. Princeton: Princeton University Press.

Forsythe, D., 2001. *Studying those who study us: an anthropologist in the world of artificial intelligence*. Palo Alto: Stanford University Press.

Funders and Founders Notes, 2016. *Who are hackers? Hackers are doers*. Available at: http://notes.fundersandfounders.com/post/50417296471/who-are-hackers-hackers-are-doers (accessed 28 January 2016).

Galloway, A.R. and Thacker E., 2007. *The exploit: a theory of networks*. Minneapolis: University of Minnesota Press.

Ghosh, R.A., 2005. Understanding free software developers: findings from the FLOSS study. In J. Feller, B. Fitzgerald, S.A. Hissam, and K.R. Lakhani (Eds.), *Perspectives on free and open source software*. Cambridge: MIT Press, pp. 23–46.

Hafner, K. and Lyon M., 1998. *Where wizards stay up late: the origins of the Internet*. New York: Simon and Schuster.

Hakken, D. and Andrews, B., 1993. *Computing myths, class realities: an ethnography of technology and working people in Sheffield, England*. Boulder: Westview Press.

Helmreich, S., 1998. *Silicon second nature: culturing artificial life in a digital world*. Berkeley: University of California Press.

Himanen, P., 2001. *The hacker ethic and the spirit of the information age*. New York: Random House.

Kelty, C., 2008. *Two bits: the cultural significance of free software*. Durham: Duke University Press.

Lallement, M., 2015. *L'âge du faire: hacking, travail, anarchie*. Paris: Seuil.

Lapsley, P., 2013. *Exploding the phone: the untold story of the teenagers and outlaws who hacked Ma Bell*. New York: Grove Press.

Latour, B., 2008. *We have never been modern*. Cambridge: Harvard University Press.

Lessig, L., 2000. *Code: and other laws of cyberspace*. New York: Basic Books.

Levy, S., 1984. *Hackers: heroes of the computer revolution*. Garden City: Anchor Press/Doubleday.

Macmillan, A., 2011. Empire, biopolitics, and communication. *Journal of Communication Inquiry*, 35 (4), pp. 356–361.

Murillo, L.F.R., 2015. Transnationality, morality, and politics of computing expertise. PhD thesis, University of California, Los Angeles.

Murillo, L.F.R. and Kelty, C., 2016. Stack of power: an analytic decomposition of hacking practices and how they might relate.

Nafus, D., 2012. 'Patches don't have gender': what is not open in open source software. *New Media & Society*, 14 (4), pp. 669–683.

O'Connor, E., 2005. Embodied knowledge: the experience of meaning and the struggle towards proficiency in glassblowing. *Ethnography*, 6 (2), pp. 183–204.
Pfaffenberger, B., 1992. Social anthropology of technology. *Annual Review of Anthropology*, 21, pp. 491–516.
.:: Phrack Magazine::., 2016. Available at: http://phrack.org/issues/65/2.html#article (accessed 28 January 2016).
Raymond, E.S., 1993. *The new hacker's dictionary*. Cambridge: MIT Press.
Raymond, E.S., 1999. *The cathedral & the bazaar: musings on Linux and open source by an accidental revolutionary*. Beijing: O'Reilly.
Raymond, E.S., 2004. *The art of Unix programming*. Boston: Addison-Wesley.
Sennett, R., 2008. *The craftsman*. New Haven: Yale University Press.
Star, S.L. and Ruhleder K., 1996. Steps toward an ecology of infrastructure: design and access for large information spaces. *Information Systems Research*, 7 (1), pp. 111–134.
Sterling, B., 1993. *The hacker crackdown: law and disorder on the electronic frontier*. New York: Bantam.
Suchman, L., 1987. *Plans and situated actions: the problem of human-machine communication*. Cambridge: Cambridge University Press.
Takhteyev, Y., 2012. *Coding places: software practice in a South American city*. Cambridge: MIT Press.
Taylor, P.A., 1999. *Hackers: crime in the digital sublime*. London: Psychology Press.
Thomas, D., 2003. *Hacker culture*. Minneapolis: University of Minnesota Press.
Turkle, S., 1984. *The second self: computers and the human spirit*. New York: Simon and Schuster.
Turkle, S., 1995. *Life on the screen: identity in the age of the Internet*. New York: Simon & Schuster.
Turner, F., 2006. *From counterculture to cyberculture: Stewart Brand, the Whole Earth Network, and the rise of digital utopianism*. Chicago: University of Chicago Press.
Winner, L., 1978. *Autonomous technology: technics-out-of-control as a theme in political thought*. Cambridge: MIT Press.
Xiang, B., 2007. *Global "body shopping": an Indian labor system in the information technology industry*. Princeton: Princeton University Press.
Zetter, K., 2014. *Countdown to zero day: Stuxnet and the launch of the world's first digital weapon*. New York: Crown Books.

6 'A brilliant copy every time!'
Aspects of a cultural proportion

Christian Schönholz

The first part of the chapter's title is copied: 'A brilliant copy every time' was a slogan of Sony Group's ads and commercials for the Minidisc in the mid-1990s (Sony 1997: 56). This 20-year-old slogan expresses two ideas central to my argument. *First*, 'A brilliant copy every time' highlights the immanent paradox of the debated concepts: if any copy were truly of the same material and sentimental value as its original, it would become fundamentally redundant to talk about originals or trying to define them. *Second*, the alleged equivalence of the two quantities, original and copy, relies on digital technologies for its realisation: only a digital copy of a music file is a truly identical one-to-one copy. This is what distinguishes digital production from its analogue predecessor; while digital technologies generate 100 per cent identical duplicates, analogue dubbing provides versions of a template.

This chapter elaborates on the assumption that the conceptual distinction Original_Copy (also related to digital objects and processes) is still powerful and influential, despite all formal objections and paradoxical anachronisms.[1] To illustrate this, I begin by expanding on the historical point of origin of this proportion, which I believe to be responsible for a certain mythologisation that continues to be relevant up to today. Albeit the proportion Original_Copy is not (or no longer) useful as an empirical method, neither demanding the abolition of the one nor praising the other is helpful. I rather propose to understand it as a part of discourses on power and value, wherein it still forms a weighty opposition against which it is difficult to argue. I further dwell on this by introducing three theoretical positions in the second section. Subsequently, I respond to current trends, discussions, and prospects regarding Original_Copy. In the concluding section, I pursue the question of how to practically make use of Original_Copy in cultural analysis.

The myth of Original_Copy

Where to start? With Benjamin (1939). Since industrialisation, the mechanical effects of modern replication technologies are the seriality and the inherent repeatability of production processes. Objects of the same production series are apparently no longer distinct and do not differ qualitatively from one another. They no longer bear the individual hallmark of their creator, as there is none anymore. This threatened the existence of the mechanisms that attributed value and significance to objects and works of art and made them distinguishable from others.

Benjamin already pointed to the fact that works of art have always been reproducible, but he differentiates between the mimicry (in terms of master and student) and technological reproduction (1939: 10). In special cases, the latter already existed long before industrialisation (he mentions the minting of coins or terra cottas, then especially letterpress printing), but for him only photography and film mark the decisive break to the age of mechanical reproduction. Another crucial effect of this change is that technological reproduction captured its own significance among the artistic manufacturing methods (1939: 11). As a consequence, the recognition of reproducibility as an artistic act in itself resulted in techniques of theoretical and practical differentiation between originals and their copies. Hence, the originals were attributed genuine and unique, while copies were identified as counterfeits and fakes, with particular impact on value discourses. This mostly unquestioned step is of immense importance for our cultural dealings with Original_Copy. As Carlo Ginzburg pointed out, the fact that there is a difference between a picture by Raphael and a copy of it is a 'cultural choice', which has never been made explicit and is anything but self-evident (Ginzburg 1979: 275). Just because this condition was accepted for paintings (and not for written texts), it could create the new social figure of the connoisseur.

The modern obsession with the original is most identifiable in the social figure of the art connoisseur, as it represents the ability to distinguish originals from counterfeits without any doubt. Art history still deals to a large extent with precisely this differentiation, in order to make unambiguous assessments of originality. However, this assessment is faced with blurred definitions, all the more since the reproduction of works of art has become a main source of income for the (in this time, newly) established type of autonomous artist from the Renaissance until now.[2]

But the diagnoses' and expertise's defining power and validity unravel whenever they come to erroneous findings or art forgers are convicted only after decades. The analogy to the detective work of a Sherlock Holmes is obvious, as Ginzburg had it (1979: 276). Edgar Wind also described the work of a connoisseur with reference to a satire of Hogarth (Wind 1968: 39). He outlined four steps that are relevant for any attempts to distinguish between Original_Copy: (1) attribution, (2) impression of accuracy, (3) details as a major indication, and (4) gesture of conviction. This long (here, only touched on) debate on genuine and false pictures, unrepeatable unique and false copies that had original status for a long time (and those which have it until the next scrutiny), justifies the persistent mythologisation of the cultural proportion Original_Copy.[3] Successful counterfeiters – and one could be of different opinions on the notion of counterfeiting – are able to meet the conditions for an original, at least for some time, and thus make a mockery of the 'cultural choice' observed by Ginzburg. Prominent counterfeiters often gain public sympathy and respect for disclosing commercial structures in art trade and giving connoisseurs the runaround, as the case of the German painter Beltracchi testifies.[4]

However, the dispute about originals and copies in the visual arts is just one area of this once so 'intimate partnership' (Täubner 2012: 39), though one that was decisive for a long time; thus, it has been transferred to other topics, in which the relationship of the two variables was even harder to articulate.

The work of the connoisseurs – which is trying to ascribe original of fake pictures to artists – was an initially economic, but more importantly an ideological pattern for other cultural fields, especially for digitised contents. The tested, but ambiguous and still unreliable concept of Original_Copy was transferred to the copyright and consequently to many forms of cultural expression such as music, literature, dramatic arts, and science, including their respective media. That is why historicising copyright is necessary to illuminate today's debate (Dommann 2014).

Today, aspects of originality, duplicates, and copies are relevant in almost all cultural classification systems. No matter whether visual artists, writers, or musicians, they all owe the copy that they can sell their intellectual property as cultural commodities, as Täubner underlines (2012: 41). But since the 1980s, accelerated and simplified digitisation processes interrupted this value chain; hence, the copy came under suspicion to be indecent and even criminal.

How appropriate and feasible this opposition of cultural originality and criminal copies is depends on the used medium: on the web, the very assumption of an original seems to be fundamentally unnecessary, notwithstanding that bootlegging and software piracy are emotive issues. Material goods, buildings, information carriers, and performative acts still raise the question of how faithful the reproduction is to an imagined original content, which is an endeavour as paradoxical as the art market.[5] Consequently, the body of literature from various disciplines is as diversified as the applications of Original_Copy; I discuss three key approaches in the following section.

Key texts and exemplary studies on Original_Copy

Due to their disciplinary origins, the art market and museums rely on exact diagnoses of an original or a fake. However, it is remarkable that the fairly common use of replicas in historical museums usually does not cause irritation to the visitors. For example, no one seems to bother that the exhibited *Treasure of Priam* in the New Museum in Berlin is only one of many replicas there. After all, the replicas are unquestionable testimonies of high manufacture skills, partly even with considerable historical value themselves. Imagine, however, the *Mona Lisa* in the Louvre was a copy, though not recognisable for the layman, and the original was stocked in the depot; or the *Bust of Nefertiti* in the Egyptian Museum Berlin was one of her numerous replicas – hard to imagine the outrage this would cause among visitors, as they expect to see the unique originals for the price they paid at the entrance. The importance of an original or copy for the representation is obviously extremely variable, negotiable, and depends on considerably more criteria than on a clearly identifiable authorship.

The specialist literature is as diversified as the subject itself. Hereinafter, I discuss three selected positions and exemplary studies that may help to understand the difficult relationship between original and copy and make it fruitful with regard to the digitisation in cultural fields.

What is innovative in seriality?

An inspiring starting point for the discussion is a contribution by Umberto Eco, in which he discusses the potential for innovation in the seriality of mass media products (Eco 1990). According to him, modern aesthetics denied that industrial products could have any innovative features, since they were regarded as repetitive variations (1990: 155). Hence, Eco offers a helpful distinction of 'the same' with regard to the practical dealing with forms of repetition in everyday life. He points out that two identical sheets of paper may be the same for the purpose of our functional needs, but they are not the same for a physicist who is interested in the molecular composition of the two sheets (1990: 158). Transferred to our topic, what is 'the same' in a digital copy depends, following Eco, more on our functional needs than on the molecules, meaning the ones and zeros and their variations, which are irrelevant for everyday dealings.

Taking examples of television series, Eco explains specific types of repetition and their own variations and intertextual references. The stories, characters, and places may repeat, but the variability no longer aims at the already known scheme only, but rather at the endless possibilities of variation (he mentions especially 'Columbo'). The endlessness is the decisive new quality, which does not depend on originality, but on the recurrence of the same and its minimum deviations in the reception that the audience or the user expects. Eco finally summarises these future aesthetics and their reading of seriality in a vivid thought experiment: in the year 3000, there is only one single episode of Columbo left; then, how would we read a part of a series, when all the rest is unknown? Eco answers that nothing prevents us from reading the serial products already like this (1990: 180).

Here, the decisive criterion is the reading of serials, not their inherent quality or originality. This understanding extends the uniqueness that is attributed to originals in tradition of an art historical term to repeatable and varying products. Innovation in seriality is not defined by its absolute newness and identifiable authorship, but by recognising minimal deviations from a known scheme through a mode of reading that searches for endless variations. Eco undertakes an early and important revaluation of established aesthetic criteria in terms of uniqueness and repeats in serial media, offering a possible solution to the dilemma of the propagated unrepeatability of originals.

Is it possible to praise copies, and if so, what for?

In the title of his widely acclaimed *Praise the Copy*, German journalist Dirk von Gehlen (2011) makes clear what it is all about: an appreciation of the category of copy in its relation to the original. Similar to Täubner, he describes the copy as the very basis of creativity and draws attention to the economically motivated criminalisation of copying practices (2011: 13).

In order that a copy could be praised, he formulates specific criteria any copy ought to meet. The first is to disclose the sources, which does not seem to be surprising in a scientific context. The second requirement is to choose a new temporal, social, and spatial context, in which the copy is to be located; the example given is the iconic Campbell Soup cans of Andy Warhol. Third and most important, a copy must contain a creative element, so that it is recognisable as a new work (2011: 20). This claim is raised especially to the mash-up culture and remixing, whose practices of transmediality come along with changes in production and distribution in media systems.

These three criteria appear likeable and convincing; copiers have to respect at least some rules if they want to copy creatively. However, von Gehlen remains (unconsciously?) caught up in a traditional concept of Original_Copy as he applies the same standard to the creative copy as to works with an original status. In addition, it is not always possible to comply with these criteria in digital contexts. Circulating files do not need any crediting of sources and the copyright of copies cannot be clearly determined. How could this be possible, when even art history is not able to identify original works beyond doubt? The second criterion is absurd for another reason, since any copy can fulfil it easily: when another person uses the copy of a copy or an original, it always creates a new spatial, social, and temporal context in its 'here and now', as we know from Benjamin (1939: 12). Thus, different contexts alone are insufficient to differentiate between copyright infringement and creative copies. Likewise, the third criterion cleaves to a traditional notion of the original. The necessity of a 'creative element' as a decisive criterion brings back a traditional concept of artwork with its identifiable creator and author again through the back door. Von Gehlen's attempt to differentiate illegal plagiarism and creative copies thus ultimately reinforces the potency of the two categories themselves, which (though unnoticed by von Gehlen) can claim again in the theoretical reflections on their reformulations and revaluations. Moreover, it is questionable whether the already problematic notion of copy gains accuracy by adding creativity as another yardstick. Who ultimately decides what is creative to a copy? What could be an uncreative copy and how should it be evaluated?

Similarly, von Gehlen even tries to redefine the concept of original. Therefore, he first takes a look to Asian martial arts, where the students' imitation of the master is considered as a method of perfection (2011: 166). This may be true, but it eventually circumscribes a relationship similar to that of student and teacher in the tradition of artistic imitation in European painting.

By the way, the subtle assumption that imitation and copy were the same is not very convincing with regard to the character of a 1:1 duplicate in digital processes. Even the best imitation always somewhat bears the hallmark, gestures, or facial expressions of the imitator and never fully resembles the imitated, otherwise it would no longer be discernible as an imitation. Digital copies lack exactly these discernible marks, although von Gehlen establishes them as a criterion. This cannot be practically accomplished: copy and paste practices do not produce something new but rather a duplicate!

Eventually, Dirk von Gehlen's attempt to praise the copy seems to be overplayed in some places, especially when he associates the copy with the right to free speech (2011: 175). It also remains unclear why references and links are equated with copies, as they are only referential techniques such as footnotes. Conversely, practices of digital copying are not automatically forms of expression (although they can assume such functions in certain contexts) but simply media practices in order to produce indiscernible duplicates. Wanting them to be more and connecting them with consistently positive connoted concepts such as creativity and civil liberties, however, is indicative of a contemporary ideal of a more democratic society with open access to all levels of culture. This is what makes the analytical substance of von Gehlen's pleading. It impressively illustrates how the proportion Original_Copy and its partly paradoxical orders fosters (yet utopian) visions of cultural production and revaluation of work and culture, particularly in digital fields and the creative industries.[6]

Can copies be dangerous?

To the same extent that von Gehlen is exemplary for a downright, though somewhat unreflective upgrading of the copy, Mercedes Bunz (2003) represents quite the opposite. She emphasises the post-avant-garde's current positivity in terms of repetition and determines the question of repetition as the political question of our time par excellence. Since these discourses on repetition follow economic imperatives, they also bring about effects of inclusion and exclusion. This is the very core of the proportion Original_Copy: it is all about the qualitative evaluation and the quantitative valorisation of cultural goods. The problem with it is that the inherent logics of this valorisation are not articulated by those who produce these goods, but depend on strategic market laws. These laws do not only govern, to some extent, the access to the totality of culture, but more importantly, they frame the circulation of cultural goods.

In times of digitised and digitalising culture, repetition and the concomitant, strong need for viable distinction between original and copy underlies the insight that intellectual property becomes, as Bunz (2003) puts it, the current productive power par excellence. But precisely this transition from (former) labour force and material goods to (today) intellectual property as a means of creation of value causes the existing capitalist logics to alter.

As Bunz (2003) puts it, intellectual property possesses an inherent potential for duplication, which means that it can be reproduced without consuming itself. When I depicture a painting, take a photograph of something or download MP3 files, I appropriate it without taking anything away.

That sounds simple, but it has immense consequences. Reproduction without consumption does not fit in with the existing system of trade in goods, which always assumes an unrepeatable transfer of possessions from one person to another (whether by sale, gift, or inheritance). This incidentally makes obvious the fallacy inherent in the term *piracy*: it cannot be robbery, since it takes nothing away but reproduces without consuming. The thereby resulting supplementary chains secure the clear division of production and appropriation.

The copyright law is still the starting point of this need for a descending logic of the original, because it is the only way to maintain the established and traditional value chains (Bunz 2003). Unlike the traditional product and its appropriation, reproduction without consumption does not close the cycle of production and appropriation but keeps it open, since it is both production and appropriation at the same time. This opening also does not run in descending order, but the other way around. Precisely this irritation of the existing value logic by repetition and appropriation constitutes the dangerous moment of the copy (dangerous for the descending logic) – which is why it can be used as a political argument.

The three approaches of Umberto Eco, Dirk von Gehlen, and Mercedes Bunz each represent different priorities in dealing with the theoretical complexity of the concept Original_Copy. Eco proposes a change of perspective with regard to innovation in serial products, thereby somewhat easing the restrictions of the traditional paradigm of originality. Dirk von Gehlen represents the perspective of cultural producers in the digitised creative industries, which no longer see themselves bound to an existing regime of Original_Copy and therefore try to challenge it by proposing other principles. Mercedes Bunz brings up the painful subject of economic laws, which also govern the proportion Original_Copy – or is it the traditional regime of Original_Copy governing the economic laws?

The perception of Original_Copy in digitised fields

It is difficult to narrow down the level of perception of the concept Original_ Copy. The two concepts are ascribed or emphasised every time a certain object's quality is to be established or denied. This is known with regard to material goods, buildings, or performances, and it apparently will continue to be a fluid and changeable quality criterion. Therefore, this is of particular interest to advertisement, as the emphasis of originality helps to distinguish the advertised product from supposedly inferior copies. The argument for Original_Copy still seems to be too strong and too convincing to give up on it, despite the insistence from numerous theoretical sides. There is no other explanation to the fact that this line of argument – despite all calls for its disposal – continues to prevail over available alternatives. Therefore, empirical investigations should track the use of this argument as a marker of quality or distinction.

Thereby, Original_Copy is at an advantage, since it can be adapted for effects of digitisation processes in cultural fields. For example, negotiations about reproducing copyrighted files are nowhere near the end. Corporations, lawyers, artists, and beneficiaries of the propagated digital freedom and unlimited access to information still argue about the interpretational sovereignty in this area. Meanwhile, copyright laws as well as the GEMA or other collecting societies assuming a traditional hierarchy between original and copy stand their ground.[7]

A particular difficulty lies in the fact that, depending on the thematic focus, a large repertoire of different terms is used synonymous to copy: plagiarism (in science), sampling, mash-up (in media production), imitation (in sports, theatre, film), remix, remake (in the creative industries and media studies), fake (in satirical or political actions), re-enactment (in historiography, which also is greatly influenced by discourses on authenticity), or quotation (in literature and journalism).[8] At this point, providing conceptual clarification and differentiation of these terms would be much more helpful than demanding general revaluations that lead nowhere in the end.

There is yet another fallacy. The Internet certainly offers endless access to an unprecedented mass of cultural material, and digital technologies constantly provide increasing possibilities to process this. But the often-imputed creativity of reworking is the exception while the digital copy in the form of a duplicate in the cloud or on the hard drive still is the rule. This simple multiplication of cultural material is historically unique but it is rarely creative on its own, though it may be a starting material for creative action. Those who consume music, movies, and photos via networked contexts neither perform a creative act nor do they contribute to a more democratic formation of opinion.[9] Simply declaring this to be a criminal act is not a solution, neither is trying to define it as the expression of a new, more democratic system of production that consistently lives off a positive and progressive self-stylisation of its protagonists.[10]

How to analyse (culture with) Original_Copy?

The question of how to make my previous considerations useful for the analysis of culture remains. Since Original_Copy is a relational and strategically applicable proportion, trying to operationalise or even to quantify it does not make sense in my opinion. Attempts to measure originality or a copy's degree of creativity, as proposed by von Gehlen, are beside the point.

Also, the previously mentioned attempts of art history to properly determine originals reach an impasse: the good copies pass the test for originality until they are unmasked by refined methods. Thus, only copies can be definitely determined, not the originals; therefore, all originals are latently at risk to be yet another copy.

Thus, approaches to empirically assess Original_Copy, such as experts' reports, might remain to be useful and valid for economic and market-oriented contexts, but they neither achieve a cultural understanding of this complex field nor do they overcome its paradoxical anachronism.

Analytical premises and approaches

I cannot trace the different methodological and conceptual discussions on cultural analysis in recent decades at full length.[11] Nevertheless, most concepts agree on the assumption that the key issue of cultural analysis is to decode the meaning and semantic relations of cultural phenomena for the participating protagonists by investigating their specific contexts. That this requires us to think in relations and not in dichotomies became commonplace in cultural studies, at least since Stefan Beck's (2008) reflections on a relational anthropology and Lindner's (2003) contribution to the character of cultural analysis. Methodologically, the analysis draws on ethnography or historical, discourse analytical or narratological approaches or combinations of these.

The Dutch cultural historian Mieke Bal (2002) delivered an approach that productively brings together the different conceptual and methodological foci and thus is very pertinent to my argument. Bal puts three notions at the heart of her concept of cultural analysis: terms, intersubjectivity, and cultural processes (Bal 2002: 9). I will adopt these notions with regard to Original_Copy.

Bal's first notion, the variable usage of terms in narratives as a primary reservoir of our cultural baggage (2002: 9), is particularly fertile for my argument on Original_Copy. Intersubjectivity means, in this case, to impart knowledge by using appropriate terms, in order to make it available outside the academic world (2002: 10). This approach is important, all the more as Original_Copy plays a decisive role in diverse fields where different groups of experts argue with distinct concepts. Finally, Bal's reference to the processuality of culture is especially important for the cultural analysis of digital phenomena in networked lifeworlds in order to not lose track of the dynamics and historicality of the phenomena under investigation.

Original_Copy on the tramp

Science requires a precise terminology to facilitate mutual understanding and interpretation as well as to enable professional discourse; but also – for the purpose of a low-threshold intersubjectivity – to circulate the results of its knowledge production within non-collegiate circles. Regarding conceptual accuracy, the fascination and the immense variability of Original_Copy feed on the very tension between the putatively accurate opposition of original and copy and the frequent failure to discern this difference in actual objects, especially with regard to serial products. The fact that both original and copy are still potent concepts in digital contexts, is therefore, following Bal, due to their mobility.

She points out that concepts are not fixed, but rather wander around individual subjects, scientists, historical periods, and geographically dispersed communities (2002: 11). Being on the tramp, they convey different meanings and contexts, which is why they must be re-evaluated after each trip in order to avoid misunderstandings.

These characteristics of cultural study terminology (Bal cites as examples hybridity, text, media, identity, and even culture) – mobility, the capacity to carry meaning, and the necessity of reassessment and audit – can be aptly transferred to Original_Copy, except for the final step; obviously, the need for an update after each excursion got forgotten all too frequently. This means that a cultural analysis of Original_Copy must attempt to reconstruct the wanderings of both terms and their respective meanings in each given field of study.

Using the example of digital piracy of music or movie files, it is necessary to start with pointing out the various narratives of previous original or copied carrier media in order to take its respective implications and references into view. Only then can it be reconstructed which interests are pursued by using the concept of copy and how these are compatible with existing cultural practices. The reconstruction of the relevant regularities of the narrative of originals and copies is necessary, since both are not essential characteristics of the objects by themselves. They rather represent a relational quality obtained by the attribution of meanings. This is why the corresponding narratives are so instructive. One of the undoubtedly most powerful narratives related to Original_Copy is that of the unrepeatable image as the very essence of an original, although it is, strictly speaking, only a narrative of pictorial science or simply a 'cultural choice' (Ginzburg 1979: 275).

Besides these analytical questions on Original_Copy, it is also useful to do research through Original_Copy. Considering the existence of a howsoever organised discourse on Original_Copy, an indicator of cultural conflicts over interpretation can serve to better understand the respective discourse and brings to the surface that Original_Copy provides access to strategic, political, and hegemonic disputes over the both material and ideological valuation of cultural phenomena. It is a material valuation insofar as it determines economic criteria (for images, file sharing, and so forth) and an ideological valuation with regard to, for example, pioneering achievements in technology or science, but also with regard to the conservation status of restored objects, whether classic cars, furniture, or buildings. Anyone claiming an original status always pursues a special interest – for or against what is to be clarified in each individual case.

Cultural analysis and copyright

The relationship Original_Copy defines one already mentioned area, in which cultural processes (particularly those that are digitally mediated) must come to terms with existing regulations and standards: the copyright. Cultural analysis often comes into conflict with connecting creative practices of appropriation, makeovers, duplications, and demands for availability and openness to the statutory provisions. The appreciation of (and partly also the own enthusiasm for) the side of production seems to prevail, which is not necessarily conducive to a discussion of legal concepts and traditions. Cultural analysis through Original_Copy must also take note of statutory regulations without defining them as an obstacle to cultural practices from the very beginning; only then may it seek to influence them. It is not enough to determine the existing conditions as anachronistic. Such a claim requires considering the aforementioned intersubjective approach, since mediating between the concepts of cultural studies and the juridical logic undoubtedly is in need of translation work and techniques of anthropological understanding.

The public debates in recent years, in which artists and scholars claimed either to ease the legal requirements (when it comes to remix practices, sharing, or GEMA) or to sharpen them (when it comes to Internet surveillance, privacy, the right to dispose of its own data), show that the juridical dimension of cultural action in digital fields is an important reference point. Due to the different paces of technological and legal development, it is doubtful that laws could always correspond to the requirements of communication media and digital life. However, I also consider it an ethical necessity for cultural analyses in these areas to take account of both sides.[12] A cultural studies perspective on Original_Copy should question the juridical logic of the partially anachronistic regulations, which necessitates adding perspectives of legal history and discourse analysis.

Finally, even more radical approaches are fertile and provide helpful perspectives for this context, as it is a fundamental question whether culture (understood in the broadest sense as traditional knowledge), collective achievements, and established values should fall under the (Western) construct of copyright anyway.[13] Does not the dynamic, communication, and exchange-based character of any cultural phenomena contradict the concept of individually owned and legally protected intellectual property?[14] Don't we always understand culture as public common property, which is, or at least should be, available to all? Recently, European Ethnology also seized on the multifaceted public controversies about cultural property in various research projects.[15] This subsequently raised both critical and productive questions about the efforts in cultural policy to conserve and preserve cultural heritage, a concept which should, following Martin Scharfe, be distrusted by cultural studies scholars, because, being a concept of political practice, it is not so much a tool for reflection rather than a vehicle of making policy (Scharfe 2009: 15).

Conclusion and perspective

The current hype about utopias of future production capabilities is indicated by buzzwords such as 'The Internet of Things' (IoT, e.g., Anderson 2013) or 'Industry 4.0'. The former is generally associated with 3D-printing technology, which various parties consider to make the differentiation between consumers and producers obsolete, although the devices are still limited in their capabilities. Taking this vision even rudimentarily seriously will put the proportion Original_Copy to the test again. Patent holders will invoke originality in order to try to maintain the dominant logic of the descending value chain, and jurisdiction will come to their assistance. As a result, the template file could actually be acknowledged as original, which could then only be realised in the form of another copy by the technical digital reproduction method. This, in turn, would have consequences for the current patterns of argumentation regarding Original_Copy, because being an original in the sense of a unique unrepeatable piece is not conceded to files; this admittedly corresponds to a traditional logic, but not to the common cultural practice of digital duplicating.

The proportion Original_Copy in all its various forms does not always deliver satisfactory, unambiguous results for analysing and understanding digital phenomena in cultural fields, as culture is not unambiguous either. Functioning as some kind of magnifying glass, it reveals a clear view of the usually hidden cultural self-evidence and the different logics of valence. This is the reason for both its strength and the necessity to first question a phenomena's relation between original and copy before other analytical steps follow. Finally, it remains only to wait with bated breath to see whether each copy will indeed be a brilliant original.

Notes

1 In the following I use the spelling Original_Copy to indicate that it is not a dualistic relationship.
2 On the importance of the signature Tietenberg (2013).
3 Concerning this see the contribution by Bolz (2006).
4 The case of Wolfgang Beltracchi is currently the most comprehensive example of art forgery. Beltracchi counterfeited lost images over the years (especially from Campendonk, Ernst, and Pechstein). Later, he even created his own works and sold them with false provenances. This frequently succeeded due to positive expertises from connoisseurs like Werner Spies, who fell for the authenticity of the fake images. Despite the undisputed facts of the commercial fraud (Beltracchi had to do six years in prison), he still receives a lot of credit and sympathy from outside the art market; see Briegleb (2012) and Keazor and Öcal (2014).
5 Also relevant in this context are the discussions about the original conditions and restored or reconstructed parts of historical buildings and their authentic state in the World Cultural Heritage; see Gisbertz (2013).
6 Ethnographic approaches to the subject of creative industries offer Huber (2012), Krämer (2014), and Löfgren (2000).

7 GEMA (German society for musical performing and mechanical reproduction) is a German performance rights organisation. It represents authors, composers, and music publishers and the rights of other copyright owners.
8 Of course, these assignments are not always obvious. It is rather a tendential designation that can be elicited out of the literature on the topics; see in the order of the above mention Reulecke (2006), Theison (2009), Fuchs (2011), Mundhenke et al. (2015), Bonz (2006), Diederichsen (2006), Binas (2004), Djordjevic and Dobusch (2014), Lessig (2008), Doll (2012), and Fischer-Lichte (2012).
9 At this point begin the discussions about modern piracy, which on many levels acts as a metaphorical description of media-transmitted counterfeiting and copying practices. The concept of pirate modernity was shaped by Sundaram (2010) for urban contexts. Another rewarding perspective on the issue of patents and pirates offers Haedecke (2011).
10 One of the first scenarios of legal repression and control of cultural creativity in networked contexts comes from the US copyright expert Lessig (2004). He also has written a column for the magazine *Wired* since 2003.
11 A useful overview on these is provided by Bennett and Frow (2008).
12 I refer to Young and Brunk (2009) for the ethical and legal dimensions of cultural appropriation and distribution.
13 Michael F. Brown gives the example of the cultural knowledge of indigenous peoples that is currently not (yet) protected by copyright: 'Can culture be copyrighted?' (Brown 1998, 2003). Whimp and Busse also show in detail the current developments in the field of intellectual, biological, and cultural property, using the example of Papua New Guinea (Whimp & Busse 2013).
14 For the relation of authorship, ownership, and legislation, see Coombe (1998).
15 Regina Bendix and Kilian Bizer identify the main challenges and prospects for the debates on cultural property with regard to material and immaterial cultural heritage (Bendix & Bizer 2010).

References

Anderson, C., 2013. *Makers. Das Internet der Dinge. Die nächste industrielle Revolution*. München: Hanser.
Bal, M., 2002. *Kulturanalyse*. Frankfurt am Main: Suhrkamp.
Beck, S., 2008. Natur | Kultur. Überlegungen zu einer relationalen Anthropologie. *Zeitschrift für Volkskunde*, 104 (2), pp. 161–199.
Bendix, R. and Bizer, K., 2010. Cultural Property als interdisziplinäre Forschungsaufgabe: Eine Einleitung. In R. Bendix, K. Bizer, and S. Groth (Eds.), *Die Konstituierung von Cultural Property: Forschungsperspektiven*. Göttingen: Universitätsverlag Göttingen, pp. 1–24.
Benjamin, W., 1939, 1963. *Das Kunstwerk im Zeitalter seiner technischen Reproduzierbarkeit Drei Studien zur Kunstsoziologie*. 3rd ed. Frankfurt am Main: Suhrkamp.
Bennett, T. and Frow, J. (Eds.), 2008. *The sage handbook of cultural analysis*. Los Angeles: Sage Publications.
Binas, S., 2004. "Echte Kopien" – Sound-Sampling in der Popmusik. In G. Fehrmann, E. Linz, E. Schumacher, and B. Weingart (Eds.), *Originalkopie. Praktiken des Sekundären*. Köln: DuMont, pp. 242–257.
Bolz, N., 2006. Der Kult des Authentischen im Zeitalter der Fälschung. In A.-K. Reulecke (Ed.), *Fälschungen. Zu Autorschaft und Beweis in Wissenschaften und Künsten*. Frankfurt am Main: Suhrkamp, pp. 406–417.

Bonz, J., 2006. Sampling. Eine postmoderne Kulturtechnik. In C. Jacke, E. Kimminich, and S.J. Schmidt (Eds.), *Kulturschutt. Über das Recycling von Theorien und Kulturen*. Bielefeld: Transcript, pp. 333–353.

Briegleb, T., 2012. Bilder unter Verdacht. *art Magazin*, 10, pp. 32–41.

Brown, M.F., 1998. Can culture be copyrighted? *Current Anthropology*, 39 (2), pp. 193–222.

Brown, M.F., 2003. *Who owns native culture?* Cambridge: Harvard University Press.

Bunz, M., 2003. Yo, Entfremdung. *De:bug*, 31 March 2003. Available at: http://de-bug.de/mag/yo-entfremdung/ (Accessed 12 August 2015).

Coombe, R.J., 1998. *The cultural life of intellectual properties: authorship, appropriation, and the law*. Durham: Duke University Press.

Diederichsen, D., 2006. Sampling und Montage. Modelle anderer Autorenschaften in der Kulturindustrie und ihre notwendige Nähe zum Diebstahl. In A-K. Reulecke (Ed.), *Fälschungen. Zu Autorschaft und Beweis in Wissenschaften und Künsten*. Frankfurt am Main: Suhrkamp, pp. 390–405.

Djordjevic, V. and Dobusch L., 2014. *Generation Remix. Zwischen Popkultur und Kunst*. Berlin: iRights Media.

Doll, M., 2012. *Fälschung und Fake. Zur diskurskritischen Dimension des Täuschens*. Berlin: Kulturverlag Kadmos.

Dommann, M., 2014. *Autoren und Apparate. Die Geschichte des Copyrights im Medienwandel*. Frankfurt am Main: Fischer.

Eco, U., 1990. Die Innovation im Seriellen. In U. Eco, *Über Spiegel und andere Phänomene*. München: dtv Verlagsgesellschaft mbH & Co. KG, pp. 155–180.

Fischer-Lichte, E., 2012. Die Wiederholung als Ereignis. Reenactment als Aneignung von Geschichte. In J. Roselt (Ed.), *Theater als Zeitmaschine. Zur performativen Praxis des Reenactments. Theater- und kulturwissenschaftliche Perspektiven*. Bielefeld: Transcript, pp. 13–52.

Fuchs, P., 2011. Jedes Original ist ein listiges Plagiat und umgekehrt – Zur Funktion des Plagiierens in der modernen Gesellschaft. In T. Rommel (Ed.), *Plagiate – Gefahr für die Wissenschaft? Eine internationale Bestandsaufnahme*. Berlin: LIT Verlag, pp. 41–52.

Ginzburg, C., 1979. Clues: roots of a scientific paradigm. *Theory and Society*, 7 (3), pp. 273–288.

Gisbertz, O., 2013. Reproduzierte Originale und originale Reproduktionen. Zur Paradoxie von Authentizität in der Architektur. In U. Daur (Ed.), *Authentizität und Wiederholung. Künstlerische und kulturelle Manifestationen eines Paradoxes*. Bielefeld: Transcript, pp. 59–80.

Haedecke, M.W., 2011. *Patente und Piraten. Geistiges Eigentum in der Krise*. München: C.H. Beck.

Huber, B., 2012. *Arbeiten in der Kreativindustrie. Eine multilokale Ethnografie der Entgrenzung von Arbeits- und Lebenswelt*. Frankfurt am Main: Campus.

Keazor, H. and Öcal, T. (Ed.), 2014. *Der Fall Beltracchi und die Folgen. Interdisziplinäre Fälschungsforschung heute*. Berlin: De Gruyter.

Krämer, H., 2014. *Die Praxis der Kreativität. Eine Ethnographie kreativer Arbeit*. Bielefeld: Transcript.

Lessig, L., 2004. *Free culture: how big media uses technology and the law to lock down culture and control creativity*. New York: Penguin Books.

Lessig, L., 2008. *Remix: making art and commerce thrive in the hybrid economy*. London: Penguin Books.

Lindner, R., 2003. Vom Wesen der Kulturanalyse. *Zeitschrift für Volkskunde*, 99 (2), pp. 177–187.

Löfgren, O., 2000. The cult of creativity. In Institut für Europäische Ethnologie der Universität Wien (Ed.), *Volkskultur und Moderne. Europäische Ethnologie zur Jahrtausendwende*. Wien: pp. 157–167.

Mundhenke, F., Ramos Arenas, F., and Wilke, T. (Eds.), 2015. Mashups. *Neue Praktiken und Ästhetiken in populären Medienkulturen*. Wiesbaden: Springer.

Reulecke, A-K., 2006. Ohne Anführungszeichen. Literatur und Plagiat. In A-K. Reulecke, *Fälschungen. Zu Autorschaft und Beweis in Wissenschaften und Künsten*. Frankfurt am Main: Suhrkamp, pp. 265–290.

Scharfe, M., 2009. Kulturelle Materialität. In K.C. Berger, M. Schindler, and I. Schneider (Eds.), *Erb.gut? Kulturelles Erbe in Wissenschaft und Gesellschaft Referate der 25. Österreichischen Volkskundetagung vom 14.-17.11.2007 in Innsbruck*. Wien: Selbstverlag des Vereins für Volkskunde, pp. 15–34.

Sony International (Eds.), 1997. *in side 7. Katalog Herbst/Winter 97/98*.

Sundaram, R., 2010. *Private modernity: Delhi's media urbanism*. Abingdon: Routledge.

Täubner, M., 2012. Hassliebe. *brand eins 07*. Geld verdienen im Netz. Schwerpunkt Digitale Wirtschaft, pp. 39–43.

Theison, P., 2009. Copy/Paste. Das Plagiat als digitaler Schatten (Heute und Morgen). In P. Theison, *Plagiat. Eine unoriginelle Literaturgeschichte*. Stuttgart: Kröner, pp. 518–538.

Tietenberg, A., 2013. Die Signatur als Authentifizierungsstrategie in der Kunst und im Autorendesign. In: U. Daur (Ed.), *Authentizität und Wiederholung. Künstlerische und kulturelle Manifestationen eines Paradoxes*. Bielefeld: Transcript, pp. 19–34.

Von Gehlen, D., 2011. *Mashup. Lob der Kopie*. Frankfurt am Main: Suhrkamp.

Whimp, K. and Busse, M. (Eds.), 2013. *Protection of intellectual, biological and cultural property in Papua New Guinea*. Canberra: Australian National University E Press.

Wind, E., 1968, 1979: Kritik des Kennertums. In E. Wind, *Kunst und Anarchie. Die Reith Lectures 1960. Durchgesehene Ausgabe mit den Zusätzen von 1968 und späteren Ergänzungen*. Frankfurt am Main: Suhrkamp, pp. 38–69.

Young, J.O. and Brunk, C.G. (Eds.), 2009. *The Ethics of cultural appropriation*. Oxford: Wiley-Blackwell.

7 The manifestation of mash-up categories

Joan Kristin Bleicher

Introduction

Western societies have experienced long periods of media history where a single media form, such as the newspaper or television, was considered the primary form of public communication (see Wilke 1999). In this context, a large number of cultural and social changes have thus far been accredited to the influence of this single primary media form. Diagnostic analysis of an information society has been based, for example, on the range of diversity in print or broadcast media (see Mattelart 2003) or, as evidenced in the work of Gerhard Schulze, on how television has influenced our progression towards an experience society (see Schulze 2005). Similar relationships can also be found when one studies cultural changes and development. The term 'mash-up culture' has matured into a creative opportunity for professional and non-professional artists to capture existing cultural and/or media content and utilise it to create a new message (see Bruns 2007, Mundhenke et al. 2015). A closer analysis in the field of aesthetics and mash-up culture not only offers interactions within a single media form, but also between music, fine arts, literature, and theatre (see Müller 1996, Helbig 1998, Bleicher 1999, Mecke & Roloff 1999). Terms such as 'mash-up' or 'remix' are often used interchangeably to describe this diverse mixture of cultural and media forms. In analysing these two terms, however, it becomes quite clear that mash-up refers to visual forms such as videos (see Stuhlmann 2011) and remix refers to musical productions (see Djordjevic & Dobusch 2014).

Media studies is able to look into both current forms of the mash-up culture and its precursors and establish a basis for how our current media and cultural trends have developed. Within this development, difficulties that have arisen and the resulting evolution of creative and artistic trends that utilise existing media become obvious. As these trends have developed, cultural perspectives on the conceptualisation of an artistic production and the criteria that are considered in the fields of innovation and authorship have shifted. When a creative reproduction of existing media is presented, what was once considered a creation can be seen instead as an interaction. In this fashion, it is possible to assess a culture through mash-ups that are created in the existing and developing cultural environment.

Current manifestations of mash-ups

The broad variety of current manifestations of mash-ups in music, art, and literature create an environment in which a contemporary analysis has to be carried out with a variety of different approaches (see Mundhenke et al. 2015: 2):

> Online Remix Culture (Angeloro 2006), Read/Write Culture (Lessig 2008), post digital remix Culture (Harley 2009). Remix-Kultur (Stalder 2009), Cut-up-Kosmos (Fahrer 2009), Sampling Culture (Navas 2012), Mashup Culture (Sonvilla-Weiss 2010), User-Generated Culture (Frank 2010), Configurable Culture (Sinnreich 2010), Referenzkultur (von Gehlen 2012), Bastard Culture, Participatory Culture (Schäfer 2011).
> (Wilke 2015: 15)

There are many different multimedia forms of mash-ups, which can be studied in contemporary literature. In his novel *House of Leaves* (2000), Mark Z. Danielewski combined personal memories, novels, non-fiction books, pseudo-documentaries, and feature films with various typographic design forms, in a form that the reader can see as a type of concrete poetry. Seth Grahame Smith created *Pride and Prejudice and Zombies*, a literary 'clash up' (Kuhn 2012: 12) of high- and entertainment-cultures (see Wilke 2015: 22f.). Rainald Goetz's drama 'Festung' (1993) utilised existing text from the media to create dialogues between characters. Walter Kempowski's 'Bloomsday '97' (1997) followed a similar principle of organisation media with its integration of 24 hours of television reception. Helene Hegemann's novel *Axolotl Roadkill* (2010), which used blog entries from a variety of sources, was both successful and subsequently marked by critics as a mash-up.

Other literary mash-ups can be seen as a deconstruction of conventional narrative values. In Thomas Meinecke's novels, such as *Musik* (2004), *Jungfrau* (2008), or *Hellblau* (2001), characters interact and recontextualise text quotations from various sources. Raul Zelik asserts that Meinecke's stories consist of discourses and raise different questions in ever-changing contexts. Furthermore, Zelik argues that, in Meinecke's books, every detail is reproduced in any form or processed and duplicated by electronic means. Therefore, the reference to electronic music is obvious: the composition of quotations creates new works, which are reproduced and hence liberated from the Benjaminian aura of ingenuity (see Zelik 2004: para. 3).

Sixty years ago, John Cage used eight different soundtracks, which played and mixed simultaneously to create what one could say was the precursor to the digital mash-up. In contemporary music, combinations are often created between existing sound recordings and a new sound. Thomas Wilke from the Department of Media Studies at the Martin Luther University in Halle offers an interesting look into the development and potential for innovation in an environment of cultural technology. In his opinion, there has always been a discussion on contents connected to traditions and existing cultural techniques, evolving new works until the innovation potential is exhausted. Upon this, according to Wilke, old content is again picked up and, with a look back, set in a new context (see Forberg 2012: para. 3).

But is the decontextualisation of the source material enough to qualify a work as a mash-up? In this context, Dirk von Gehlen refers to the following characteristics of creative copies that qualify it as an original work: credibility; a new temporal, social, and spatial context; and an independent artistic achievement.

Thomas Mann spoke of a higher form of transcription, which one could say Thomas Meinecke updated in his work. The disclosure of one's sources became necessary. One good example of this is Hegemann's *Axolotl Roadkill*, in which the sources are revealed in the appendix of the book and therefore outside of the diegeses of the text.

Christian Kortmann focuses on the paradoxical, satirical orientation of many mash-up productions. In his opinion, satirical combinations are the most popular expression of the mash-up principle: the less the elements fit together, the better. This principle is also known from the visual arts as collage, from literature as the cut-up method, and from pop culture as bastard mix. Kortmann also observes that Internet users are constantly performing remixes, and reinterpreting and recoding art works. This leads to the parodic disclosure of stereotype storytelling, like for instance, integrating solemn music into the movie trailer of Sacha Baron Cohen's comedy 'Borat' that makes it seem like a social drama from Kazakhstan (Kortmann 2009: para. 3). Kortmann thus illustrates the focus on the creation of new meanings through cross-border combinations of existing media and cultural contents and the aesthetics and the resulting changes in the concepts of artistic production (Frank 1987, Zima 1995). Literary mash-ups are not only a wide range of multimedia covers but also, as the examples showed, parodic, repetitive, and constructivist forms of expression.

Mash-ups in documentary film

Through its historical development, film has supplied various models for dealing with existing image materials in visual media. In 'A Movie' (1958), Bruce Conner edited a collage of newsreel and film clips together to create self-reflexive conflicts and offer a perspective on current events and culture (see Wilke 2015: 21). Since the 1970s, demonstrations have utilised documentaries as a communication tool. These documentaries were often compilation films using a mash-up of image material of different film-makers and teams to create a documentation of circumstance and events (Brunow 2015). 'Von der Revolte zur Revolution' (1968/1969) combines image material from various members of the Hamburger Filmkoop, such as Helmuth Costard, Gerd Meissner, Martin Grasmann, Fritz Strohecker, and Kurt Rosenthal (director). The film uses image material from anti-Springer Demonstrations, rallies on the 1st of May, and at the Bonn demonstrations against the German Emergency Acts. In the 1980s, a number of video teams compiled video footage to document the port strikes in Hamburg (Wees 1993, Brunow 2015, Kügle 2015).

Various forms of creative interaction with film material from different sources can be found in both non-fiction and fiction movies (for example, in horror films such as 'The Blair Witch Project'). The German online lexicon of film terms describes the documentary genre of found footage film from the production perspective as movies or parts of movies composed of found footage, which was shot by other film-makers in different contexts. Found footage can stem from all kinds of sources and comprises archive photos, film garbage, film rests, or home movies. In any case, the footage originates from other contexts and receives its new meaning through its integration in new footage. The online lexicon of film terms especially emphasises that the art-house film genre makes use of found footage, whereas its formal-aesthetic quality often outranks the content. The content rather dominates in the compilation movie, which is also composed of found footage originally produced for another purpose (see Lexikon der Filmbegriffe 2012: para. 1).

'Mocracy: Neverland in me' (von Borries 2013) helped to establish the current methods that are used to incorporate found documentary footage in films with its utilisation of original radio tone, such as the live coverage of a funeral or film clips with narration that reaches into the area of the mockumentary (see Roscoe & Hight 2001). Existing resources were incorporated, such as the radio drama 'Meine Tonbänder sind mein Widerstand' by Thomas von Steinecker (BR 2007). In a similar way, prototypical documentary mash-ups combine essay film and found footage film, blending picture and sound from different sources to offer a perspective on a topic. In this fashion, film-maker Chris Marker offered a glimpse into human memory by using different film material with the theme of cultural differences in how we recall what has occurred (see Scherer 2001). Marker actively mixed his own footage with external footage and photos from Japan and Africa, narrated by a woman's voice reading poetically composed letters she has received from a fictitious cameraman who has written about his experiences. Through this narration, the viewer is offered a unique perspective on a diverse variety of cultural interactions. This interaction with the viewer is created with the medium of film itself, wherein each scene has not only its own conditions, but also the elements of time and rhythm, melancholy and recollection. The sound that accompanies each scene finds a balance between the intonation of serious analysis and detail-loving curiosity and playfulness. In his contribution to the *Metzler Filmlexikon*,

Brunow describes Marker's method of mixing different image material as follows:

> he combines pictures of a rocket being launched with pictures of emus on the Ile-de-France, pictures of the last lighthouse using kerosene on the island Sal with pictures of the fortress tower that once protected Joan of Arc. Other pictures, freed from the burden of meaning, seem to explode into pure colour and motion. Marker is said to take found footage from different contexts and to act unlimited in the way of recombination. Despite lacking a coherent concept, Marker's pictures are not considered to be arbitrary; on the contrary, Brunow argues that Marker does justice to the pictures by keeping their secrets.
> (see Brunow 1995: 181; trans. author)

Interestingly, the same structural principle is utilised in the film's dialogue and sound. Brunow describes how the role of a commentator is to accompany the pictures, to anticipate them, or even to follow different associations than the sequence of images suggests. Thus, the nature of the commentator remains ambivalent: it is not a singular author sharing his or her thoughts. The line between the person reporting and the person quoting is blurred. This approach is considered to prevent a uniform perspective on the level of speech (see Brunow 1995: 181).

Comparable to a literary essay, the film-maker maintains a subjective approach within a structure of cinematic reflection. Long before film existed, literary essayist Michel Eyquem de Montaigne made an interesting observation on the degree of truth or misconception that lies in one's individual perspective. As he considers his own text a record of different and changing events, as well as unfinished and sometimes conflicting thoughts, he admits to be self-contradictory at times. Nonetheless, he argues that, in doing so, it is impossible for him to 'contradict the truth' (see Michel Eyquem de Montaigne 1998: 398).

This subjective fragmentation of thought and perception and how they affect one's associative interpretation of events is a model for the construction of reality in Chris Marker's documentary film 'Sans Soleil' (1983). This associative model of perception is contrary to the linear, chronological, or causal models of reality, which are the dominant methods for information transfer in media. In effect, we interact with media with a varying range of complexity, in a variety of ways: the direct observation of live broadcasts, the linear structure of a news report, the causal structure of periodicals, and the associative structure of a documentary.

A further form of mash-ups can be described as first-person documentaries and offer a subjective, autobiographical perspective, mostly using amateur films and private family photos. These include David Sieveking's 'Vergiss mein nicht' (2012) or Agnès Vardas's 'The Beaches of Agnes' (2009) (see Weißer-Gleißberg 2014). The biographies and portraits of the protagonists are the end product of narrative structures, which are created in post-production, combining both professional and non-professional image material.

In the field of documentary films, found footage films are created with a comparable procedure to mash-ups, utilising a combination of image material from various sources (Hausheer & Settele 1992). Media artist Candice Breitz argues that the creative process is not about originating and animating, but rather about recycling, translating, interpreting, in short, a process of reanimating materials and languages that pre-exist one's own practice as an artist (Djordjevic & Dobusch 2014: 25). This process of translating, interpreting, and reanimating is clearly displayed in Christian von Borrie's 'Mocracy: Neverland in me' (2012), where the conjunctures and dialogues between fiction and reality form through switches between film scenes, original footage filmed in Dubai, and war movies. But there are also mash-ups or polylogue from Michael Jackson videos, profiles of dictators, images from news events, and surrealistic architectural shots from Kazakhstan and Kosovo. These mash-ups create not only analogies between film scenes and architecture, but almost more significantly, between the images that stars develop and maintain in popular culture and the media self-representation of political leaders. This applies for example to Michael Jackson, Vladimir Putin, and the Kazakhstani President Nursultan Nazarbayev. Democracy is expressed by laying images of American election posters on Kosovar houses as a recontextualisation. Through this, the director is able to enhance the reflection of documentary, film, and acoustic conventions. Key factors are shown such as overlaid lettering on the screen and media-theory quotes from Friedrich Kittler are read by passers-by into a microphone in several locations such as Berlin (Kügle 2015). What is also particularly interesting is the acoustic aesthetics of the synchronous usage of music and asynchronous use of language. Primarily in the post-production, only image sequences were produced with the sound on and then the sound was cut. From the very beginning, the film develops a rhythm of its own through the intermittent playing of the German National Anthem in sync with the symphony 'Freude, schöner Götterfunken'.

Film-maker Christian von Borries reported, in a public discussion about political documentary films in Germany (29 July 2013), that the only material that was filmed for the production of 'Mocracy: Neverland in me' was the material that could not be found online. In this context, the crew travelled to both Kazakhstan and Kosovo. The triangular relationship between these countries with Germany offered an opportunity for film viewers to take a critical look at the situation in their own country from a new perspective.

Correlations between the images, the language (such as quotes from media-theory publications), locations, and focal points (individualism, mass phenomena, and growing tension in both localisation and globalisation) offered the viewers themselves a partial context in which they could decide what the text 'Film is psychoanalysis' at the beginning of the film implies. A clarification of meaning of the images was deliberately omitted. The central theoretical concepts were, however, integrated in the images, offering viewers the opportunity to reflect and associate these images within the surrounding context. Media theorist Marshall McLuhan had already discussed the relationship between the media connections and cognitive processes of the spectators in 1964 (McLuhan 1964: 63).

Quotations, allusions, and small-scale mash-up forms

The small-scale forms of mash-ups can be seen as manifestations of intertextuality and intermediality, quotes, and allusions (Helbig 1998).[1] There can be different quote forms and functions such as quotes from a genre, a programme, a direct quote, an image, or a movement.

Genre quotes integrate a confrontation with stereotypes within established genre conventions. Director Wolfgang Petersen was able to accomplish this in his 'Tatort'-episode 'Jagdrevier' (1973) by including traditional action elements of the Western genre such as a showdown in typical locations – the village pub serves as the saloon – or camera settings such as the empty village road. Through this integration of typical elements of the Western genre in the genre of the TV crime thriller, the director extends the range of substantive and formal narration.

Quotes from a programme are frequently found in the context of television parodies, such as the series 'Kalkofes Mattscheibe' (Tele5). Short extracts from TV shows are first shown in their original version and then parodied by comedian Oliver Kalkofe.[2] Quotes from people or roles focus, for the most part, on stereotypes that the characters want to portray, such as Else Kling ('Lindenstraße'), who resembles the chattering cleaning woman from the TV series 'Family Hesselbach' (HR). In the 'Tatort' episode 'Jagdrevier', Dieter Brodschella, the protagonist, uses quotes to reinforce the role of a Western outlaw (see Bleicher 2005b). Direct quotes integrate text excerpts or known sayings from other sources into the current script. For example, 'I think this is the beginning of a beautiful friendship' from 'Casablanca' has been used in many movies and television series (such as 'The Simpsons'), thereby creating and integrating an allusion to its historical film precursor.

Image quotes are made, for example, by incorporating a clip from a film or TV series or by the playing of image sequences in the background (such as on a TV screen in the room). Image quotes can be a palimpsest of different integrations into a film or TV plot: the image of the new episode or film is deposited on the image of a previous episode or the older films, yet the background image remains visible. This blending of images blurs the boundaries between fact and fiction. One example is the integration of a self-produced 'Tagesschau' news story based on a fictitious event in a 'Tatort' series episode. Comparable allusions are often found in TV movies, based on established conventions of feature films. Another example is the TV movie 'Hangover in High Heels' (SAT.1 2015), wherein a variety of image and action elements, such as getting tattooed while under the influence of alcohol, mirror the cinematographic film series 'The Hangover'.

Dirk von Gehlen claims that football player Messi copied a goal previously scored by Diego Maradona, and shows the two goals as mirror images of each other.[3]

These detailed ways in which media uses different quote forms is also a basis of the current trend towards fragmented story telling. Paul Booth (2012) has observed an adaptation to the structure of the online services, such as microblogs (such as Tumblr), in current forms of fragmented and complex narrative series. Most short web series can be seen in this context as a mash-up of television series, amateur videos, and/or video blogs elements (see Fagerjord 2010).

The general departure from the demand for innovation in various forms of mash-ups can be associated with a departure from established linear order models. Instead of innovating new forms of production, it is acceptable to innovate new ways to network, link, and utilise existing forms. This has also resulted in a detachment from the demand for innovation of the traditional avant-garde, instead working within existing cultural hierarchies (see Twitchell 1992: 253–274).

Cultural and social contexts of current mash-ups

The multitude of different descriptions that have been surfacing for the cultural and economic contexts of the current mash-up phenomena indicate ongoing changes of existing concepts and production procedures. Convergence terms describe the fusion of different media forms (see Jenkins 2006). Economists and media journalism researchers have often used the term crossmediality to describe the recycling of the same content in different media contexts (see Jakubetz 2011).

Henry Jenkins has looked into various aspects of current intermedial references and production processes and defined it as a Convergence Culture (Jenkins 2006), which can be compared to the traditional definition of a remix culture in the following way.

> Generally speaking, remix culture can be defined as the global activity consisting of the creative and efficient exchange of information made possible by digital technologies that is supported by the practice of cut/copy and paste. The concept of Remix often referenced in popular culture derives from the model of music remixes which were produced around the late 1960s and early 1970s in New York City, an activity with roots in Jamaica's music. Today, Remix (the activity of taking samples from pre-existing materials to combine them into new forms according to personal taste) has been extended to other areas of culture, including the visual arts; it plays a vital role in mass communication, especially on the Internet.
>
> (Remix Theory: para. 2)

Convergence is considered to be a key term for cultural and media changes that take place in the mash-up phenomena. Henry Jenkins welcomes the readers to his popular scientific study *Convergence Culture* with the words:

> Welcome to convergence culture, where old and new media collide, where grassroots and corporate media intersect, where the power of the media and the power of the media consumer interact in unpredictable ways.
>
> (Jenkins 2006: 2)

The quote refers to fundamental changes in existing media systems. Digitalisation technology enables the production and distribution of content of different origins in different media forms through the usage of binary codes. This technical innovation has rapidly expanded the spectrum of both the distribution and further processing of existing media services (cf. Claussen 2014) and has allowed the innovation of different forms of intermediality (Rajewsky 2002). These disappearing limitations and regulations of the mash-up culture not only change the media system, but also the production and distribution contexts, media planning models, available media forms, the individual reception of media, and the collective impact of media. They thus form a relevant subject area which can be researched in communication studies.

One central aspect of the cultural changes has been the increased interactivity of users in the observable circulation processes (Schade & Tholen 1999):

> This circulation of media content – across different media systems, competing media economies, and national boarders – depends heavily on consumer's active participation.
>
> (Jenkins 2006: 3)

Musician Greg Gillis spoke from a similar perspective of a 'remix and appropriation culture' (Djordjevic & Dobusch 2014: 33).

Production aspects of mash-ups

Interactivity is a basis for the dissolving of borders between professional and amateur productions that help to increase the exposure of the social conflict between economics and the strategies of counter-public information. Jenkins notes: 'Corporate convergence coexist with grass root convergence' (Jenkins 2006: 18). Media scholar Mirko Tobias Schäfer refers to the context of the amateur culture, present among other ways, through the user-generated content on the Internet (Schäfer 2011). Mash-ups have changed from a method of professional creative expression to an expression of fan culture, according to Henry Jenkins.

The manifestation of mash-up categories   141

On YouTube, for example, one can find many fan videos compiled from existing movies or series in which the fans desire to bring the plots together: in the mash-up online video 'Buffy Meets Edward', the protagonist of a popular television series is in a conflict-filled relationship with the protagonist of a popular cinema film (see Stuhlmann 2011). A montage from the series and movie scenes creates common dialogues, views, and action sequences, the beginning and the dramatic end of a relationship in high school. Claussen sees this form of video editing as the manifestation of a diverse fan culture in which fan videos are created to express an enthusiasm that existing media alone cannot convey (Claussen 2014: 89).

Although such mash-ups are mainly seen as a hallmark of amateur and fan culture (Bruns 2007), there are also attempts to market mash-ups with economic interests in mind. Both professional and non-professional entertainment products are used for this crossmedia marketing. Media scholar Dan Harries notes that the film and television industry see the potential to connect through the Internet with wide, globally dispersed audiences and communities: '(...) so they continue to develop entertainment products that combine the entertainment industries' proven ability to "entertain" with the Internet's ability to "connect"' (Harries 2002: 172).

Henry Jenkins emphasises the variety of interests held by participants within the Convergence Culture (2006: 19): while producers try to expand their markets and revenue, participants try to increase the number of recipients with whom they can interact.[4] This interaction is the primary constitutive element of the social web, which has also established its position as an important distribution medium for mash-ups (see Bleicher 2014).

In his contribution to the collection *Generation Remix*, the producer of Mashup Germany emphasises that, in addition to the production and the technical aspects of mash-ups, there is also specific software for mash-up production (Wessel 2014: 126). The production process itself involves much more than just the remix. Christian Forberg divides the production process of musical mash-ups in two main steps. First, one uses any number of new music tracks, placing them on the respective audio track number. Then, one adds any number of electronic sounds, music, and noises by picking them to pieces and inserting them in the desired position to create one single file for the mash-up (see Forberg 2012: para. 1).

Authors in publications such as the anthology *Generation Remix* often discuss how remix productions are traditionally in the artistic avant-garde. The artists themselves take the songs and videos that they find interesting and use them as material within the creative process. When they use a re-sync strategy as a form of remix for example, they separate the picture and soundtracks, synchronise the picture with music from a completely different context, and thereby change the context of both, often creating a parody of sorts (Claussen 2014: 87).

Forms of visual mixing and rearranging

In a similar way, mash-ups are created from current movies and TV series (see Claussen 2014: 88). An example of this is the YouTube Video 'Virales Marketing im Todesstern', where the character Dodokay in a scene from 'Star Wars' discusses the potential of viral marketing using a Swabian dialect. In the rearrangement of scenes, the separation of image and soundtrack is not always entirely necessary: according to Claussen, Pogo's Pulp Fiction cutup 'Lead Breakfast' serves as an example, in which the audiovisual entity of individual elements from the original movie plays a crucial role for authenticity. The context of the original movie remains complete, whereas the film sequences are arranged in the style of a groovy music video (see Claussen 2014: 91).

Despite frequent analyses on the importance of both identifiability and immutability of the original, the organiser of the 2012 conference 'Mashup. Theorie – Ästhetik – Praxis' in Leipzig focused on the processes of mixing different mash-ups. 'Mash-up' has become the common term for auditory, visual, or audiovisual 'mashed' new arrangements, collages, and bricolages in different art forms (e.g., in music, video, computer games, contemporary media, art, and architecture). Wilke emphasised that what all of the mentioned techniques have in common is that they join heterogeneous elements to a (seemingly) new piece or at least give it a new quality. Moreover, mash-ups are often self-referential insofar as they reveal their own mediality and make it a subject of discussion (see Wilke 2012).

Candice Breitz emphasised in an interview with Lawrence Lessig that artists in general utilise the work of their predecessors to deal with a contemporary, current perspective (Lessig 2014: 25). This diagnosis implies a symbiosis of creative production and critical, interactive reception of the previous linear understandings of the communication processes. Thus, mash-ups are a particular challenge for media-related research.

Research perspectives on media interactions

Previous studies have focussed for the most part, along with the technical fundamentals of mash-ups, on the changed relations of production and reception in media systems. As digitalisation technology has developed and become more common, new forms of distribution have directly influenced and incited change in established cultural and media participant constellations, which have been described with a number of coined terms such as prosumers or produsage (see Bruns 2007; von Gehlen 2012: 50). Formerly, passive recipients have become active commentators and producers of content, a concept to which researchers have given names such as audience participation (see Schmidt & Loosen 2014) or Amateurkultur (amateur culture) (Reichert 2008). Jan Torge Claussen describes this change using the example of YouTube in accordance with Geert Lovink's redefinition of the 'social'. The 'social' or rather the processes of exchanging, evaluating, commenting, collecting, and sharing media content are considered the core of contemporary video practice, not only a side effect of audiovisual media, as Geert Lovink remarks in his essay 'Online-Videoästhetik oder die Kunst des Datenbankschauens' (see Claussen 2014: 85).

An important yet limiting resolution within the mash-up culture lies in the field of old and new media aesthetics and design principles. In their influential study on Remediation Phenomenons (2000), media researchers Jay David Bolter and Richard Grusin look into this integration of analogue cultural forms of expression in the digital culture. Interactions between literary texts, arts, and the media as an integral part of cultural-historical developments have long been a research focus. Literary scholar Gerard Genette (1993) has stated in his palimpsest study that there is a broad range of different forms of literary intertextuality. This literary model of the overlaying of text palimpsest acts as a formal precursor as the image overlays in earlier mash-up videos such as 'Take this dance'.[5]

In his study, Genette also emphasises the dependence on and the importance of intertextual references. Michel Riffaterres describes the phenomenon of *intertextuality* from the perspective of the recipient insofar as the 'intertextual' quality depends on whether or not the reader perceives the references between one work and other works preceding or succeeding it (see Genette 1993: 11). The intertextual references between different media that one can describe as intermediality rely on the recipient's ability to identify the reference.

Further research foci are the historical precursors to the current mash-up phenomena and a conceptual assessment of different cultural and media interactions. Media scholar Yvonne Spielmann stresses the long tradition of artistic interaction in relation to the history of the concept of intermediality. The terms 'intermedia', 'intermedio', and 'intermezzo' describe dramatic interludes between acts, common from the 15th to the 17th centuries, or music pieces, dances, vocal numbers, pantomimes, acrobatic elements, and farces performed during the breaks of a comedy (see Spielmann 1998).

Literary scholar Irina Rajewsky researched the relationship between film and literature and distinguishing the differences between the inter- and intramediality categories (2002). She defines intermediality as phenomena that cross media boundaries and involve at least two distinctly perceptible media (Rajewsky 2002: 12f.). Furthermore, she describes transmediality as the occurrence of the same subject, the implementation of aesthetics, or a certain type of discourse in various media, without the possibility or relevance to determine the contact source media.

Generally describing a parody, intramedial references are created in the procedure of constituting meaning in a medial product by referring to a product (single reference) or to one or more subsystems of the same media or its own media qua system (system reference) (see Rajewsky 2002: 76). Intramedial references are based on literary texts as contact-taking products and designate the relation between one text and another individual text (or other texts) or one or many semiotic systems without crossing media boundaries (Rajewsky 2002: 71). In intramedial references, Rajewsky distinguishes the difference between individual references in the form of quotes from other media or system references such as system mentions or discussions of the reference system. In system contamination principles of foreign media, systems are integrated and then translated in the context of the existing system. These different forms of intermedia references offer structural models for current manifestations of mash-up culture.

Studies on border resolutions of literary and journalistic mash-ups

A further focus of research is on the manifestation of resolving limitations between cultural and media forms. These include reducing limitations in the areas of authorship and online-based literary forms, among others, which is also referred to as digital literature (see Simanowski 2001). The network structure of the Internet is reproduced through the networking of authors, their writing projects, and their texts. Roberto Simanowski defines digital literature as an artistic form of expression whose existence relies on digital media as a foundation, because it is characterised by at least one of the following specific criteria of digital media: interactivity, intermediality, and orchestration (Simanowski 2001: 4).[6] Furthermore, Simanowski defines interactivity as 'the participation of the recipient in the construction of the works' (Simanowski 2001: 5).

In turn, terms such as cross- and transmediality refer to different forms of the multiple use of content in different media contexts (cf. Ryan 1992, Jakubetz 2011). One can also find economic forms of reusability: online newspapers such as the *Huffington Post* can be seen as a mash-up blog of print media, where citizen journalism is bundled together and offered without the respective authors receiving any payment.

It has already become clear, in the field of remediation research of online design (Burgess & Green 2009), that these forms of digital culture integrate existing artistic traditional standards and rules. The continuing development and adaptation of accepted standards in the Fine Arts has created a bricolage of postmodern literature and cinematic installation, which can also be seen as a mash-up of artistic or political forms of expression in combination with existing media material as a form of recycling and re-updating (Stahl 2015: 216). Possible additions and enhancements are made by utilising additional picture material, voice-overs, synchronisations, or re-enactments.

Studies on the relationship between the original and the resulting production in mash-ups

In this context, the combinations of previous and current productions can be classified by looking into the relationship between the original and the end product. Jan Torge Claussen emphasises the fact that for YouTube users, the ability to collect and share videos is limited. Videos have been repeatedly uploaded and remain almost unchanged (Claussen 2014: 86). Roman Marek created the term video clone for this form of identical repetition (Marek 2013: 82ff.).

In the case of other types of creative processing, the question is: which elements are combined, in what way, and with which aim? Mash, as the English term for mixing, is similar to bricolage in postmodern literature and sampling and remixing in music (DJs could associate this with club appearances with two turntables and two different records). Remediation in the area of web design refers to the handling of the Fine Arts. Based on their features, Eduardo Navas differentiates two forms of mash-ups in the music and software areas:

> Regressive mashups, which mix two different songs into one song.
> Reflexive mashups, which mix different forms of information e.g. in the form of news feeds or maps.
> (Navas 2010: 157f.)

Navas assigns mash-up forms the parent category 'remix', which is then segmented as follows: extended, selective, reflexive, and regenerative remixes (2010: 159). This model is an example which shows that mash-ups are similar to artistic remixes and can be categorised in a similar fashion to the collage or bricolage phenomena.

Different definitions include different aspects of interactions and relationships between media and art. These interactions can ask for the identification of the respective aspect of media or a cultural starting point. Media theorist Stefan Sonvilla-Weiss emphasises the importance of the original in distinguishing between mash-up and remix. Collages, montages, sampling, or remixing can create a new product through changes, recombination, manipulations, and copying, whose roots remain recognisable. A mash-up distinguishes itself as follows: 'the original format remains the same and can be retraced as the original form and content, although recombined in different new designs and contexts' (Sonvilla-Weiss 2010: 9). Dirk von Gehlen refers to mash-ups as a modified, fluid state in comparison to traditional cultural offers (von Gehlen 2013). Lawrence Lessig looks at mash-ups from a legal perspective and sees 'samples that are remixed without the approval of the original artists' (Lessig 2014: 30). This principle creates a situation that exceeds and actively changes existing power structures and the hierarchies of cultural productions.

Studies on traditions within the mash-up culture

A wide variety of volumes and studies look into current cultural traditions in the mash-up phenomena. Eduardo Navas, for example, includes references to cultural and historical developments in his book on remix theory. The introduction states:

> *Remix Theory: The Aesthetics of Sampling* is an analysis of Remix in art, music, and new media. Navas argues that Remix, as a form of discourse, affects culture in ways that go beyond the basic recombination of material. His investigation locates the roots of Remix in early forms of mechanical reproduction, in seven stages, beginning in the nineteenth century with the development of the photo camera and the phonograph, leading to contemporary remix culture. This book places particular emphasis on the rise of Remix in music during the 1970s and '80s in relation to art and media at the beginning of the twenty-first century. Navas argues that Remix is a type of binder, a cultural glue – a virus – that informs and supports contemporary culture.
>
> (Remix Theory 2012: para. 1)

One substantial influence from the field of the visual arts is the collage work of the Dadaists in the 1920s (Bruns 2010: 24). According to Georg Fischer, the artistic-intellectual avant-garde of the 1920s used techniques that are comparable to present digital techniques (Fischer 2014: 69). Collages of modern art have already provided templates for a broad variety of experimental animation films (see Herbst 2012).

Further publications describe mash-up developments in both music and in film (see Stuhlmann 2011). Mixing and sampling have been a consistent element in popular music since the 1970s and can be seen as the predecessor to mash-ups. Through the development of hip-hop (originating in the 1970s and 1980s) and of drum and bass music (originating in the 1990s), repetition and editing of sound fragments have found their way into popular music (see Fischer 2014: 73).

Terry Riley's experimental sound study 'You're no good' is considered to be one of the first remixes (Fischer 2014: 71). Fischer points out that there was a distinct motivation to create a unique, innovative, avant-garde work from existing materials. New work must deviate from the familiar to be considered original, surprising. Such a deviation must be significant but not too radical to still keep a connection between the new and the familiar work (Fischer 2014: 72). These modifications can also be seen in a second example. Matthijs Vlot combined a series of short film fragments to recreate the lyrics to Lionel Richie's 'Hello', while playing the instrumental track in the background (see Wilke 2015: 29f.).

One can use terms from literature studies such as quote or bricolage to refer to existing traditions and creative cultural interactions, theories, and forms of expression. Various bricolage definitions portray the dealing with the history of culture as a kind of stone quarry from which writers can mine their materials and write text freely and without time commitments. Fundamentally, this would include both the open form of artistic expression and the dialogical principle of communication between literary texts. Based on

Bakhtin's concept of the dialogic nature of novels, Julia Kristeva develops an open understanding of texts by considering intertextuality as the generator of text structures. Arguing that the constitution of meaning in texts makes them fundamentally connectable resembles Roland Barthes's concept of texts (see Krewani 2001: 22f.).

Jan Torge Claussen refers to it as a remix-culture, noting that copy, cut, and paste are not only standard functions of computer programmes, but are deeply rooted in our culture (Claussen 2014: 79). Furthermore, in several other texts, there is a variety of synonyms used for the terms mash-up and remix. Claussen emphasises that terms such as collage, mash-up, and remix denote the reuse of different media objects (Claussen 2014: 81). Mash-up as a concept has become an umbrella term, which has a constantly evolving list of terms, ideas, and concepts that it encompasses.

These discussions regarding different cultural and historical precursors and current mash-up developments imply that modern innovation is no longer the primary concern, and people are looking to find creative ways of dealing with and utilising existing material in a postmodern era. Progress is still evident, however, as new ways to network existing material are created and the content and the ways in which it is represented changes. But even in this context, one can see it as a form of innovation. Lawyer Till Kreutzer describes this ability to innovate as the creation of new and original works through cultural techniques such as mashing, remixing, and sampling, exploiting the synergy between existing and new contents. According to him, the new works have their own expression as a whole and address the recipient in a different way than single elements would (see Kreutzer 2014: 44).

The crisis of originality that is often associated with the boom of mash-ups, as described by Dirk von Gehlen in his anthology of mash-ups, goes back in cultural history to times where often a lack of content and forms of expression innovation was criticised. Von Gehlen's crisis concept offers an exemplary view into a creative environment where, when an artist creates a work by interacting with existing materials, it is often seen as plagiarism or falsification. Film mash-ups have been particularly problematic as far as legal and copyright issues are concerned, and this ongoing debate is quite evident when one looks at the number of YouTube videos that are currently blocked in Germany by GEMA.[7]

Issues, methods, and foci of mash-up analyses

Mash-up phenomena are currently being discussed and studied by a number of different scholarly disciplines and are therefore analysed using a variety of different methods. One example is Gebhard Rusch at the University of Siegen, who describes his research foci on his project on mash-ups from a sociological perspective as follows:

> The project focuses on socio-media processes in prosumer media-cultures across the World Wide Web. The term "presumption" refers

not only to the latest mashup media but also to a new way of communicating and handling media: Freely configurable mashup media that are generated on an ad hoc basis are taking on a whole new form across all the traditional categories for the production, distribution and reception of stabilised social and media structures. This produces prosumer mashup cultures and is driven by the enforced technological process of digitalisation and the mass-market rendering of information by means of digital editors, aggregators and other software tools, from RSS feeds to news & entertainment pipes. The result is the ongoing fusion of socio-media production, distribution and reception contexts within a cross-media world of "digital presumption". At this point, mashup cultures leave the multimedia of "Web 2.0" – including all the individual media they simulate – behind them. Prosumer activity is no longer a question of expressing interest in the original information by traditional means; nor does it deal with the multimedia networking of individual snippets of media: The principle driving force is the cross-media hybridity of mashups. This throws up new perspectives on media history, as well as new theoretical and empirical challenges for research that treats the mashup media as both indicator and symptom of the next leap forward in the mediatization of our society.

(Mashup-Media 2010: para. 1)

The ways in which different cultural and media forms of expression interact in mash-ups are the subject of study in cultural studies and media research. Research in the humanities has looked into the relationship between artistic forms of expression and media texts and offered a number of different terms to describe this interaction.

Early on, Julia Kristeva looked into the intertextuality between different literary works. Gerard Genette examined interactions between literary texts in the areas of motifs, characters, and quotes in his study of literary palimpsest. With the term transtextuality, he described the manifestation and hidden connections between one text and others (see Genette 1993: 9). This can take on the form of dialogues between different texts from different phases in the history of literature.

Christopher Dallach extended Genette's theory in his review *Eheversprechen und Leichenschmaus* with literary, musical, and media mash-ups. In this context, he was able to discuss the development of pop music mash-up and its transition into the development of the DJ culture in the 1990s, in which amateurs and DJs started to blend different artists' songs into 'bastard pop' collages; this technique corresponds to the literary cut up, promoted by William S. Burroughs (see Dallach 2010: para. 5).

Shortly after the success of *Pride and Prejudice and Zombies*, Christopher Dallach observed that a whole series of mash-up novels were being created and marketed in the United States, such as *Sense and Sensibility and Seamonsters* and *Android Karenina* (ibid: para. 6).

This form of mash-up literature helped motivate Gerard Genette to utilise the online research concept of hypertext to observe and document the direct and indirect transformation of the text (Genette 1993: 18). Genette's definition can also be applied to other forms of artistic expression. The German band Tocotronic used this concept for their song 'Jenseits des Kanals' (1999) with Gustav Flaubert's novel *Bouvard und Pecuchet*. Machinima, in turn, uses real-time graphics to reproduce imitations of films in online game environments (see Schwinghammer 2008). Literature and media scholars such as Marie Laure Ryan (1992) research transmediality regarding the selective distribution of different sub-areas of a narrative to various media distribution platforms. The plot of the US series 'Lost' has not only been used for television, but also for online videos and computer games. Within this transmedial narration, one experiences interactions with the characters and the plot layout. However, the degree of interactivity integrated in the use of transmedial narrations can greatly differ. While YouTube videos of the 'Lost' series explain and clarify events and interactions in the plot, the players in computer or online games are able to explore the locations of the series interactively and immerse themselves in the diegesis of the plot.

Immersion and narrative complexity are closely connected in this specific form of transmedia diegesis and the concept was already implemented by Hollywood with commercially oriented goals.

> Some of the earliest attempts by Hollywood to put 'original' entertainment on to the Internet were efforts to create new media properties that had this 'added feature' of limited interactivity. Sites such as The Spot, Eon 4, and The Pyramid fostered a spectatorship defined through actions of 'narrative exploitation' in which the user explored multilinear options of a story and literally 'read' media content.
>
> (Harries 2002: 175)

The narrations make larger worlds and contexts available as participants explore the initial context of the media. This model is also reflected in current developments in transmedia storytelling.

Levels of examination and methods of mash-up analyses

In order to develop a media-scientific methodology of analysis of mash-up media, this section uses established foci of analysis and methods for visual mass media, with an emphasis on fiction, documentation, and entertainment.[8] For some time now, the emphasis has been either on the relationship to the meaning of the original or to the method in which the media forms were combined. A study of the various mash-up forms should not be based exclusively on methods established in media studies, but should also include methods of analysis that are used to study literature. This would imply that one would not only analyse what the mash-up offers the viewer or what type of mash-up it is, but would also delve into the ways that it can be aligned with other works. The methods of analysis in media studies focus on formal and substantive aspects of the development of meaning, as well as on the complex dimensions of the respective subject of study and how this can affect the interpretation.

Many current supply-oriented analyses of intermedial forms focus on the following question: what is combined, how is it combined, who combined it, and with what objective? Which new meaning has been generated by this combination of existing content and aesthetics? The possible levels of analysis include many different phenomena and their potential to be clarified to the viewer. One can quickly see how the research methods, including the culture, have to be considered and recognised in media studies: the largest variable is the integration of the analysis in social, cultural, and historical contexts. Mash-ups and their key issues operate as a cultural form of expression that the media offers or that political movements choose to employ.

A media and cultural-historical perspective offer formal and thematic models of intermedial and intertextual references: for example, collage, bricolage, cut-up technology, or found footage films. This not only allows for different models of the combination of forms of expression and content, but also one to determine changes in the construction of the message.

The range of the mash-up concept is closely linked to an analysis of the participants. Will you choose to state your objective and identity or will you remain anonymous? Are you going to work with professional or non-professional participants? Will there be noticeable objectives or intentions, such as political education about hidden power structures, or will you remain playfully disconnected from a direct objective? Has this method of presenting media been criticised? What are the thematic, dramaturgical, and formal preferences that are recognised by the respective participants in the production? Due to a focus on intermediality, there are a few factors that all mash-ups have to take into account. Will viewers recognise the original work and its author? Which material, from which media source, in which cultural areas or period of time will be combined? To determine who has the rights to the original work, the usage of paratext from existing records and contact with those who possess the authorship can also be economically beneficial.

One possible form of production analysis is the observation of or the record keeping for one's own mash-up production. The following questions focus on production analysis: how will we go about the processes of selecting, copying, or cutting and pasting? In which ways can the production be influenced by planned, specified plot developments within the media? Is there an ironic and/or factual reflection on medial stereotypes?

One of the central points considered in researching works in media studies is the identification of sources: which genres or motifs are linked to the origin of the media and how? From what cultural-historical perspective are the sources used?

Traditionally, the first focus of media studies is to analyse the plot structure of a film or television production, followed by the structure and the arrangement of events and content. Is there any evidence of linear narration or an associative structure evident in the plot structure? Which visual, textual, and thematic references and relationships (e.g., the encounter with a film and television character) take place? In what way are action structures and dramatic experience dimensions linked together? This includes the analysis of focal points such as tension, emotionalisation strategies, and rhetoric (e.g., the argumentation of the language and pictures in a demonstration documentation or the construction of verbal arguments in interview documentation).

A particularly relevant aspect of mash-ups that can be studied lies in the field of post-production. What type of montage is the mash-up based on? How do images and sound material transitions influence the potential for creating illusions and reflectivity? What kind of space, visual axis, and/or dialogue design arises through the diverse range of clips? Mounting and collage are also musical principles of the DJ culture, and the remix can be seen as an equivalent to a mash-up of moving images. Another result of the post-production process is the time structure of mash-ups. Here, the plot can be linear or non-linear. A specific rhythm creates the respective length of sequences, allowing the musical soundtracks to be customised.

By analysing how quotes are incorporated and represented, one is able to analyse the quote forms, their remuneration, and their semantic dimensions (see Bleicher 2002). An interesting aspect to look into is the temporal origin of the edited and mixed material: what time or cultural-historical eras are brought into connection with the main theme of the mash-up?

Picture and language are essential elements that allow a meaning to be integrated into a mash-up. In what way will the existing language material be mixed, replaced, complemented, or contrasted with added comments, monologues, or dialogues? How are the semantics and image shapes created and related (motif, image design, light, colour)? Is a relation created between the images by internal visual references and/or by narrative voiceovers?

The reception a mash-up gains is an established research topic in media studies audience research. Each individual reception of the film can be assessed based on scientific or ethnological approaches through techniques such as monitoring or performing a subjective survey of participants. An empirical analysis of clicks on an Internet site can determine the frequency of use of mash-ups.

This empirical survey forms a basis on which one can further analyse comparable mash-ups. Frequently, mash-ups that are widespread across the Internet experience a snowball effect, resulting in many new versions (see Claussen 2014: 90). This instigation of the snowball effect is both a cause and an effect of the close link between reception and distribution in the mash-up culture (see Groscurth 2011).

One can also employ this analytical model by changing what questions are asked and applying it in reverse order, i.e., starting from the reception and the presentation of the mash-ups and, from the results of the analysis, reconstruct the initial intentions and context. Another possible area of analysis could be the question of possible copyright violations within the respective mash-ups. For example, various interviews in the *Generation Remix* collection (2014) discuss the fact that many mash-up artists choose not to use material if they know that it will not be easy to gain the rights. Without knowledge of the exact context of these productions, however, the analysis of this aspect would not be fruitful and would not be implemented.

Further analysis of comparable media forms also includes research on literary or musical mash-ups. The use of existing media is, however, a legal challenge in view of copyright regulations and arrangements.

Problematic issues with copyright

One focus of current discourses is copyright in connection with the issues of lack of financing and profiling of artistic productions. With the campaign 'Recht auf Remix', artists try to actively highlight the potential of their work and to instigate changes in current copyright laws (see Fischer 2014: 76). The evaluation of this process varies between the poles of creative freedom and allegations of plagiarism or piracy. In the early stages of art history, a deceptively real copy was considered to be a performance of the artists handicraft. This assessment still influenced the art market. 'Beltracchi – die Kunst der Fälschung' (2014) is a documentary about a couple who are art forgers and can serve as an example for the current controversy about the definition of a copy. As soon as the word piracy comes up, a portion of digital culture has been criminalised.

Lawyers describe mash-ups as 'transformative works' (Kreutzer 2014: 64f.) that have their own meaning (in contrast to copies). Although mash-ups make use of foreign material, they are addressed to other target groups, due to their form of expression and intellectual aesthetic effect that differs significantly from incorporated works (see Kreutzer 2014: 64f.) From this definition, Kreutzer is able to deduce that a transformative work normally does not do any harm to the economic usage of artistic work (see Kreutzer 2014: 65). Nevertheless, he stresses that a transformative work that is too close to the original work can do harm to it. In this context, the European copyright law prioritises the protection of the author and the economic and personal interests they have invested in their work (Kreutzer 2014: 60). It refers to four individual regulations of the European Copyright Codes, a reorganisation of the European copyright law suggests:

> Article 5.1 Uses with minimal economic significance,
> Article 5.2 Uses for the purpose of freedom of expression and information,

Article 5.3 Uses permitted to promote social, political, and cultural objectives,
Article 5.4 Uses for the purpose of enhancing competition.

(Kreutzer 2014: 53)

The concrete rules include an allowance for caricatures, parodies, or pastiches and suggest a free of charge limitation on copyright for quotations (Kreuzer 2014: 54). In the United States, the same concept can be found in Article 107 of the US Copyright Act (USCA) application (Kreutzer 2014: 55). In the guidelines for fair use at the Center of Social Media, it is stated that the creation of a new work can also be performed by recombining and arranging existing elements (works or part of works), given it produces a new meaning. These guidelines explicitly include mash-ups, remixes, and collages of music videos (see Kreutzer 2014: 57).

In contrast to Germany, the United States set on incentives for the development of creativity: culturally enriching creative performances should be rewarded in terms of an exclusive right limited in time and content (Kreutzer 2014: 59). Generally, transformative uses of media ensure that the best interests of authors, readers, and third parties who utilise the media are protected (Kreutzer 2014: 62). The further development of legal regulation is faced with the challenge of taking these very different constellations of interests into account and at the same time adjusting to the continuous change of style in various media and cultural areas. Thomas Wilke emphasises that issues concerning the reliability of sources, the safety, and, by that, the verifiable originality play a part in the data-based social structural change that must not be underestimated (Wilke 2015: 18). Simultaneously, both the supply of media and its range of functionality are rapidly expanding.

Conclusion

This chapter shows various examples of the creative potential of the mash-up phenomena, which is not based on the demand for innovation of postmodernism, but much more on the intermedial interaction of existing content and forms of representation. What was once the traditional author with a copyright on their intellectual property has become a DJ, who creates a new work by remixing existing works and adding their own perspective to the content. Eduardo Navas refers not only to the economic aspects but also to the social movements these digital cultural production forms have instigated:

> At the beginning of the twenty-first century, it is evident that the Regenerative Remix is defining the next economic shift. Remix culture is experiencing a moment in which greater freedom of expression is mashed up against increasingly efficient forms of analysis and control.
>
> (Navas 2010)

Thus, an analysis of mash-ups can provide key elements and findings for changes, not only in the areas of production and reception, but also in the general economic, social, and aesthetic changes of digital culture.

Notes

1 For the similarities and differences of these terms see Rajewsky 2002.
2 See Bleicher 2005a.
3 Messidona. [video] Available at: www.youtube.com/watch?v=RFRhtWwn9T8.
4 Changing media will also pose a risk, as users who migrate from television to the Internet may not return to the old medium.
5 See www.takethisdance.com.
6 In German, 'Inszenierung'.
7 GEMA is a German collecting society defending the copyright of its members like composers and music production companies.
8 Peter Zimmermann commented on the topic of political mash-ups in documentary film during the annual conference of the GFM 2006 in Stuttgart.

References

Bleicher, J., 1999. *Fernsehen als Mythos. Poetik eines narrativen Erkenntnissystems.* Opladen: Verlag für Sozialwissenschaften.
Bleicher, J., 2002. Intermedialität im postmodernen Film. In J. Eder (Ed.), *Oberflächenrausch. Postmoderne und Postklassik im Kino der 90er Jahre.* Hamburg: LIT Verlag, pp. 97–112.
Bleicher, J., 2005a. Fernsehkritik im Fernsehen. In R. Weiß (Ed.), *Zur Kritik der Medienkritik: Wie Zeitungen das Fernsehen beobachten.* Berlin: Vistas, pp. 127–146.
Bleicher, J., 2005b. Jagdrevier. In K. Hickethier and K. Schumann (Eds.), *Kriminalfilm.* Stuttgart: Reclam, pp. 231–233.
Bleicher, J., 2014. Ökonomie, Technik, Entwicklung und Angebotsschwerpunkte des *Social Web* als Herausforderung für die Medienwissenschaft. In I. Nord and S. Luthe (Eds.), *Social Media, christliche Religiosität und Kirche. Studien zur Praktischen Theologie mit religionspädagogischem Schwerpunkt.* PopKult No. 14. Jena: Garamond, pp. 29–44.
Booth, P., 2012. *Time on TV: temporal displacement and mashup television.* New York: Peter Lang Publishing.
Bolter, D.J. and Grusin, R., 2000. *Remediation. Understanding new media.* Cambridge: MIT Press.
Brunow, J., 1995. Sans Soleil. In M. Töteberg (Ed.), *Metzler Filmlexikon.* Stuttgart, Weimar: J.B. Metzler, pp. 181–183.
Brunow, D., 2015. *Remediating transcultural memory: documentary filmmaking as archival intervention.* Berlin: De Gruyter.
Bruns, A., 2007. *Produsage: towards a broader framework for user-led content creation.* Available at: http://produsage.org/files/Produsage%20(Creativity%20and%20Cognition%202007).pdf.
Bruns, A., 2010. News produsage in a pro-am mediasphere: why citizen journalism matters. In G. Meikle and G. Redden (Eds.), *News Online: Transformations and Continuities.* London: Palgrave Macmillan, pp. 8–24.

Burgess, J. and Green, J., 2009. *YouTube: online video and participatory culture*. Cambridge: Polity.

Claussen, J.T., 2014. Remixing Youtube – Über DJ-Kultur, Videoklone und ReSync-Attacken. In V. Djordjevic and L. Dobusch (Eds.), *Generation Remix. Zwischen Popkultur und Kunst*. Berlin: iRights Media, pp. 79–98

Dallach, C., 2010. Mash-up-Bestseller: Eheversprechen und Leichenschmaus. *Spiegel Online*, 5 July 2010. Available at: www.spiegel.de/kultur/literatur/mash-up-bestseller-eheversprechen-und-leichenschmaus-a-704341.html.

De Montaigne, M., 1998. Über das Bereuen. In M. De Montaigne, *Essais. Erste moderne Gesamtübersetzung von Hans Stilett*. Frankfurt am Main: Die andere Bibliothek, Sonderband.

Djordjevic, V. and Dobusch, L. (Eds.), 2014. *Generation Remix. Zwischen Popkultur und Kunst*. Berlin: iRights Media.

Fagerjord, A., 2010. After convergence: YouTube and remix culture. In J. Hunsinger, L. Klastrup, and M. Allen (Eds.), *International handbook of Internet research*. Dordrecht: Springer, pp. 187–200.

Fischer, G., 2014. Von Jägern und Samplern. Eine kurze Geschichte des Remix in der Musik. In V. Djordjevic and L. Dobusch (Eds.), *Generation Remix. Zwischen Popkultur und Kunst*. Berlin: iRights Media, pp. 69–78.

Forberg, C., 2012. Schwerpunktthema: Mash-up – neue Technik oder neue Kulturtechnik? *Deutschlandfunk*, 14 June 2012. Available at: www.deutschlandfunk.de/schwerpunktthema-mash-up-neue-technik-oder-neue.1148.de.html?dram:article_id=208911.

Frank, P., 1987. *Intermedia. Die Verschmelzung der Künste*. Bern: Benteli.

Genette, G., 1993. *Palimpseste. Die Literatur auf zweite Stufe*. Frankfurt am Main: Suhrkamp.

Groscurth, H., 2011. Mashup Medien. Überlegungen zu ihren distributionsmedialen Eigenschaften. In S. Abresch, C. Schaumburg, and M. Wolf (Eds.), *We take what you made. Zur Theorie und Praxis digitaler Nutz(nieß)ung*. Siegen: Universitätsverlag Siegen, pp. 19–42.

Harries, D. (Ed.), 2002. *The new media book*. London: BFI Publishing.

Hausheer, C. and Settele, C., 1992. *Found footage film*. Luzern: Viper/zyklop.

Helbig, J. (Ed.), 1998. *Intermedialität. Theorie und Praxis eines interdisziplinären Forschungsgebietes*. Berlin: Turnshare.

Herbst, H., 2012. *Früher als wir noch nicht postmodern waren*. Offenbach: Hochschule für Gestaltung.

Jakubetz, C., 2011. *Crossmedia*. 2nd ed. Konstanz: UVK.

Jenkins, H., 2006. *Convergence culture: where old and new media collide*. New York: New York University Press.

Kortmann, C., 2009. Barack Obama spricht Schwäbisch. *Zeit-Online*, 23 April 2009. Available at www.zeit.de/2009/18/Mashup.

Kreutzer, T., 2014. Remix Culture und Urheberrecht. In V. Djordjevic and L. Dobusch (Eds.), *Generation Remix. Zwischen Popkultur und Kunst*. Berlin: iRights Media, pp. 43–68.

Krewani, A., 2001. *Hybride Formen: New British Cinema – Television Drama – Hypermedia*. Trier: WVT Wissenschaftlicher Verlag.

Kügle, M., 2015. Synchrone Synthesen von Aussprache und Schrift in der Kinetischen Typographie. In F. Mundhenke, F.R. Arenas, and T. Wilke (Eds.), *Mashups. Neue

Praktiken und Ästhetiken in populären Medienkulturen. Wiesbaden: Springer, pp. 179–194.

Kuhn, D., 2012. Pack die Axt aus, Präsident. Jane Austens Jungfrauen metzeln Zombies, Abraham Lincoln jagt Vampire. Die "Mash-Up"-Romane, ein Phänomen des amerikanischen Buchmarkts, erreichen das Kino – und stellen Fragen nach dem Erfolg dieser besonderen Form des Culture Clash. *Süddeutsche Zeitung*, 12 July 2012.

Lessig, L., 2014. Lawrence Lessig's "Remix". In V. Djordjevic and L. Dobusch (Eds.), *Generation Remix. Zwischen Popkultur und Kunst.* Berlin: iRights Media, pp. 15–42.

Lexikon der Filmbegriffe, 2012. *Found footage film.* Kiel: Christian-Albrechts-Universität zu Kiel. Available at: http://filmlexikon.uni-kiel.de/index.php?action=lexikon&tag=det&id=6751.

Marek, R., 2013. *Understanding YouTube. Über die Faszination eines Mediums.* New York: Transcript.

Mashup-Media, 2010. Mash-up media. Perspectives of the Netzculture in the 21. Century. [press release] Available at: www.mediatisiertewelten.de/en/projects/1st-funding-period-2010-2012/mash-up-media.html.

Mattelart, A., 2003. *Kleine Geschichte der Informationsgesellschaft.* Berlin: Avinus Verlag.

McLuhan, M., 1964. *Understanding media.* New York: McGraw-Hill.

Mecke, J. and Roloff, V. (Eds.), 1999. *Kino-/(Ro)Mania: Intermedialität zwischen Film und Literatur.* Tübingen: Stauffenburg.

Müller, J.E., 1996. *Intermedialität: Formen moderner kultureller Kommunikation.* Münster: Nodus.

Mundhenke, F., Ramos Arenas, F., and Wilke, T. (Eds.), 2015. *Mashups. Neue Praktiken und Ästhetiken in populären Medienkulturen.* Wiesbaden: Springer.

Navas, E., 2010. Regressive and reflexive mashups in sampling culture. In S. Sonvilla-Weiss (Ed.), *Mashup cultures.* Wien: Springer, pp. 157–177.

Rajewsky, I.O., 2002. *Intermedialität.* Tübingen: UTB.

Reichert, R., 2008. *Amateure im Netz. Selbstmanagement und Wissenstechnik im Web 2.0.* Bielefeld: Transcript.

Remix Theory, n.d. Remix defined. Available at: http://remixtheory.net/?page_id=3.

Remix Theory, 2012. The book. Available at: http://remixtheory.net/?page_id=491.

Roscoe, J. and Hight, C., 2001. *Faking it: mock-documentary and the subversion of factuality.* Manchester: Manchester University Press.

Ryan, M-L., 1992. *Possible worlds, artificial intelligence and narrative theory.* Bloomington: Indiana University Press.

Schade, S. and Tholen, G.C. (Ed.), 1999. *Konfigurationen: Zwischen Kunst und Medien.* München: Wilhelm Fink.

Schäfer, M.T., 2011. *Bastard culture! How user participation transforms cultural production.* Amsterdam: Amsterdam University Press.

Scherer, C., 2001. *Ivens, Marker, Godard, Jarman. Erinnerung im Essayfilm.* München: Wilhelm Fink Verlag.

Schmidt, J-H. and Loosen, W., 2014. Both sides of the story: assessing audience participation in journalism through the concept of inclusion distance. *Digital Journalism*, 3 (2), pp. 259–278.

Schulze, G., 2005. *Die Erlebnisgesellschaft: Kultursoziologie der Gegenwart.* 2nd ed. Frankfurt am Main: Campus.

Schwinghammer, A., 2008. Reporting/narrating/storytelling – anthropological explorations into machinima and its neighbours. In Y. Gächter, H. Ortner, C. Schwarz, and A. Wiesinger (Eds.), *Erzählen – Reflexionen im Zeitalter der Digitalisierung – Storytelling – Reflections in the Age of Digitalization*. Innsbruck: Innsbruck University Press.

Simanowski, R., 2001. Autorschaften in den digitalen Medien. Eine Einleitung. *Text und Kritik,* Heft 152, pp. 3–21.

Sonvilla-Weiss, S., 2010. Introduction: mashups, remix practices and the recombination of existing digital content. In S. Sonvilla-Weiss (Ed.), *Mashup cultures*. Wien: Springer, pp. 1–3.

Spielmann, Y., 1998. *Intermedialität: Das System Peter Greenway*. München: Wilhelm Fink.

Stahl, H., 2015. Die Verhältnisse zum Klingen bringen. Mashup und Sound bei Hubert Fichte, Rolf Dieter Brinkmann und Jörg Fauser. In F. Mundhenke, F. Ramos Arenas, and T. Wilke (Eds.), *Mashups. Neue Praktiken und Ästhetiken in populären Medienkulturen*. Wiesbaden: Springer, pp. 215–226.

Stuhlmann, A., 2011. Kleine Geschichte des Mashups. In J. Schumacher and A. Stuhlmann (Eds.), *Videoportale: Broadcast Yourself? Versprechen und Enttäuschung*. Hamburger Hefte zur Medienkultur, No. 12. Hamburg: IMK, pp. 103–117.

Twitchell, J.B., 1992. *Carnival culture: the trashing of taste in America*. New York: Columbia University Press.

von Borries, C., 2013. Mocracy: Neverland in me. [video] Available at: https://video.yandex.ru/users/masseundmacht/view/1/?ncrnd=2592.

Von Gehlen, D., 2012. *Mashup Lob der Kopie*. 2nd ed. Frankfurt am Main: Suhrkamp.

Von Gehlen, D., 2013. *Eine neue Version ist verfügbar. Update: Wie die Digitalisierung Kunst und Kultur verändert*. Berlin: Metrolit.

Wees, W.C., 1993. *Recycled images: the art and politics of found footage films*. New York City: Anthology Film Archives.

Weißer-Gleißberg, J., 2014. *Mein Film über mich. Regisseure vor und hinter der Kamera*. Berlin: Avinus Verlag.

Wessel, D., 2014. #1 Wessel David a.k.a. Ben Stiller "der größte Generationenkonflikt seit der 68er Bewegung". In V. Djordjevic and L. Dobusch (Eds.), *Generation Remix. Zwischen Popkultur und Kunst*. Berlin: iRights Media, pp. 125–133.

Wilke, J., 1999. *Mediengeschichte der Bundesrepublik Deutschland*. Köln: Böhlau.

Wilke, T., 2012. *Mashup. Theorie – Ästhetik – Praxis*. Conference of the Zentrums für Wissenschaft und Forschung Leipzig | Medien e.V., Leipzig, Germany, 08–09 June 2012. Available at: www.auditive-medienkulturen.de/2012/04/12/tagung-mashup-theorie-asthetik-praxis-in-leipzig/.

Wilke, T., 2015. Kombiniere! Variiere! Transformiere! Mashups als performative Diskurssubjekte in populären Medienkulturen. In F. Mundhenke, F. Ramos Arenas, and T. Wilke (Eds.), *Mashups. Neue Praktiken und Ästhetiken in populären Medienkulturen*. Wiesbaden: Springer, pp. 11–43.

Zelik, R., 2004. Im Diskurs-Cyberspace. Thomas Meineckes neuer Roman Musik. *Telepolis*, 6 September 2004.

Zima, P. (Ed.), 1995. *Literatur intermedial: musik, malerei, photographie, film*. Darmstadt: Turnshare.

8 Big Data
Katharina E. Kinder-Kurlanda

Introduction: Big Data as a digitisation phenomenon

Analyses based on *Big Data* are receiving an ever-increasing amount of attention in various domains. The potential within these analyses is usually seen in the investigation of broader connections on a large scale. The Internet has increased overall availability of data, such as user-generated content (e.g., 'tweets' on Twitter or, for example, pertinent metadata such as the time and origin/location of these messages). Concomitant to this is the burgeoning need for increased capacities for storage and calculation seemingly making possible the analysis of even global societal processes. Specifically, social media data allow research on the basis of information at a scale thus far unimagined within the social sciences (Savage & Burrows 2007). User-generated content promises insights into *in situ* behaviours that were previously difficult, if impossible, to observe, and offers the opportunity for continuous and almost simultaneous research of topics and discourses, minus the burdensome instruments of survey and interview: 'The detailed knowledge and insights that before could only be reached about a few people can now be reached about many more people' (Manovich 2011: 463). New data sources and computer-intensive methods enable the study of complex phenomena such as global social networks 'in ways never before imagined or possible' (Ruppert 2013: 269). The methodologically problematic aspect of many Big Data analyses has already been pointed out (and I will give details on this point later on), with particular emphasis on the lack of attention given to the complex conditions of origin of the data the analyses are based upon. Nevertheless, an examination of Big Data's potential is worthwhile, especially from the perspective of a qualitative culture analysis, and even more so when interdisciplinary approaches are utilised.

Both the emergence of Big Data analyses and their effects need to be considered. The algorithms utilised in computing Big Data have entered various everyday contexts and are an important dimension of the digitisation of everyday spaces in which we all live and work. More and more decisions in the most diverse domains are influenced by Big Data. An increasing number of decisions, perhaps even the majority in some areas, are now being based on Big Data analyses, with examples to be found in very different contexts, including online marketing and the control of industrial processes. With this increasing importance of Big Data and algorithms, the users of social media platforms have started to respond and are beginning to expect the results of Big Data analyses and algorithmic calculations, to react to them, and even to attempt their manipulation. For example, an overview of the products that other customers with similar attributes to oneself have decided to buy is now often expected; while on social media platforms various practices are emerging that aim to directly influence ranking algorithms in a targeted fashion; and various discourses are emerging, which are focused on the threat to privacy constituted by this data being ubiquitously generated, collected, and analysed. Big Data has entered the digitised everyday, and it simultaneously offers researchers new possibilities of observing it.

Big Data is, however, much the same as many phenomena that gain traction and actuality in the scientific discourses about digital media, being both over- and under-defined depending on the actors using the term in whichever context and with what interests or which domain happens to be the centre of attention. Big Data thus means something different in the context of business models, research, or surveillance. This chapter does not aim to offer a comprehensive or authoritative definition of Big Data, but is rather an attempt at pointing out what can be summarised under the term, and to critically evaluate the potential contribution of Big Data to a culture analysis of digitisation. The following will demonstrate how to highlight and make visible the influence Big Data is having on the everyday practices and meanings of actors in various digitised spaces. The focus is on capturing everyday life situations in digitised spaces which are of interest to cultural anthropologists or researchers from neighbouring disciplines engaged in culture analysis.

The development of Big Data

Definition and potential

Manovich condensed the definition of Big Data as it is understood in computer science as follows: 'Big Data is a term applied to data sets whose size is beyond the ability of commonly used software tools to capture, manage, and process the data within a tolerable elapsed time' (Manovich 2011: 460). This definition already alludes to a change that has occurred over time: amounts of data that were seen as Big Data a few years ago may today be computable on much smaller devices. What, in fact, can be classified as Big Data thus depends on the specific historic context, as calculations based on data, which formerly required a supercomputer, can now be performed on a home PC equipped with standard, easily obtainable software packages. A simultaneous change that has occurred is where large amounts of data are generated: the daily data volume of social media platforms alone (which have, after all, only existed for a decade or so) comprises millions of data entries. Data volume is thus not the only defining characteristic of Big Data. According to boyd and Crawford (2012), Big Data rather refers to the possibility of searching, aggregating, and sorting through large data sets to assemble and place in relation to one another. Other authors have also pointed out that the term Big Data refers to more attributes of such data than just size (e.g., Tinati et al. 2014b). Big Data promises to allow timely analyses and prognoses otherwise impossible to achieve using traditional methods of collection and investigation: currently, transpiring events, for example, are immediately reflected in social media platforms, can be observed and analysed simultaneously while they are happening, and also offer the opportunity to analyse even global interconnections on the basis of the large amounts of data being generated:

> The proportionality of social media offers information (in principle at least) on 'whole' populations, rather than sub-sets; the information is dynamic – captured in real time and over time; and social media provide data on what people say and do 'in the wild', rather than what they say they do in response to researchers' questionnaires and interviews.
> (Tinati et al. 2014b)

Applications

Currently, various scientific interdisciplinary fields of research are starting to establish themselves with Big Data analyses. New disciplines in the area of Big Data are starting to emerge. Tiropanis et al. (2015) offer an overview of Web Science, Internet Science, and Network Science, all aimed at establishing disciplines in the area of Big Data. Other efforts include Computational Social Science or the Digital Humanities. Big Data analyses have gained particular traction outside academic research among social media platforms and other businesses generating their own data such as customer data, for example. Policy makers and intelligence services are also utilising Big Data, in addition to non-governmental institutions and interest groups as well as activists. Various types of Big Data are increasingly at the centre of attention for climate data, genome databases, stock exchange data, and sensor data. These are but some examples of the uses of this new source of information. Internet data and, in particular, social media Big Data are at the centre of this chapter, both in the sense of user-generated content and in the sense of interaction data. Topics examined include analyses of numbers of 'likes' on Facebook, tweets characterised by hashtags, or search terms input into Google.

There are various ways of gaining access to Big Data from social media platforms: some platforms provide researchers with interfaces for collecting data. For example, Twitter offers the possibility to download Twitter messages (tweets) with its API (Application Programming Interface).[1] Different searches of actual tweets can be performed – for example, one can search for all tweets containing a specific hashtag that is used by users to denote a specific topic or other classification. Other possibilities range from manually collecting data to buying complete data sets from official resellers of social media data (see Weller 2014b).

Methods

Big Data research usually denotes novel types of analyses of large amounts of data that are available as the result of various digitisation phenomena and not of behaviour on the Internet employing traditional methods, such as interviews or surveys. In particular, researchers work with user-generated content, mainly social media data, e.g., from Twitter or Facebook, and so-called process-generated data employing new methods. These can be roughly divided into two types: those analysing content and those analysing networks or relations between users.

Content analyses are employed to uncover – via automated counts and comparisons of terms – overviews and connections in large text corpora. Content analyses are done in several ways, most of which are based on word frequencies. 'Sentiment analysis' allows for mapping what can be considered the atmosphere, how people are feeling about something, mostly with the aim to arrive at prognoses for future events, such as election results and the like. 'Event detection' allows continuous analysis of social media content via algorithms with the goal of identifying certain events, such as natural disasters. An example is the following Twitter posts for their relation to an event such as an earthquake (cf. Sakaki et al. 2010).

In network analyses, the focus is not so much on the content, but rather on the structure of interactions. The aim is to find and analyse patterns and connections between content and users of social media. Networks between actors on social media platforms such as Facebook are drawn, then analysed with regard to the strength of certain ties, focal points, and patterns of distribution to answer questions such as who knows whom and who people interact with (e.g. Catanese et al. 2011). In addition to the two approaches content analysis and network analysis, analyses of interaction data can be found, for example, of 'likes' on Facebook. In contrast to content analysis, these analyses are possible across various languages (Mestyán et al. 2013). There are also many other possibilities for analysing Big Data with various disciplines exploring and utilising different approaches. An overview can be found in Weller (2014a).

Critique

Research about those who are researching with Big Data or social media data has shown that these researchers themselves often problematise how Big Data analyses can be made to work. Big Data research currently can be seen as a fragmented field of very different estimations of Big Data's potential (Kinder-Kurlanda & Weller 2014). Various strategies are being employed to deal with the research-related ethical uncertainties arising from such a new and uncharted situation (Kinder-Kurlanda & Ehrwein Nihan 2013, Weller & Kinder-Kurlanda 2014). Interdisciplinary collaborations are being formed to face the challenges posed by new methods, large data volumes, and the technical skills required to collect and analyse this data (Kinder-Kurlanda & Weller 2014). More and more researchers from the social sciences and humanities, who are faced with the rise of new, computer-intensive methods and a (seemingly) universal primacy of countability and quantification, are beginning to question their own epistemological and ontological assumptions:

> The emergence of Big Data from social media has had an impact in the study of human behaviour similar to the introduction of the microscope or the telescope in the fields of biology and astronomy: it has produced a qualitative shift in the scale, scope and depth of possible analysis.
> (Tufekci 2014a: 505)

The question is how the potential of analysis possibilities with Big Data can be taken up and recombined within the framework of existing research approaches and how shifts in knowledge are to be assessed.

Big Data as a challenge for social science methods

Big Data 'creates a radical shift in how we think about research' (boyd & Crawford 2012: 665) and means a fundamental change at the levels of epistemology and ethics (Lazer et al. 2009). Berry (2011) writes of the emergence of a new ontological 'epoch'. Ruppert et al. (2013) criticise the rhetoric of an 'epochal change', while also encouraging a rethinking of the theoretical, methodological, and epistemological foundations of social science methods in the face of Big Data. In their text *Critical Questions for Big Data*, boyd and Crawford (2012) reacted with critical interventions to a rising overestimation of the potential of Big Data to illuminate and explain social interconnections. For example, Anderson (2008) argued that Big Data meant the advent of an 'end of theory' and that it had the ability to substitute scientific methods and theories by the sheer volume of data available for analysis. boyd and Crawford point out, however, that Big Data is a phenomenon arising out of the combination of new technical possibilities, analysis methods that allow for recognising patterns in large data sets, and a certain mythology. This mythology concerns the fact that there is a 'widespread belief that large datasets offer a higher form of intelligence and knowledge that can generate insights that were previously impossible, with the aura of truth, objectivity, and accuracy' (boyd & Crawford 2012: 663). In fact, since Anderson (2008) postulated an 'end of theory', many critical voices have pointed out various issues that make Big Data methods unreliable and may, in fact, lead to questionable results (e.g., Frické 2014, Lazer et al. 2014). Frické argues that 'data-driven science is a chimera' and that, although in Big Data samples may be bigger, the possibilities for testing better, and results obtainable continuously rather than being interspersed at certain points in time, scientific working nevertheless needs problems, thoughts, and theories: 'If anything, science needs more theories and less data' (Frické 2014: 10). In particular, boyd and Crawford's interventions were 'instrumental in moving scholarly engagement with "Big Data" from unquestioned acceptance of the data evangelism of interested parties towards a more critical, considered stance' (Bruns 2013: para. 3). Nevertheless, Big Data analyses still harbour the danger of lending research a 'veneer of scientificity', which the studies do not necessarily deserve (Bruns 2013). For Big Data analyses of social media to achieve methodological validity, a few conditions must be met: data collection must follow sound research questions and not opportunistically be guided by data availability; and better documentation of applied methods in publications is required to allow reproducibility of results (Bruns 2013).

In general, the difficulties in Big Data research that need to be overcome lie mainly in three areas: first, in the methodology, especially with regard to the various contexts of origin of the data; second, in questions of research ethics, particularly with respect to the fact that informed consent of social media content producers is usually missing; and third (this point is related to the first two), in the role of social media platform providers and other owners of Internet data and the limited ways in which they offer data access to researchers.

Which data is made available for research by social media platform providers and in which way determines which analyses are possible (Giglietto et al. 2012). Additionally, the platform provider's strategy and the technical possibilities of the API provided to researchers determine the research potential (Puschmann & Burgess 2013). For example, the Twitter API restricts access to past data and the completeness of provided data sets is often uncertain. Also unclear is whether tweets returned from a specific search request represent a random sample or a data set cut-off. When conducting analyses, researchers are frequently occupied with feasibility to a degree that analyses almost necessarily are conducted independently of existing traditions and discourses around epistemology and hermeneutics. Existing theories on the researched topic (e.g., communication, gender inequality, or voting behaviour) may also be ignored – a situation which realises the hypothesis about the end of theory in the form of numerous questionable research enterprises. The question whether and how these data reflect reality at all and which ideologies and decisions are the bases of their production is often not posed.

'Raw data is an oxymoron'

In relation to the term 'raw data' from Levi-Strauss's analysis of the 'raw' (as the natural) and the 'cooked' (as the social), Bowker (2013) points out that the assumption of a natural state overlooks that the natural is also always already social. For example, the hierarchical setup of early databases was oriented at the existing hierarchical structure of organisations.

Based on the understanding that 'raw data is an oxymoron' (Bowker 2013), various authors have pointed out that data used for Big Data analyses are not 'neutral' or 'raw' but always shaped by the specific context of their origin and thus come with specific ballast. In particular, user-generated content on social media platforms such as Facebook or Twitter can emerge out of very diverse usages and be created with very different intentions.

What is actually being measured or counted and what is not also depends on several decisions made when designing the research approach. In the Big Data field, such decisions, for example, pertain to which keywords were chosen to generate a given data set. Also, decisions made in the design of the software and hardware systems generating the data, which will later become available for research, likewise influence which data and format are at the researcher's disposal. User-generated content, but also automatic or process-generated data are strongly dependent on technical capacities in terms of their features, which, in turn, can be traced back to the possibility and decision space available during the developmental process. This process is influenced by the interests and preconceptions of various, disparate stakeholders. The data cannot be considered independently from the sociotechnical settings, which promote, restrict, or otherwise influence its emergence.

The related literature lists a variety of additional methodological challenges. For example, Karpf (2012) and Bruns (2013) point out that Internet research always only describes a unique state, as platforms, user groups, and habits change so quickly that it is hard for research to keep pace. Content collected at a certain point in time on a particular platform, for instance, can suddenly be unavailable because the platform provider changed its functionality. Likewise, another case may be where statements on the behaviour of Facebook users based on analyses of their privacy settings are at some point related to settings possibilities, which are no longer available at the time of publication.

For analysis at the Big Data level, it is still difficult to establish transparency of the mode of data collection necessary for replicability and for confirming validity of research. Additionally, finding adaptive methods, which fit this ephemeral and transient field is problematic, as it really necessitates particularity and focused engagement with details over holistic, overarching approaches.

Biased representation within the data, and bias problems in general, are frequently cited. Quantitative social science analyses strive for the control of bias within the data. As data is always only available for a small part of the surveyed population, attempts are made to reproduce a so-called 'representative sample' of the overall population. To be able to make statistical statements about a data set, both the origin and the shortcomings of the data have to be known. Researchers must be able to evaluate the bias in the analyses, i.e., to be able to judge the extent to which the data is truly representative of the statements and opinions of the total surveyed population. In the literature on Big Data (e.g., Ruths & Pfeffer 2014, Tufekci 2014a), various sources of bias in social media data are mentioned. Bias exists on two levels: user-generated content appears on a platform which is not representative of the statements of those not using the platform, and the available data sets are not necessarily representative for all activities on the platform itself.

The representativeness of the total population is not given, as the composition of the users is commonly unclear, as is how they are related to any other strata of people (for example, the inhabitants of Germany) (Ruths & Pfeffer 2014). It is also frequently questionable whether the available data sets are truly representative for the users of the platform itself. Numbers and sociodemographic details on the users of social media platforms are often not easily accessible for users, as these data are commercially valuable and are not necessarily made available for research by the platform providers.

Other sources of bias in social media data stem from things like different platforms, respectively, attract different types of users, for example, or that automated posts or spambots can comprise a huge share of the content. Additionally, because of the aforementioned restrictions of APIs, it is unclear how complete a given data set is or which sampling methods were utilised by the platform provider.

The overemphasis of the platform Twitter in research is also problematic. Problems include the easy accessibility of the data via the API provided by Twitter, the shortness and conciseness of the messages, and the strict structural regulations of the content. Results from Twitter analyses are often generalised without taking into account the Twitter-specific structuring, restricting, or promoting user-behaviour features (Tufekci 2014a). The specific features of Twitter lead to it often being used while on the go, for example, when good connectivity is lacking or other activities require a lot of attention, such as during protests or events (Tufekci 2014a). Twitter is therefore more suitable than other platforms for creating bridges between otherwise weakly connected communities, as Tufekci (2014a) states: it has been determined that links shortened using the online service Bitly and posted on Twitter during the Arab Spring played an essential role as a carrier of information from the Arabic uprisings to the outside world, fulfilling a bridging function (Aday et al. 2012, cited according to Tufekci 2014a). However, this does not mean that social media served as a bridge in general, that social media functioned mainly as a bridge in the studied situation, nor, that Twitter had mainly a bridge function (as only tweets containing a Bitly link were studied). It is more about user needs and specific opportunities for action provided in a selection of cases coinciding, which promoted a specific mechanism, in this case that of building bridges (Tufekci 2014a).

Sampling bias when using Twitter data also originates in the frequently used method of selecting Twitter messages via terms used within the messages themselves, which should facilitate categorisation and detection of the tweets, denoted by a hashtag '#'. If only those tweets containing a specific hashtag are used for a study, selection via a dependent variable – i.e., namely the hashtag – takes place mostly without the necessary caution. A selection via a dependent variable occurs when selection for the inclusion into the sample depends on the studied variable itself – i.e., in the case of hashtags, if only those tweets are studied that contain the hashtag. Such samples have specific limits concerning their analytical power. For example, analyses of events, which took place as a result of the coincidence of several conditions, do not take into account those cases in which the event did not take place under the same conditions. With hashtag data sets, users made a deliberate choice to state the hashtag in the tweet – this is a case of self-selection. Self-selected samples can exhibit specific correlations that are not generalisable. Furthermore, hashtags are chosen with very diverse motivations (e.g., as an expression of sympathy with a specific cause or as an indicator for the adherence to a specific context, such as a conference), so that samples based on hashtag choice may have very diverse characteristics (Tufekci 2014a).

A further problem of research solely based on social media data is that a statement in the form of a tweet can be made for the most disparate motivations, reasons, and with varied intentions, and is often characterised by an entirely specific context to which it is applied. An analysis of tweets alone cannot meet the requirements of this complexity and must be conducted with prudence and caution, while also accounting for the fact that, to some extent, it is only the shadows on the cave wall that can be observed. Social media data depend on interactions, such as when individual, specific actions are registered and counted. The number of clicks, likes, and so forth one does, for example, do not give information on how many users saw a specific message and did not execute a measurable action. The results of the analysis are therefore inevitably always limited. Additionally, the same action(s) (clicks, likes) can have different, or even opposite meanings: 'It is clear that a retweet is information exposure and/or reaction; however, after that, its meaning could range from affirmation to denunciation to sarcasm to approval to disgust' (Tufekci 2014a: 510).

Nonetheless, by considering social media data and bigger connections (facilitated by computer intensive methods), interesting analyses of the behaviour of the platform users become possible, as they allow analysing interactions as behaviour of individuals in digitised societies. Likewise, hashtag data sets can provide interesting insights into the conversation of user groups that gather in connection to a specific hashtag.

Research ethics

Big Data research often poses ethical dilemmas for researchers. While analyses are usually conducted so that the entirety of users is in the foreground, not single individuals, the risk is high that single individuals can be reidentified via cited content or sociodemographic details in publications

(Zimmer 2010). This is especially problematic as informed consent of the users of a social media platform is normally not given and users may not even be aware that they are being studied (Hutton & Henderson 2015). Researchers who work on the basis of user-generated content must deal with the problem that users have not given their informed consent to the research. They are reliant on making individual decisions about how to handle the data and about which methods to apply to protect the privacy of the platform users depending on the research topic and studied groups or individuals (Weller & Kinder-Kurlanda 2014).

Difficulties with Big Data access

The necessity of possessing specific technical skills (e.g., programming knowledge) for gaining access to Big Data restricts the number of persons who have access to it. Data access, however, requires more than sufficient and appropriate technical understanding. Computational social sciences, data mining, and data analytics are often conducted by researchers outside of the academic field who work for platform providers or, especially in the United States or Great Britain, in national security services (Lazer et al. 2009). Data access, especially beyond what is offered via APIs, is easier in the academic sector when researchers are embedded in accordingly big or influential research groups or institutions that have the possibility to negotiate special agreements with platform providers or to buy data. On the topic of access, established attention economies, which are characterised by already existing inequalities in international academic activities, are thus reproduced or, in fact, reinforced. (Weller & Kinder-Kurlanda 2015). Ruths and Pfeffer speak of a 'rise of "embedded researchers" (researchers who have special relationships with providers that give them elevated access to platform-specific data, algorithms, and resources)' (2014: 1063), causing a fissure in the research landscape, an 'environment of information asymmetry' (Tufekci 2014b).

There are various ways to collect social media data for research; however, the bigger and more comprehensive the data sets ought to be, the more difficult the data access for an individual researcher. The difficulties of gaining access also affect the quality of the research. Data is not usually allowed to be forwarded because of the regulations of the platform providers, so, for example, it becomes impossible to verify research results (Bruns 2013).

Recent developments and new trends: experimentation and interdisciplinarity

Increasing exploration has been suggested as a way to solve the methodological problems of Big Data analyses ('multi-method, multi-platform analyses' (Tufekci 2014a), 'experimentation' (Ruppert 2013), and 'kludginess' (Karpf 2012) are required) and also for testing interdisciplinary approaches (Lazer et al. 2009, boyd & Crawford 2012, Karpf 2012, Kinder-Kurlanda & Weller 2014,

Tufekci 2014a). Tinati et al. (2014b) already developed tools for addressing both the problem of 'Internet time', i.e., the ephemeral nature of content and settings on platforms such as Twitter, and the heterogeneity of the varied (sub-) groups, which gather on the platforms. Here, various methods are combined in the sense of interdisciplinary experimentation (Tinati et al. 2014a). To avoid a revival of the trench warfare between quantitative and qualitative approaches (Ruppert 2013), the social sciences have to meet the challenges posed by the new data and methods and need to become involved in interdisciplinary collaborations (Savage & Burrows 2007, Ruppert 2013, Kinder-Kurlanda & Weller 2014). The technical skills and the programming knowledge of the computer scientists could complement the more thorough knowledge of social scientific methods and analytical interpretations (Tinati et al. 2014b).

An ANT for Big Data?

Thus, without considering the composition of a sample or, put more simply: knowledge of who the actual respondents are, the size of the data set is mostly meaningless (boyd & Crawford 2012). Here, a view on the context of usage – or, from a cultural anthropological perspective, the embeddedness of technically mediated action in the context of the everyday – can help. Above all, a modelling of the various actors, structures, and materialities as sociotechnical structures in the sense of an Actor-Network-Theory (ANT), which takes into account the results of Big Data analyses, is necessary. Such an ANT model would allow one to consider the platforms as mediators for emerging social media-specific cultural techniques (in the sense of meanings and practises in digitised spaces).

Many studies analyse the behaviour of users of a specific platform or a specific subgroup of users, who, for example, share similar interests. In this way, complex networks of human and non-human actors and structures are more straightforwardly displayed and problems of representativeness and inclusion, respectively, are more easily managed analytically, as the users are mostly less heterogeneous and their characteristics are less complicated to assess.

Not all practices of social media users are observable via a mere analysis of user-generated content. The study of users' everyday routines, of the domestication of technology in various contexts of everyday life (Berker 2011), and of the complex networks of content, actors, and technology, as is done in cultural anthropology, are available for interdisciplinary cooperation, since they could fill exactly this gap. Furthermore, they enable an examination of the complexity of practices of the generation of Big Data analyses, the consequences of analyses' results, and, in turn, the emerging practices that are aimed at making use of Big Data analyses, of resistance, or of conformity. The fact that users are increasingly aware that they generate data traces can also be taken into account. For example, Germany has seen numerous scandals related to the data collection practices of the intelligence agencies BND and NSA, which are shaping the image of social media and personal digital devices. Concerns about a loss of the private sphere have led to strongly negative reactions towards the information gathering being done by the Google Streetview car. The following illustrates especially interesting phenomena of feedback loops and complex practices of appropriation via several examples for the analysis of digitalisation phenomena in daily life.

Examples of actor-networks and feedback loops

Social media platform designers adapt their products continuously based on the results of various analyses. There are feedback loops, as the users likewise adapt their practices on the basis of their own experiences with the platform or, for example, because of media events. Such practices lead to data containing an element of recursiveness and 'cultures of circulation' will emerge (Beer & Burrows 2013). The involvement of users via the provided possibilities of interaction means that users are assigned an essential role in the systems of feedback-loops. At the same time, the possibilities of the users are nonetheless restricted to exactly these provided possibilities of interaction with the platform. The complexity of motivations and explanations are omitted. Anything that cannot be expressed within the interaction is not considered. The 'why' and 'how exactly' vanishes behind the interactive settings and restrictions of the interface and explanations are left to the designers' imagination.

A further example of feedback loops is the currently observable trend in Big Data analyses, to apply learning algorithms, respectively, to motivate users to function as teachers for algorithms. Cheng et al. (2015), for example, filter web content for whether it is generated by so-called 'trolls' – i.e., users who deliberately post provoking or insulting messages to trigger a confrontational escalation of discussions in forums. Cheng et al. let their filter algorithm 'learn' the characteristics of trolls from past contributions to discussions. In this approach, which is typical for Big Data analyses, the algorithm continuously observes events, analyses them, 'learns' from them, and adapts to the changing conditions. Another example for learning algorithms would be games where users are asked for evaluations of a specific situation or web content. Nguyen et al. (2015) presented such a system where Twitter users were asked to guess other users' age and gender on the basis of a single tweet. On the website, users could have their own tweets analysed and see whether gender and age corresponded, and thus be informed whether or not their tweets were 'typical'. As they are provided with this possibility for interaction, users can influence analysis results, but they cannot change the algorithm: if it is designed to reproduce the duality of gender attribution, there is no possibility to evade this inscribed ideology.

Active 'social analytics'

Thus, various factors come together in Big Data analysis: strongly context-dependent user behaviour, platform-specific structuring and defaults, samples pre-structured by data availability, and algorithmic feasibility all converge. This combination enables the development of dynamic information about the events happening at a specific moment in time on the platform, and this information can then be reflected back to the user. In a feedback loop, the insights, which become available from the Big Data analyses, then in turn influence user behaviour, on which the analyses are based. Thus, users are individually, and in various compositions, engaged in different practices of self-reflection. They modify their behaviour because they are aware of being observed, and this modification of their behaviour then has an effect on the results of Big Data analyses. Users utilise the results and try to influence analysis results. Tufekci (2014a) studied such feedback loops on Twitter in the context of the Gezi protests. So-called trends (or trending topics on Twitter), seen on the Twitter start page of the individualised streams on the left side, are an example. Currently, the most often used hashtags are found in this list. This so-called 'trending' can be commercially utilised by platform providers, for example, to determine prices for advertising space. Metadiscourses about trending emerge among the users, for example, on which meme is more successful than others. Users thus engage in various practices of self-reflection and discourses on trending. As trending generates great attention for a specific topic, numerous practices of deliberate manipulation of trends have been established that are more or less professionally used by individual users, organisations, or enterprises. Lobbying right up to the deployment of bot networks, which push content for often opaque reasons, is being used:

> Humans understand, evaluate and respond to the same metrics that big data researchers are measuring. For example, political activists, (…) often undertake deliberate attempts to make a Hashtag 'trend'.
> (Tufekci 2014a: 513)

Users often show a detailed understanding of Twitter's trending algorithm, which, though not published, is regarded to be understood to a large extent through reverse engineering. Campaigns with the aim to let a hashtag trend are not only used by activists as a means to gain attention, but, for example, also by politicians: Tufekci describes how the mayor of Ankara announced the publication of a specific hashtag in advance and then posted it on a specific date. Subsequently the hashtag was retweeted by thousands of Twitter users (Tufekci 2014a). Not publishing a hashtag until a predefined moment leads then to the generation of a maximum spike, to which the algorithm reacts. Tufekci mentions additional user's practices being applied to utilise algorithms. Links or references between individual tweets, for example, are not recognisable for analyses if users post screenshots of the original tweets instead of linking to them, or if they message about other users without stating their Twitter names, i.e., to make fun of them or to criticise them:

> Such behaviors, aimed at avoiding detection, amplifying a signal, or other goals, by deliberate gaming of algorithm and metrics, should be expected in all analyses of human social media.
> (Tufekci 2014a: 513)

Couldry and Powell (2014) have investigated different domains in which users draw on Big Data analysis results and found that too little attention is given to the consideration of the power of algorithms: the agency of Big Data (in the sense of the consequences of analysis results for user behaviour) is neglected as well as algorithmic reflexivity. According to Couldry and Powell, Big Data exhibits its effectiveness mostly in the area of governance and decision making and alters both their point of reference and their mechanisms. On the basis of uninterrupted flow of information and surveillance, recurring interventions are substituted with consistent management practices:

> Both management and government increasingly are becoming predicated upon the continuous gathering and analysis of dynamically collected, individual-level data about what people are, do and say ('Big Data').
> (Couldry & Powell 2014: 1)

This vision of government and control structures constantly and uninterruptedly supplied with new and current information is nevertheless not yet realised. For example, the question of which information, for which decision(s), at which point in time and space is actually the right information and how it should get to where it is required, surely is not realised; rather, this information is currently based on different, fragmented infrastructures that happen to allow the collection, analysis, and presentation of results of continuously running analyses (Couldry & Powell 2014).

The bigger role algorithms increasingly play in various areas has been analysed by Lash (2007) as a form of 'power through the algorithm': 'A society of ubiquitous media means a society in which power is increasingly in the algorithm' (Lash 2007: 17). This analysis has been criticised by Couldry and Powell, as it leaves no space for agency or reflexivity on the part of the 'smaller' actors. The establishing cultures of data collection have to be analysed in a mode that brings the agency and reflexivity of individual actors and the various ways in which they construct and implement power and participation to the foreground. Thus, an examination of what social actors and groups of actors undertake under the conditions of algorithmic power in diverse contexts is required (Couldry & Powell 2014). How do actors and groups of actors react in various situations to the existence of Big Data analyses and to the permanent availability of new, dynamic results?

Web citizens are aware that they are part of a process of production and analysis of data and that individual and collective behaviour or actions have effects on algorithms, which then in turn become visible in the results of algorithmic computations (Couldry & Powell 2014). Studies of the usage social actors make of Big Data analyses to achieve their aims are necessary (Couldry & Powell 2014). Beer (2009) mentions three areas for such empirical studies: the organisations of the social media providers, the software infrastructure of the web itself, and how these two affect each other in the everyday lives of those using (or not using) specific web-applications. Couldry and Powell (2014) call this a 'social analytics' approach, which considers itself to be a decidedly sociological approach of how Big Data analyses are utilised by social actors. This approach reveals how specific actors reflect their online presence and are being involved in social processes of reflection, surveillance, and observation, as well as in iterative adjustment. However, often only minimal adjustments are possible for the users (Couldry & Powell 2014). Couldry and Powell see special potential for sociological analyses in those cases, in which the aims of the actors are at odds with the interpretations of their actions, which are generated by the analyses.

The proximity of the objects of the data analysis to or distance between them and the aims and practices of the social actors shape the degree of tension and reflexivity, which exists in connection with the implementation of analytics. At the one end of the spectrum, there are analyses that directly promote other mechanisms of exercise of power (e.g., performance management); at the other end, there are cases in which it is about the redefinition of the aims and performance of the organisation, without directly influencing the evaluation and the management of individuals. In the former case, social analytics needs an investigation of changing and reinforcing power structures; while in the latter case, the investigation of social analytics would rather offer a phenomenology of how social actors and organisations present and observe themselves (Couldry & Powell 2014). As an example for how actors can make use of Big Data analyses for themselves (beyond total surveillance or market analyses conducted by social media platform providers on the basis of complete data sets or user-generated content), Couldry and Powell note that there are activists who lay out strategies in various areas for setting up alternative economies of information. Projects in the area of environmental protection and sustainability are an example, for which various data sources are used for making the provenance of products more visible, or there are others that ask Internet users to collect data and contribute to making environmental problems visible.

Conclusion

The availability of Big Data enables new analyses that, despite some challenges, allow for new perspectives on digitisation phenomena. Making use of Big Data requires a critical investigation of quantitative methods and of issues of validity and explanatory power, in addition to the willingness to acquire certain technical skills. However, the emergence of Big Data also opens up many possibilities for qualitative analyses of phenomena including the everyday lives of the various actors who conduct Big Data analyses, the contexts within which actors utilise the results of analyses, how Big Data affects everyday contexts, and of the diverse practices of utilising, reacting, and resisting. The research area of Big Data is all the more interesting because Big Data mostly means active analysing. However, to facilitate such studies a theory is required of the distribution and fragmentation of the various actors and agencies in the generation of digital traces and in feedback loops of analyses: 'Who are the subjects of digital data and devices?' (Latour et al. 2012).

The combination of quantitative Big Data analyses with the instruments of qualitative research offers special potential. Nevertheless, only with the cooperation of the designers of the algorithms will it be possible to find ways to allow users significant feedback, which account for the complexity of the usage situations and consider individuals' everyday world of meanings and practices (Beck 1997). Combinations of design perspectives and qualitative research approaches could enable more open and flexible feedback loops, while also, as Bowker (2013) demands, allowing for conceptualising new visions of the future on the basis of hermeneutic approaches.

Further resources

The Web Science Institute Southampton offers critical evaluations of Big Data's potential from a sociological perspective and gives an overview of various interdisciplinary projects. A link to the Southampton Web Observatory is provided where web data and visualisations can be shared (www.southampton.ac.uk/wsi/research/index.page).

The Digital Methods Initiative provides various tools for collecting and analysing Internet data, e.g., for searching Google for the frequency of occurrence of specific pictures or for searching within comments on web sites (https://wiki.digitalmethods.net/Dmi/ToolDatabase).

The Oxford Internet Institute provides information on methods and research projects along with descriptions of tools (such as for visualisation) (www.oii.ox.ac.uk/blogs/).

Various Summer Schools offer introductory and advanced courses on social media research methods and also cover other types of Internet data. An example is the Digital Methods Summer School in Amsterdam (https://wiki.digitalmethods.net/Dmi/SummerSchool2015).

Note

1 Available at: https://dev.twitter.com/streaming/overview.

References

Aday, S., Farrell, H., Lynch, M., Sides, J., & Freelon, D., 2012. New media and conflict after the Arab Spring. *United States Institute of Peace, 80.* https://www.files.ethz.ch/isn/150696/PW80.pdf

Anderson, C., 2008. The end of theory: will the data deluge make the scientific method obsolete? *Edge.* Available at: www.edge.org/3rd_culture/anderson08/anderson08_index.html.

Beck, S., 1997. *Umgang mit Technik. Kulturelle Praxen und kulturwissenschaftliche. Forschungsrezepte.* Berlin: Akademie Verlag.

Beer, D., 2009. Power through the algorithm? Participatory web cultures and the technological unconscious. *New Media & Society,* 11 (6), pp. 985–1002.

Beer, D. and Burrows, R., 2013. Popular culture, digital archives and the new social life of data. *Theory, Culture & Society,* 30 (4), pp. 47–71.

Berker, T., 2011. Domesticating spaces: sociotechnical studies and the built environment. *Space and Culture,* 14 (3), pp. 259–268.

Berry, D. M., 2011. The computational turn: thinking about the digital humanities. *Culture Machine,* 12, pp. 1–22.

Bowker, G., 2013. Data flakes: an afterword to "raw data" is an oxymoron. In L. Gitelman, (Ed.), *"Raw data" is an oxymoron.* Cambridge, London: MIT Press, pp. 167–171.

boyd, d. and Crawford, K., 2012. Critical questions for Big Data. *Information, Communication & Society,* 15 (5), pp. 662–679.

Bruns, A., 2013. Faster than the speed of print: reconciling "Big Data" social media analysis and academic scholarship. *First Monday,* 18 (10), Available at: http://firstmonday.org/ojs/index.php/fm/article/view/4879.

Catanese, S.A., De Meo, P., Ferrara, E., Fiumara, G., and Provetti, A., 2011. Crawling Facebook for social network analysis purposes. In *WIMS' 11: Proceedings of the International Conference on Web Intelligence, Mining and Semantics.* Sogndal: ACM.

Cheng, J., Danescu-Niculescu-Mizil, C., and Leskovec, J., 2015. Antisocial behavior in online discussion communities. In *ICWSM: Proceedings of the 9th International AAAI Conference on Web and Social Media.*

Couldry, N. and Powell, A., 2014. Big Data from the bottom up. *Big Data & Society,* 1 (2), pp. 1–5.

Frické, M., 2014. Big Data and its epistemology. *Journal of the Association for Information Science and Technology,* 66 (4), pp. 651–661.

Giglietto, F., Rossi, L., and Bennato, D., 2012. The open laboratory: limits and possibilities of using Facebook, Twitter, and YouTube as a research data source. *Journal of Technology in Human Services,* 30 (3–4), pp. 145–159.

Hutton, L. and Henderson, T., 2015. "I didn't sign up for this!": informed consent in social network research. In *ICWSM: Proceedings of the 9th International AAAI Conference on Web and Social Media.* pp. 178–187.

Karpf, D., 2012. Social science research methods in Internet time. *Information, Communication & Society,* 15 (5), pp. 639–661.

Kinder-Kurlanda, K.E. and Ehrwein Nihan, C., 2013. Ethically intelligent? A framework for exploring human resource management challenges of intelligent working environments. In A. van Berlo, K. Hallenborg, J.M.C. Rodríguez, D.I. Tapia, and P. Novais (Eds.), *Ambient intelligence – Software and applications.*

4th International Symposium on Ambient Intelligence (ISAmI 2013). Advances in intelligent and soft computing. Switzerland: Springer, pp. 213–220.

Kinder-Kurlanda, K.E. and Weller, K., 2014. "I always feel it must be great to be a hacker!" The role of interdisciplinary work in social media research. In WebSci '14: Proceedings of the 2014 ACM conference on web science. New York: ACM, pp. 91–98.

Lash, S., 2007. Power after hegemony. Cultural studies in mutation? Theory, Culture & Society, 24 (3), pp. 55–78.

Latour, B., Jensen, P., Venturini, T., Grauwin, S., and Boullier, D., 2012. The whole is always smaller than its parts: a digital test of Gabriel Tarde's monads. British Journal of Sociology, 63 (4), pp. 590–615.

Lazer, D., Kennedy, R., King, G., and Vespignani, A., 2014. The parable of Google flu: traps in Big Data analysis. Science, 343 (6176), pp. 1203–1205.

Lazer, D., Pentland, A., Adamic, L., Aral, S., Barabasi, A.L., Brewer, D., Christakis, N., Contractor, N., Fowler, J., Gutmann, M., Jebara, T., King, G., Macy, M., Roy, D., and Van Alstyne, M., 2009. Life in the network: the coming of age of computational social science. Science, 323 (5915), pp. 721–723.

Manovich, L., 2011. Trending: the promises and the challenges of big social data. In M.K. Gold (Ed.), Debates in the digital humanities. Minneapolis: The University of Minnesota Press, pp. 460–475.

Mestyán, M., Yasseri, T., and Kertész, J., 2013. Early prediction of movie box office success based on Wikipedia activity Big Data. PLoS ONE, 8 (8). Available at: http://journals.plos.org/plosone/article?id=10.1371/journal.pone.0071226.

Nguyen, D., Gravel, R., Trieschnigg, D., and Meder, T., 2015. "How old do you think I am?" A study of language and age in Twitter. In ICWSM'14: Proceedings of the 8th International AAAI Conference on Weblogs and Social Media. Ann Arbor: AAAI Press, pp. 439–448.

Puschmann, C. and Burgess, J., 2013. The politics of Twitter data. HIIG Discussion Paper Series, No. 2013-01.

Ruppert, E., 2013. Rethinking empirical social sciences. Dialogues in Human Geography, 3 (3), pp. 268–273.

Ruppert, E., Law, J., and Savage, M., 2013. Reassembling social science methods: the challenge of digital devices. Theory, Culture & Society, 30 (4), pp. 22–46.

Ruths, D. and Pfeffer, J., 2014. Social media for large studies of behaviour. Science, 346 (621), pp. 1063–1064.

Sakaki, T., Makoto O., and Yutaka M., 2010. Earthquake shakes Twitter users: real-time event detection by social sensors. In Proceedings of the 19th International Conference on World Wide Web. ACM, pp. 851–860.

Savage, M. and Burrows, R., 2007. The coming crisis of empirical sociology. Sociology, 41 (5), pp. 885–899.

Tiropanis, T., Hall, W., Crowcroft, J., Contractor, N., and Tassiulas, L., 2015. Network science, web science, and Internet science. Communications of the ACM, 58 (8), pp. 76–82.

Tinati, R., Halford, S., Carr, L., and Pope, C., 2014a. Big Data: methodological challenges and approaches for sociological analysis. Sociology, 48 (4), pp. 663–681.

Tinati, R., Philippe, O., Pope, C., Carr, L., and Halford, S., 2014b. Challenging Social Media Analytics: web science perspectives. In WebSci '14: Proceedings of the 2014 ACM Conference on Web Science. New York: ACM, pp. 177–181.

Tufekci, Z., 2014a. Big questions for social media Big Data: representativeness, validity and other methodological pitfalls. In *ICWSM'14: Proceedings of the 8th International AAAI Conference on Weblogs and Social Media*. Ann Arbor: AAAI Press, pp. 505–514

Tufekci, Z., 2014b. Engineering the public: Big data, surveillance and computational politics. *First Monday*, 19 (7). Available at: http://firstmonday.org/article/view/4901/4097 (accessed 06 February 2016).

Weller, K., 2014a. What do we get from Twitter – and what not? A close look at Twitter research in the social sciences. *Knowledge Organization*, 41 (3), pp. 238–248.

Weller, K., 2014b. Twitter und Wahlen: Zwischen 140 Zeichen und Milliarden von tweets. In R. Reichert (Ed.), *Big Data: Analysen zum digitalen Wandel von Wissen, Macht und Ökonomie*. Bielefeld: Transcript, pp. 239–257.

Weller, K. and Kinder-Kurlanda, K. E., 2014. *"I love thinking about ethics!" Perspectives on ethics in social media research*. Presentation at the conference Internet Research (IR15), Daegu, South Korea, 22–24 October 2014.

Zimmer, M., 2010. "But the data is already public": on the ethics of research in Facebook. *Ethics and Information Technology*, 12 (4), pp. 313–325.

Part III
Approaching the world digitally

9 From GUI to No-UI
Locating the interface for the Internet of Things

Nishant Shah

The contemporary digital moment is the moment of the Graphical User Interface (GUI). When Douglas Engelbart introduced the world to the mouse with his oN-Line System (NLS) that developed the manipulation of text-based hyperlinks, it was a new way of imagining the computer (Terranova 2009). The computer, till then, before it shrank in size following the inevitability of Moore's Law (1965) and the development of integrated circuits, was a behemoth. It was not an interface that you accessed the digital with, but it was the digital complex within which you lived.[1] The large mainframes of the mid-20th century were huge boxes of designed circuitry that were managed and occupied by human bodies that had to manipulate the components and the infrastructure in order for automatized computing to take place. These computers were also prone to breaking and constant repair, reminding us that computing was not a seamless process but required constant human interaction and intervention. As John Markoff (2005) points out in his counterculture history of the computer, the computer was an imagined whole; it was obviously fragmented, larger than life, impossible to imagine, and it was a triumph of physics and engineering if it functioned for more than a few minutes at a time without something collapsing. The early mainframe of the computer did not have a way of visualising the Von Neumann architecture (1945) that it was premised on or a way of explaining, to the lay user, the complex mechanism that was at work in the digital circuitry.[2] In fact, the computer was represented not by visuals or images but by architectural plans and blueprints that resembled physical engineering and modelling processes that became the abstractions through which the computer (and all its living and non-living components) could be controlled and governed towards efficient work.

Engelbart's NLS, further extended to graphics by the researchers at Xerox Palo Alto Research Center (PARC), with Alan Kay as the visionary leader, produced the first instance of the GUI as the primary interface for the Xerox Alto computer in 1972, producing an entirely new imagination of the computer (Johnson et al. 1989). The computer was no longer an open block that the user had to walk through, looking at exposed circuitry and inexplicable surges of light and electricity. Instead, it became a black box, completely hiding not only its physical computational components but also its machine language protocols, which were disguised by the seductive, animated, graphical buttons, arrows, and check boxes that became the default vocabulary of the personal computer – often acronymised as WIMP – introducing Windows, Icons, Menus, and Pointing device as the new metaphors through which the computer had to be imagined.[3] This imagination of the GUI as the front end of the computer became the common idiom through which desktop environments got popularised by Microsoft and Apple, thus quickly erasing other interfaces that preceded the GUI.[4,5] The GUI has a significant role to play in the democratisation of the computer as a device that could be accessed and used by 'children of all ages' (Kay 1973) and becoming a visible, ubiquitous, toy-like presence that transitioned from being a sinister machine to a companion gadget.[6]

The GUI, as our single point contact with the digital, draws from this history of interface design, so much so that it has become a salient part of the feature set and descriptions of new iterative gadgets.[7] Even as portable computing devices develop a touch-and-scroll language involving gestures rather than tactile input of the keyboard and mouse era, the GUI persists as the most dominant interface that brings together the mystified world of accelerated computing and the irrational realm of human intention. The GUI has not only become the singular interface that is identified by the screen, but it has also become the only mode by which the rest of our interfaces are designed and understood. The GUI has, in fact, spilled over the single screen units that were being developed at Xerox 'Star' at the turn of the century, and has slowly become the backdrop of modern life and its computational interactions, offering the 'messiness of everyday life' as an operational space that houses the 'ubiquitous computing of the present' (Dourish & Bell 2011: 4). As Helen Grace (2014) marks in her study of camera phones in Hong Kong, the GUI has become 'mundane', producing a 'great body of ordinary work that is ignored' and producing 'patterns, regularities, series, dynamic sequencing within which we might establish repetition', which needs to be seen as 'the rhythm of a beat that constitutes life, maintains it and guarantees its reproduction' (2014: 10). Similarly, Audrey Yue and Sun Jung (2011), looking at the urban landscapes of Incheon and Melbourne, argue that the GUI of the large urban screen, which are both 'public amenities and cinematic screens', become the new interfaces that bring together 'multiple stakeholders' (2011: 15) and ideas of urban regeneration, transcultural consumption, and processes of social belonging.

The persistence of the logic of the GUI enables it to move beyond just bringing together the human and the computational. It becomes a hub that connects multiple stakeholders, intentions, logics, and logistics, all seamlessly and smoothly connected under the slow glow of pulsating screens. The GUI has become such a naturalised metaphor that we forget to see it. It has become something that we see through, look at, but don't really see. As Wendy Chun (2013) argues, exploring the ways in which the transparent surfaces of the GUI hide the complex mechanisms of power and control in the age of the fibre optics, the more our machines become transparent, the more they become opaque. Chun argues that the graphical interface can only be understood as an 'opaque metaphor' – where it hides more than it shows and it shows only itself, even as it makes itself invisible and impenetrable. Resonating with Chun's thesis, Larissa Hjorth et al. point out that the GUI, embodied in the iPhone, has become a 'symbol, culture, a set of material practices around contemporary convergent mobile media, as well as a particular form of proprietary platform' (2012: 2). In her essay in the book, looking at female Australian iPhone workers, Hjorth further argues that the GUI 'participates in the relationship between public and private, work and leisure boundaries' (2012: 194), thus presenting the GUI glamorised in forms like the iPhone, as the aesthetic, political, social, economic, and cultural interface that enables techno-social practice and forms.

John Seely Brown and Mark Weiser (1996) have very concisely postulated, 'if the computers are everywhere they better stay out of the way' (1996: 3). With the rise of ubiquitous computing, we are already reaching a stage where we have shifted from keeping track of our computers to being tracked by them. As computing devices start getting embedded in everything around us, there is a questioning of the GUI as the site and location where intersections of human design and technological intention can be studied. There is an emerging narrative premised on the promise of the 'Internet of Things' that threatens to destabilise the centrality that has been afforded to the GUI as the reigning interface metaphor of our times. As Paul Daugherty et al. (2015) write on the *Harvard Business Review* blog, the new interface for the Internet of Things is going to be gestures. They argue that gestures are habitual, instinctive, and intuitive, and do not require the kind of learning that tactile and haptic media devices like the keyboard or the touchscreen demand. For Daugherty et al., the GUI is already ancient and not suitable for a connected environment where the ubiquity of computing devices means that the human subject could no longer be expected to process and keep track of all the information flows. The GUI was invented when we had imagined the human subject as the central reader of digital information.

The loss of the GUI is not just about finding new sites of interfaces or more intuitive design. If Engelbert's production of the GUI was in the service of making human-centric computing, then the disappearing GUI is indicative of a new world order of computing. As David Talbot (2014) argues in the *MIT Technology Review*, in the world of Big Data and predictive algorithm, the central 'readers' of our data societies are machines that do not require GUI to process the information. The human subject only needs visualisation of the final outcomes and decisions rather than of processes and mechanics, making the GUI nothing more than a projection site and denuding it of its importance as the primary interface of the digital.

The Internet of Things, as Galen Gruman (2015) points out, is made of 'headless devices' which no longer bear the imperative or pretence of making themselves transparent to the human eye, satisfied in their closed, opaque, and mysterious conversation between themselves. Gruman identifies the Internet of Things as consisting of three paths: machine to machine, smart systems, and ad hoc connectedness. He sees this as a mistake and insists on the need for a human readable interface for the digital. However, Mike Iacobucci (2015) writes in *Wired*, we are already witnessing a shift from the Interface of Things to Interfacing with things – the interface as a verb, as a series of negotiations and interactions with the digital objects, as opposed to the interface as a noun wedded to the fixed location and static surfaces of the GUI locked digital gadgets.

Most contemporary understanding of digital cultures and processes of digitisation is contingent upon the primacy and the immediacy of the GUI. Yet, as the brief overview shows us, the GUI is not only recent in the history of computing but it might also be the shortest-lived interface phase of computing. At the same time, we have seen that the production of the GUI played a significant role in the democratisation of computing and the rise of the World Wide Web and its ambitions of creating a connected, equal society. The GUI, even as we process its imminent decentering, has produced a subject of computing that has social, political, and cultural implications beyond usage, penetration, access, and adoption. The disappearance of the GUI or the rise of the new interface poses significant challenges to the study of digitisation for scholars and researchers as well as developers and coders. While the future of the GUI still seems to be up in the air, the contradictions it embodies and the resistance that it generates should not stall our interventions into understanding the Interface in our practices of digitisation.

This chapter looks at the precarious nature of the GUI to propose that the interface needs to be understood as more than the screen that has been so central to the disciplines and fields of Computer-Human-Interaction design and development. It seeks to disarticulate the idea of an interface by locating it in three different histories, narratives, and imaginations, drawing from Post-Colonial computing, Feminist epistemology of science and technology, and cybernetics studies to show how the Interface can be studied and decoded using different entry points and locations of access. In these three propositions, it also illustrates how we might be able to re-conceptualise the form, format, and function of the Interface outside of its quotidian touch-scroll-click-access transactional functionalities that the GUI is reduced to. It also further examines how this separation of the Interface from the GUI – this transition to the No-UI – offers us new modes of understanding protocols of power and realms of regulation that are not easily addressed by the omniscient presence of the GUI.

The body: the form of the user

Perhaps the most problematic conception that the GUI brought forward was the imagination of the User as the subject of computing. The figure of the user carries with it different connotations of infantilisation, alienation, and homogenisation that we need to unpack. The emergence of the GUI produces the subject of digitalisation as necessarily in a state of ignorance. Implicit in Allan Kay's remarks of building computing for 'children of all ages' and the strong influence of Piaget's theories of child development in early Silicon Valley experiments to make the computer accessible, is the idea that the 'user' is to remain unaware of what happens behind the seductive screen of the computer. Mercedes Bunz (2015), in her currently unfolding work, has argued that the GUI has persistently reinforced a visual aesthetic which promotes a juvenile articulation that conveys information and knowledge through simplified narratives and comic iconography, thus often rendering the complex, dark, or problematic ideas as legible and acceptable conditions.[8] Bunz looks at the playful, colourful, primary shapes and forms, and cute animals that form such a huge part of the Web 2.0 to suggest that this idea of the Internet as playground is dangerous because it acquits the user of all responsibilities while concentrating them on a few key central actors who remain hidden in the burgeoning disneyfication of the web.

This resonates with the thesis where Wendy Chun (2008) argues that the production of the GUI was a powerful way by which we produced 'transparent machines' that made the user transparent, while persistently hiding the mechanics of control and regulation that emerge in the age of fibre optics. The user was markedly not the developer, the coder, the technologist, or the systems administrator who designed the logics and was in control of the logistics of digitisation. Thus, while the emergence of the GUI inspired new generations of digital subjects to engage with computing as building blocks to their daily informational and affective practices, their interaction remained controlled by and limited to the graphical interface. Alexander Galloway (2006), for instance, in his eponymous work on Protocols shows how the hidden messages that travel as layers that wrap our human information traffic, define and design the routes of information transfer as well as the possibilities of our usage. Galloway argues that the seeming transparency of the GUI makes invisible the hidden mechanisms of limits and logic that override the user interaction and desire and make it conform to the designs that have been hard-coded by the proprietary companies and the developers who shape our digital interactions.

I have argued elsewhere (Shah 2015a) that, in examining the imagined user of our digital practices, the user is a template identity. It is premised on scale rather than on diversity. The idea of the user builds a homogeneous demography that celebrates difference but not diversity. In that essay, I have shown how the user is an identity that is only afforded to those who can make themselves legible, intelligible, and accessible to the demands and transactions of the digital networks. Through a historical and temporal journey, I have shown that the user is not necessarily a figure with agency, but constructed out of the protocols of penalisation, punishment, and pathologisation, culpable in its actions and restricted in its scope by the leaky networks and sneaky interfaces that collect data and share identifiers beyond our human capacities to compute or control this distribution.

However, the way the GUI arrests our imaginations and polices the possibilities of the human subject of digitisation can be resisted and questioned by dislocating the centrality of the visual interface as our single point access to digital technologies. One of the most insightful works that takes this challenge is by the feminist material historian of technology Jennifer S. Light, whose long-form essay "When Computers Were Women" (1999) seeks to make visible women's labour and presence in the development of computing, which has been obliterated from the current histories of digital making. In her essay, Light shows how, in the mainframe days of computing, the systems were not only run by women, but these women, with fresh degrees in Maths and Physics, were employed as computers to manually process code in multiple parallel batches. Light also uncovers pictures that show the woman in the computer and the woman *as* the computer, in those early days of digitisation. For Light, the woman's body was the first interface of mainframe computing. The women as computers were eventually abstracted only as counting bodies, thus discounting their gendered presence and making them invisible in the history of computing. The women computers were the points of access to the code, they were the physical workers who sustained the different components of the system, and relayed information and data between the nodes of that digital network.

Yet, the traditional histories of digitisation and computing have long forgotten these women and postulated the digital as the domain of the masculine triumph over technology and science. As Light further shows, when the mainframe shrinks, as the computer becomes personal, it is the body of the woman that gets replaced by automatic algorithms as well as visual interfaces that mimic the affective labour that was often associated with the women in the mainframes. Melissa Gregg (2013), in her book *Work's Intimacy,* shows how the body has been used as an interface in the development of assembly line techniques as well. Gregg proposes that the body of the female assembly line worker was also the body that bore the burdens of efficiency, productivity, development, and innovation, which are machine characteristics but can only be performed by and inscribed on these gendered bodies. This subject of digitisation that Light and Gregg conceptualise is no longer a user. Instead, it is an interface that is made invisible in its transparent presence, so that we can see through it and focus on the preconceived ideas of machine labour, as well as in the way by which it allows us the first interaction and visualisation of a machine that is otherwise too large and cumbersome to appeal to human narratives.

Positioning the body as an interface produces a new imagination of the digital. Especially with the growing discourse on Internet of Things and Big Data regimes, there is a way by which the human body is thought as redundant or incidental to the new data empires. The GUI replaced the body to also hide the gendered, sexual, and racial discrimination and violence that

From GUI to No-UI 185

are enabled and designed within the visual logic of our interfaces. Mapping the subjects of digitisation, not as users, but as interfaces opens up the study of digitisation to new social, cultural, and political interventions, which are not just about the digital but also about the material contexts within which the digital exists.

The feedback loop: of formats greater than graphics

One of the most basic functions that the Graphical User Interface embodied was the performance of interaction between the human and the computer. The earliest instances of the computing machines were transparently opaque. In the absence of software as we understand it or a lack of human language programming, to work with a computer was also to work on the computer. In the older instances like the ENIAC, this meant a constant configuration of vacuum tube devices to perform complex mathematical equations.[9] The computer was a piece of hardware and the tasks that it performed were transparent in the assembly of circuits and the flow of electricity through the system. The computer operators followed the whirring of the software and the clanking of the machinery to see it perform its tasks. The computer, in other words, was a tool, and much like we do not expect the hammer to light up on impact with a nail, the computer was not supposed to demonstrate its calculations. The computer had no 'inside' that needed to be demystified or indeed, decoded.

The production of the GUI was a moment where the computer became opaquely transparent. The GUI, as a 'graphical and visual representation of, and interaction with programs, data, and objects on the display of a system' (Plocher et al. 2012: 175) was an ergonomic need that emerged with the mass commercial availability and adoption of the computer. As the computer became more personal and shrunk in size, it postulated a non-specialised user who was to interact with but not understand the computing process. It was the moment when the computer stopped being a computing device and became a black box through which other things could be done; there was the production of the computer as a 'thinking machine'. It was given an interiority. This interiority was a pre-assembled hardware design that remained opaque, mystical, and unknown to the end-user. The shrinking of circuits and the production of portable computers available in an opaque case that hid the very processes of computing, along with the rise of human language programming and mass usage of computers for different purposes, necessitated the presence of a Graphical User Interface.

The GUI was a way by which the complex processes of the computer, the normative nature of pre-programmed algorithms, the restricted variables that determined the stable state of the computer, and the controlled nature of code, were made invisible behind animated point and move gestures that would allow the newly enfranchised user to initiate a pretend conversation with the computer. As our computing devices became more opaque, alienating their inner mechanisms from the user, it became necessary to produce a GUI that ensured the user that even when nothing was happening, something was at work. The light bulb that lit up with the assurance of a bright idea, the hour glass that showed that the lag was merely the computer thinking, the cursor that blinked expectantly on a blank document were all ways by which the computer became both transparent and opaque. The user received feedback, without which it would be impossible to communicate with the computer. This was the moment when What You See is What You Get (WYSIWYG) became an established acronym that insisted that there was a direct connection and correlation between the input and the output of computing. The WYSIWYG paradigm established a relationship between human input gestures and the presented GUI outputs, hiding the different layers of computing, control, restraint, and policing that the computer hides. The same principle informs another acronym that was foundational to the understanding of an 'error' in computing – GIGO, standing in for garbage in, garbage out, reinforced the idea that the human had complete control over the computing device, where an error in the output was because of the human error in inputting bad or corrupt data into the computing system. The computer itself was infallible, and when faced with a BSOD (Blue Screen of Death) or a Sad Mac, the error was not with the transparent machine but the opaque user whose irrationalities and neglect result in system crashes.

Sherry Turkle (1998), one of the earliest and most prolific digital anthropologists, in her examination of children interacting with 'alive toys', examines how the computers were designed to give immediate feedback that ensured authority, efficiency, and standardisation. Recounting the anecdote of a tic-tac-toe game where the user was playing against an algorithm that was designed to generally play towards a draw, Turkle points out that the game was also programmed to occasionally lose, providing impetus for the user to continue playing. However, this programmed error was invisible in the placid interface of the game that showed no logic behind why, without reason, sometimes it lost a game, even though the user had tried the same move multiple times before and had not won using that strategy. The Graphical User Interface, for Turkle, offers the possibility of a strategy, of a hack, of a cheat, which would allow the user to override and 'fool' the machine, not knowing that the only hacks that the machine entertains are the hacks that the machine allows.

The GUI was a feedback loop. It was a mechanism by which the user was ensured of the godlike centrality in the interactions with the computer, while hiding the amount of control that protocols and algorithms exercised in the pre-packaged computing devices. The GUI hid more than it showed.[10] Once we recognise the feedback function of the GUI that we experience in our quotidian digital practices, from the ubiquitous Like button to the swipe and scroll gestures, we realise that the interface is not about the visual, but that it is located in the feedback mechanisms which hide structures of power and discrimination, on the one hand, and processes of extra human communication that our connected devices perform, on the other hand. Especially in the world of Big Data, it is obvious that the seductive feedback of our sensuous artefacts hides the sinister workings of data harvesting, profiling, customised targeting, surveillance, and de-privatisation of the individual. One of the most telling examples of this seduction of the GUI is in Amazon's Mechanical Turk. Mechanical Turk is a platform that seamlessly connects those with work to adjunct, distributed task-wagers across the world, who are paid minimal costs to perform repetitive, fruitless, unskilled, and exploitative work. As Lilly Irani points out in her work on the Turkopticon (2013), examining the precarious conditions of life and labour that the Mechanical Turk operators are often put into, the interface of the platform constantly produces a sense of movement, of urgency, of connectivity, and traffic. The mechanical nature of the work and the accompanied exploitative labour conditions are glossed over as each completed task gets rewarded, and the task-wager is made to feel a sense of accomplishment through the interface that hides the fact that they are not afforded their dues for the work that they do. The interface focuses on making the processes visible, tracking tasks, and seeing them to efficient completion. The surface visibility of tasks ensures the invisibility of precarious workers, thus glossing over the neoliberal condition of work and making the precariousness completely opaque.

The robotics artist and researcher Kelly Dobson disrupts this narrative of the GUI and its capacity to hide in her body of work. One of her earlier projects, called Blendie, is about a mixer grinder that does not have a graphical or a tactile user interface that makes the grinder run. Instead, it requires the human user to make the noise that they would want Blendie to make while it would be churning and grinding. This noise gets translated by voice recognition circuits and gets converted into electricity regulation that powers Blendie to do the quotidian task which we expect our machines to do, and hence otherwise would not pay attention to. In her more recent work, Dobson produces neurotic companion robots called OMO. OMO is a series of machines that 'have expressive engaging behaviours, strength of character, negotiative egos and neurotic propensities' (Dobson 2012). As the patent description for OMO suggests, it is 'an object that interacts with a user at a visceral level'.[11] OMO responds, through touch, pulses, heat, and vibration, to its companion, and it learns to alter its behaviour through recognising patterns of the human it is interacting with. It disrupts the seamless feedback of the GUI by laying bare the idea that the human, in order to successfully and meaningfully interact with OMO, must learn to produce new gestures, movements, motions, and sensations for the gratification of receiving predictive responses. As I have argued elsewhere:

> This is not very different from the ways in which we change our lives so that our quotidian practices match the filtered realities of Instagram, or

our wit operates in byte-sized tweets, or our relationships get mapped and networked entirely through the rubrics offered by Facebook.

(Shah 2015b: 2)

Recognising this role that the feedback plays in shaping our behaviour and crafting our subjectivities is crucial to understand the interface as a site of political contestation and argumentation. With the Internet of Things, the loss of the GUI also has to be understood as a sign of how the new connected devices do not have the human user as the intended recipient of information.

Our machines speak with each other. Going back to the first order of cybernetics where scientists like John C. Lilly (1967) were postulating a world where human communication would be carried out by extra-human agents, we are finally realising the dream of machines that speak to and learn from each other. As we find our place in the age of perpetual information overload, we depend more and more on our machines to process, parse, and present the information that is produced and distributed faster than human cognition. Locating the interface, not on the surface and the visible, but in the mechanisms and intentions of feedback, opens up a new way of thinking about the computational relationships and subjectivities that have been kept outside the fold of digital cultures and critique. It comes closer to questioning ideas of agency, action, intention, and programme, in the ways suggested by Bruno Latour (2005), who proposes thinking of an Actor-Network-Theory that concentrates on 'actants' as hybrid units that configure and conspire to perform actions and transactions within our social, political, and computational systems. For Latour, focusing on the feedback between the human and the prosthetic is the beginning of a map of visible, invisible, implicit, and complicit actors which might be shaping and drafting the course of action which goes beyond just the visible performance of it. It allows us to be suspicious of the performative nature of computed systems and ask what they hide when they perform transparency through feedback loops, thus opening up new avenues for thinking about mechanisms of control and possibilities of freedom in our connected realms.

The contradictory object: towards the function of the interface

Within design and Computer-Human-Interaction (CHI) studies, the interface is often seen as the moment of interaction, rather than the space of interaction. This dislocation of the interface from spatiality, and locating it in time and practice, has interesting consequences. It suggests that the interface is not a predefined location, but something that gets activated as an interface because of an encounter between two distinct entities. The interface, thus conceived, is not the site where the norm unfolds and plays itself out. Instead, identifying something as an interface unpacks the different interests and intentions that are invested in the construction of the interface.

This idea of the interface as ex-potentia, as almost there, draws its parallels in computing hardware and software studies. The sociologist and network theorist Duncan Watts (2003) has pointed out in his work on 'the small world phenomenon' that networks form patterns of clustering that are not the democratic, free-floating space of every node connecting with every other node, but mimic patterns of organisation and proximity aimed at efficient flow of traffic. For Watts, the digital has to be located in the contradiction between probability and possibility; or between the dichotomies of logic and mathematics. As a logical, self-contained system, the computer acts as a self-referential space of probability, where it produces descriptions of what is happening based on patterns of the occurrence of that event. Which means that at any point of encounter, it is possible that any combination of moment and location might become an interface. However, given the programming limitations that have been designed into the computer, it only highlights and posits a few of the sites as spaces which can be located as the interface.

However, the computer is also a mathematical system and computes possibilities as well. This suggests that within the computing machine there is no part or component that is unknown, and every time-space continuum can become an interface on random interference and encounter patterns. Even though it favours the probable, it is capable of predicting the possible, which is not contingent upon pattern recognition but on computing the fluctuating and changing nature of variables. The emergence of the computer in this duality, as both a probable and a possible machine, allows us to think of the computer itself as an interface that brings these two contradictory systems together and offers new modes of conceptualising the future and describing the past. In other words, we are looking not at the interface of the computer, but at the computer as an interface that brings together several contrary forces, which are reconciled but not resolved.

Understanding these impulses that design the computer and the connected web as a large network dislocates the interface from being a site that connects the interior with the exterior, or the biological with the technological. Instead, the interface emerges as the porous boundary between contradictory spaces, ideas, concepts, and actions that the digital web embodies. It helps us revisit the world of binaries in a different manner. For instance, the cliché idea around binaries is that they are absolute contradictions that are represented in 0 and 1. The something-nothing interpretation of binary systems belies the running of electricity through the computer and indeed maps the troughs and crests of movements of electrons that power the system. However, the computer has to be seen as an interface that brings the 0 and 1 together. Interestingly, the bringing together of these two apparent contradictions has a further nuance where the 0 and the 1 are not predetermined. The regulation of electricity to run an efficient system is controlled by the computer, so that different locations within the system, dependent on the computational task at hand, are assigned the 0 and 1 value.

In other words, there are no hard-coded values within the computational system and thus, in its bringing together of the two values, the computer also becomes an activation system that invokes these values based on an intention. The role of the interface, following this physical computation logic, is not to visualise the 0–1 but to in fact produce them as a part of its computational process. The physical computer is thus freed of the ontological or teleological burdens. It is not something that we interface with and its role is not to produce interfaces. Instead, the computer as an interface object defines the various components that formulate the computing system and assigns values based on the logic of efficiency and circulation of traffic.

The network social scientist Albert-Laszlo Barabasi (2002) describes this phenomenon as the basis of the web as we understand it. On the web, which is a scaled-up model of our physical computation device, the nodes within a network are not predefined. Each entity within that network can be activated as a node in order for transfer of traffic as information packets to be enabled. When the network is in idle time or there are redundant resources, in order to reduce the expense of energy, different nodes are made to lurk, ready to be activated but not defined as nodes yet. Similarly, Tim Berners-Lee and Mark Fischetti (2000), in his original proposal for the World Wide Web, shows us that within the new connected environments, there is no distinction between a document and a link. Within the digital storage systems, the document and link both are the same kind of object. Based on the query that is presented to the system, different documents are activated to be links, tracing a path between the two entities seeking an encounter, thus forming temporary interfaces.

This tenuous nature of the interface, the understanding of interface as temporary, to imagine them as formed through encounter as opposed to being the sites where encounters happen, is the third trope that this essay has to offer to the study of digital interfaces. It opens up for us the idea that, just because the GUI is receding and the new No-UI objects are becoming common, this does not mean that we let go of the idea of interface. Instead, we conceptualise the new things in the Internet of Things as interfaces that embody different contradictory values, while being both the values simultaneously. Donna Haraway (1991), when trying to make a case for hybrid cyborg existence, argued for thinking about the human and the technological as being in a relationship of irony. Irony, for Haraway, is a mode where the cyborg, an ironical subject, is neither human nor technological, but also human and technological. We might extrapolate that framework from Haraway and suggest that the interface is a cyborg function. It brings together the contradictions of the biological and the prosthetic, the 0 and the 1, the something and the nothing together. However, in this bringing together, it does not separate, distribute, or disburse. Instead, it behaves as a probable and a possible system. It becomes an interface because it can be both and none of those two absolute values that it inhabits. It brings together the two things even as it separates them.

Identifying this duality that is at the heart of our devices and translating it into its social, cultural, political, and human presence dislocates the focus from decoding the interface to investigating the role of the interface. It suggests that the 'interfacization' – the making of something into an interface – is not just a description of a moment or an encounter. It signals a strategic, specific encounter that brings together contradictions that are simultaneous and distributed. The production of an interface is a move towards dually unpacking these contradictions, realising that they are intertwined in quantum principles of interference and separation, and that the interface, like a metaphysical simile, yokes them together, often with violence. It allows us to look at the various structures of contradictory and contrary forces that the interface is able to reconcile and recalibrate in its very production.

Facing the interface: towards the future of digitalisation

In this chapter, I have been interested primarily in looking at the shapes that digitalisation might take, given the new conditions of technosociality that we inhabit. Instead of taking the contemporary moment of the digital as the basis for looking towards the future, I have suggested that the current period is one of transition. Looking at the Graphical User Interface, which is extremely popular and almost inseparable from our imagination of the digital, I have shown that the GUI is neither the inescapable nor the default mode of the digital. The emergence of the GUI was historically necessitated by the new design of the computer from the mainframe to a personal device, just as it was also produced by certain intersections of cultural, economic, and subjective powers. While the GUI has grown in its scope with the increase in computing power and reduction in processing time, it is evident that it has had a short history in the longer temporality of the digital. As we transition into the Internet of Things, where the human-computer interaction is no longer the only form of interaction that can be imagined within future digitality, it is safe to predict that the GUI is going to be dislocated from the centrality it occupies right now.

In fact, with the Internet of Things we are looking at a future where connected machines of Big Data storage, retrieval, and transfer shall become the primary consumers and distributors of data and creators of informational meanings that would be mapped on our individual and collective bodies. As the computing devices grow smaller and more hidden, penetrating our bodies and surveying us from superhuman summits, the GUI, which was introduced as a way of making the computer acceptable for a user who is removed from its hardware mechanics, is slowly going to lose its centrality and primacy. However, I have shown that the loss of the GUI is not the same as the disappearance of the interface. Just as the emergence of the portable personal computing device shifted the idea of the interface from the computer as an interface to the interface of the computer, marking in the shift, the production of new sociopolitical subjectivities, so the transition into the Internet of Things is going to find new locations of the interface.

Drawing from three layers of discourse and development, and in examining the alternative locations of the interface that resist, question, critique, and reconfigure our restricted imagination of the interface as graphical in nature, I offer three approaches to understand the interface as a key to decrypt the intertwining of the digital and the social in this contemporary moment of transition:

The first approach is to locate the bodies that are being invoked, subsumed, hidden, and interpellated in the technosocial condition. The insistence is to think about the body not merely as a user that engages with the interface. Instead, we think of the body as an interface that allows the questioning of discursive and historical relationships that perpetuate the construction of the body as discretely separated and intentionally interacting with the digital for transactional and actionable processes. Instead, we go to the complex processes by which digitalisation has sought to replace, augment, and compete with the body – thus reading the presence or 'making invisible' of the body as a reifying of different intersections of mechanics and machine logics with human intentions and aspirations. The location of the body as an interface makes us aware that the regulatory structures, analytic modes, and infrastructure aspirations of technology need to be understood as the forces that shape the interface rather than are being described by the interface. The body, in its material practice, questions the GUI as the default form, and lays bare the idea that interfaces are not just representations of a reality but are generative forces that prescribe, persuade, and position new meanings and values, which further inform the regulatory intent of the technological.

The second approach is to locate the interface in feedback loops, which are both graphically and structurally inherent to the orders of cybernetics that govern the systems of computation. Recognising the feedback loop as performing the function of an interface dislocates the fetishisation of the visual, the haptic, and the responsive qualities of efficient interfaces, and instead looks at the complex processes of power, control, discrimination, and exploitation that are often made invisible in the seductive visuality of our rich GUIs. The feedback loop as an interface looks beyond the formats of interface, which get located in the conversations about design, usage, efficiency, speed, and so forth. It suggests that the formats that we need to pay attention to are the ones that question the WYSIWYG transparency of the interface and instead pay attention to the systems and structures that engender the enterprise of the interface.

The third approach suggested in this essay is to understand the function of the interface by locating it in the processes of contradictions that the digital reconciles. Shifting the site of the interface from space to time, and building upon network sciences, computational theory, and software studies, we can argue that the basic architecture of the digital and the impulses that govern our connected environments have the capacity to be simultaneous and distributed. Understanding the interface as a contradictory object enables a mapping and critique of the forces that activate a particular moment of encounter as an interface, thus questioning the obviousness or the naturalness with which the GUI is accepted in existing digital cultures narrative.

In all these different discourses, I wanted to propose that we need to shift our attention in digitalisation from the description of the interface to the role that the interface plays in organising and governing our societies. In building these new ideas of the forms, formats, and functions of the interface, I argue that the interface is not only an aesthetic and design process but is the site for political contestation and manipulation. We need to see the production of interface as an instance of regulation that seeks to make legible and intelligible the bodies, the processes, and the institutional structures that operationalise their control through the interface. Often, the interface is seen as a connector, as a seamless rendering of interoperability, and as a way by which the web becomes accessible to us. However, in these critical locations of the interface, the interface actually forces us to become accessible to the different power institutions and brokers that use the moment of the interface to exercise their control over our individual and collective bodies and organisations. Retaining the interface as a site but expanding its locations can create new, interdisciplinary, methodological frameworks that can also help us understand the moment of digitalisation, not merely as a technological transition but as deeply intertwined with the reconfiguration of social structures and individual identities in the age of ubiquitous computing. This chapter hopes to open up the scope and intersections of digitalisation beyond the literal study of form and format of the digital. Instead, it offers three approaches that examine the form, format, and function of the interface as ways by which digitalisation needs to be understood as symptomatic of larger systems that offer and organise the conditions of control and freedom, of probable and possible, and of protocol and power.

Notes

1 Moore's Law is the observation that, over the history of computing hardware, the number of transistors in a dense integrated circuit has doubled approximately every two years, based on the proposal made by Gordon E. Moore (1965).
2 In his Introduction to *The First Draft Report on the EDVAC*, John von Neumann described a design architecture for an electronic digital computer that contained a central processing unit, which included an arithmetic logic unit and processor registers, a control unit consisting of an instruction register and programme counter, a memory to store both data and instructions, an external mass storage capacity, and input and output mechanisms. Neumann's architecture thus gave the computer form and the possibility of embodying an interface. Available at: https://web.archive.org/web/20130314123032/http://qss.stanford.edu/~godfrey/vonNeumann/vnedvac.pdf.

3 David Smith (1975), at PARC, under the supervision of Alan Kay, had written a thesis where he first introduced these four concepts that fundamentally inform the ways in which our graphical user interfaces are defined.
4 One of the most noteworthy figures in the design of these new icons and interface interactions is Susan Kare, who was responsible for producing some of the most iconic graphics for Apple's Macintosh.
5 It is worth remembering that, even before the GUI, the computer had different interfaces that displayed the outputs and allowed for results and processes to be tracked. Console line interfaces or command line interfaces, which still get used for non-GUI machines, quickly got replaced by the GUI as a more intuitive form of Computer-Human-Interaction.
6 Alan Kay's provocative idea of the computer as a machine that could be developed so that children of all ages could use it signals the production of a GUI because, for Kay, the idea of the children was not limited to people who are young. Instead, he was proposing that every user of the computer has to be understood as childlike and hence, we need to build interfaces that will allow this infantile user to interact with these new devices, without too much training,.
7 Take, for example, Apple's marketing of its phones as having a 'retina display', making the richness of the UI and the density of pixelation a premium feature of its new products.
8 Bunz's current work was first presented at a keynote titled 'Well, hello there! Talking to technical interfaces', at the symposium on digital technologies and their affect, organised by the Dutch Art Institute, in Eindhoven, The Netherlands. Available at: http://dutchartinstitute.eu/page/7263/september-16---did-you-feel-it---a-symposium-on-digital-interfaces-and-their-a.
9 German media theorist Friedrich Kittler (1995) argues in his essay "There Is No Software" that software is a concealment. It is a construction that restrains the possibilities of computing and restricts the capacities of the user. He sees software as limiting and restrictive, and programmability as a force that enables this concealment, and is, instead of an advantage, considered an indictment. His argument about software also resonates with the interface and, just like Kittler's argument that software cannot be given the responsibility of programming our futures, the interface can also not be depended on to either be the site of prediction or location of description for our technologically mediated lives.
10 Alexander Galloway (2006) argues that protocols need to be understand as the visible opaque instances where data is shown and the metadata, where actual control resides, becomes invisible and only for machines to know and follow.
11 The description of the patent in the US Patents registry reads, 'Objects that interact with a user at a visceral level when the object comes within the user's personal environment. The objects detect a user's visceral behaviour, for example breathing pattern or perspiration. In response to the visceral behaviour the object simulates a behaviour of a living entity such as breathing, or produces an output to which the user responds viscerally, such as an electric field. The form of the output or simulated behaviour is determined by the visceral behaviour. The output or simulated behaviour may be modified to guide the user's visceral behaviour, for example by first synchronising simulated breathing to the user's breathing and then slowing down while the user's breathing is entrained to the simulated breathing. One such object has a skin that is warm to the touch, and simulates breathing with a breathing sound. Another such object produces electric fields like electric fields of the heart. A further such object simulates a purring sound in response to the user's breathing: the form of the purring also depends on sounds in the environment.' Available at: http://www.google.de/patents/US20100112537.

References

Barabasi, A-L., 2002. *Linked: the new science of networks*. Cambridge: Perseus Publishing.
Berners-Lee, T. and Fischetti, M., 2000. *Weaving the web: the original design and the ultimate destiny of the world wide web*. New York: Harper Collins.
Brown, J.S. and Weiser, M., 1996. *The coming age of technology*. Available at: www.cs.ucsb.edu/~ebelding/courses/284/papers/calm.pdf (accessed 2 October 2015).
Bunz, M., 2015. Well, hello there! Talking to technical interfaces. Keynote at *Did You Feel It? A Symposium on Digital Interfaces and Their Affects*. Dutch Art Institute, Eindhoven, The Netherlands, 16 September 2015. Available at: http://dutchartinstitute.eu/page/7263/september-16—-did-you-feel-it—-a-symposium-on-digital-interfaces-and-their-a (accessed 12 January 2016).
Chun, W.H.K., 2008. *Control and freedom: power and paranoia in the age of fibre-optics*. Cambridge: MIT Press.
Chun, W.H.K., 2013. *Programmed visions: software and memory*. Cambridge: MIT Press.
Daugherty, P., Schybergson, O., and Wilson, H.J., 2015. Gestures will be the Interface for the Internet of Things. *Harvard Business Review*. Available at: https://hbr.org/2015/07/gestures-will-be-the-interface-for-the-internet-of-things (accessed 10 September 2015).
Dobson, K., 2012. *Machine therapy*. Available at: http://web.media.mit.edu/~monster/ (accessed 20 February 2016).
Dourish, P. and Bell, G., 2011. *Divining the future: mess and mythology in ubiquitous computing*. Cambridge: MIT Press.
Galloway, A., 2006. *Protocol: how control exists after decentralization*. Cambridge: MIT Press.
Grace, H., 2014. *Culture, aesthetics and affect in ubiquitous media: the prosaic image*. New York: Routledge.
Gregg, M., 2013. *Work's intimacy*. Cambridge: Polity Press.
Gruman, G., 2015. *What the 'Internet of Things' really means*. Available at: http://core0.staticworld.net/assets/media-resource/16434/ifw_dd_internet_of_things.pdf (accessed 3 September 2015).
Haraway, D.J., 1991. *Simians, cyborgs, and women: the reinvention of nature*. New York: Routledge.
Hjorth, L., 2012. iPersonal: a case-study of the politics of the personal. In L. Hjorth, J. Burgess, and I. Richardson (Eds.), *Studying the iPhone: cultural technologies, mobile communication, and the iPhone*. New York: Routledge, pp. 190–212.
Hjorth, L., Burgess, J., and Richardson, I. (Eds.), 2012. *Studying the iPhone: cultural technologies, mobile communication, and the iPhone*. New York: Routledge.
Iacobucci, M., 2015. The Interface of Things: a universal remote for your life. *Wired*. Available at: www.wired.com/insights/2015/01/the-interface-of-things-a-universal-remote-for-your-life/ (accessed 3 September 2015).
Irani, L., 2013. Turkopticon: interrupting worker invisibility in Amazon Mechanical Turk. In *Proceedings of the SIGCHO Conference on Human Factors in Computing Systems*. New York, ACM, pp. 611–620.
Johnson, J., Roberts, T.L., Smith, D.C., Irby, C., Beard, M., and Mackey, K., 1989. *The Xerox star: a retrospective*. Available at: www.digibarn.com/friends/curbow/star/retrospect (accessed 20 February 2016).

Kay, A.C., 1973. A personal computer for children of all ages. *Xerox Palo Alto Research Center.* www.mprove.de/diplom/gui/kay72.html (accessed 20 February 2016).

Kittler, F., 1995. There is no software. In J. Johnston (Ed.), *Literature, media, information systems: essays.* Amsterdam: Overseas Publishers Association, pp. 147–155.

Latour, B., 2005. *Reassembling the social: an introduction to actor-network-theory.* London: Oxford University Press.

Light, J.S., 1999. When computers were women. *Technology and Culture,* 40 (2), pp. 455–483.

Lilly, J.C., 1967. *The mind of the dolphin: a non-human intelligence.* New York: Double Day.

Markoff, J., 2005. *What the dormouse said: how the sixties counterculture shaped the personal computer industry.* New York: Penguin Books.

Moore, G.E., 1965. *Cramming more components on to integrated circuits.* Available at: www.cs.utexas.edu/~fussell/courses/cs352h/papers/moore.pdf (accessed 20 February 2016).

Plocher, T., Rau, P-L.P., and Choon, Y-Y., 2012. Cross cultural design. In G. Salvendy (Ed.), *Handbook of human factors and ergonomics.* New Jersey: Wiley and Sons, pp. 172–185.

Shah, N., 2015a. Of heathens, perverts and stalkers: the imagined learner in MOOCs. In S. Bayne (Ed.), *The Europa world of learning.* London: Routledge, pp. 21–25.

Shah, N., 2015b. When machines speak to each other: unpacking the 'social' in 'social media'. *Social Media + Society,* 1 (1), pp. 1–3. Available at: http://sms.sagepub.com/content/1/1/2056305115580338.full (accessed 10 February 2016).

Smith, D., 1975. *Pygmalion: a creative programing environment.* PhD Thesis, Stanford University.

Talbot, D., 2014. An easy interface for the Internet of Things. *MIT Technology Review.* Available at: www.technologyreview.com/news/526006/an-easy-interface-for-the-internet-of-things/ (accessed 10 September 2015).

Terranova, T., 2009. Together forever. In J. B. Slater and P.v.M. Broekman (Eds.), *Proud to be flesh: a mute magazine anthology of cultural politics after the net.* London: Mute Publishing, pp. 244–245.

Turkle, S., 1998. Cyborg babies and cy-dough-plasm: ideas about self and life in cultures of simulation. In R. Davis-Floyd and J. Dummit (Eds.), *Cyborg babies: from techno-sex to techno-tots.* New York: Routledge, pp. 29–63.

Von Neumann, J., 1945. The first draft report on the EDVAC (Electronic Discrete Variable Automatic Calculator). Contract no. W-670-ORD-4926, between the United States Army Ordinance Department and the University of Pennsylvania Moore School of Electrical Engineering, University of Pennsylvania, June 30.

Watts, D.J., 2003. *Small worlds: the dynamics of networks between order and randomness.* USA: Princeton University Press.

Yue, A. and Jung, S., 2011. Urban screens and transcultural consumption between South Korea and Australia. In D.Y. Jin (Ed.), *Global media convergence and cultural transformation: emerging social patterns and characteristics.* Philadelphia: IGI Global, pp. 15–36.

10 Ubiquitous computing and the Internet of Things

Katharina E. Kinder-Kurlanda and Daniel Boos

Introduction: ubiquitous computing and the Internet of Things as digitisation phenomena

Producing digital data has become part of almost all everyday practices. Data is being captured by various digital devices and systems; whether we are at work, moving through public spaces, or at home – our actions produce digital data traces. Often, we are not necessarily even aware of this fact. Also, what happens to the data that is being generated is often uncertain or may change over time. Dataveillance and other surveillance concepts have been suggested (e.g., Lyon 2007, van Dijck 2014) to describe the phenomenon of ubiquitous technology and data. Changes to our societies are occurring at a global level even though capacities for technology and data capture are not evenly distributed. New expressions of power arise in which 'often illegible mechanisms of extraction, communication, and control ... effectively exile persons from their own behaviour while producing new markets of behavioural prediction and modification' (Zuboff 2015). However, there is not necessarily always a clear-cut or even explicit strategy 'behind' a specific data collection effort; a collection strategy may also change over time. Often, contingencies arise once the data is 'there' – in the sense that the data is contingently found to be useful for purposes other or in addition to those originally intended, possibly in combination with other data.

The ubiquity of data collection devices is part of a vision that has been called 'ubiquitous computing' (ubicomp) or, more recently, an 'Internet of Things' in which, increasingly, every 'thing' either is a computer, has one attached to it, or at least in some way is connected to the Internet. Ubicomp, as a concept, drove a large wave of prototypes and development in computer science and engineering about a decade ago; today, while we effectively live in an age where computing has indeed become almost ubiquitous, the term is used less often. Several technological developments have contributed to enabling the extension of the Internet outside of traditional computers and to computerising everyday environments. Among the most important developments are technologies that allow identifying and localising objects in time and space. A very widely used example is the GPS localisation feature in smart mobile phones that allows them to track their users across space and time, thus generating digital data traces that can, for example, be visualised on a map. Many other technological developments have also contributed to the emergence of an Internet of Things, in particular ever smaller and more powerful sensors and other hardware. Beyond hardware, the development of ubicomp is also driven by software-based developments, such as the spread of online social networks and by the way in which the borders between personal and public communication are increasingly becoming blurred.

The fact that data is being collected 'everywhere' by computing technology that has become 'ubiquitous' opens up many new possibilities for communication, participation, optimisation, or surveillance. The ubiquity of technology and the Internet is starting to blur the difference between being online and being offline. We are getting used to being informed and connected everywhere we go. Various actors navigate different everyday digital environments, try to make use of them, or try to resist them. Studying their motivations, actions, and beliefs in everyday situations can lead to new insights about how ubiquitous computing or an Internet of Things is starting to change societies.

One area where such technologies can be studied is business organisations. Many current development efforts in computer science and engineering aim to contribute to improving business processes through new forms of automation and machine-to-machine communication. An example of this is the German government initiative 'Industry 4.0':

> Combining embedded systems with business application software leads to entirely new business models and considerable potential for optimising production and logistics. At the same time, Industry 4.0 facilitates more resource-conserving production, greater individualisation and a perfect fit of products at mass-production prices.
> (Federal ministry of education and research n.d.: para. 5)[1]

Ethnography as a method for studying organisations has a long tradition (see Emmett & Morgan 1982) and was widely used to inform the design of new ubiquitous systems in the first decade of the 21st century. Ubicomp technologies within organisations and the changes that occur in the workplaces of those interacting with these technologies have been analysed in several studies (e.g., Stanford 2002, Kinder & Kortuem 2008, Boos et al. 2012, Schulz-Schaeffer & Meister 2015). Within business organisations, ubicomp technologies promised to have several benefits. They were often believed to increase efficiency and accuracy in general; to facilitate better control and supervision over operations; to speed up the handling of goods; to enhance products by making them smart; to allow avoiding human error, e.g. in record keeping; to reduce the amount of tedious, administrative tasks for people through automation; and last but not least, to allow the developing of new services (Fleisch & Tellkamp 2006).

Stanford (2002) studied, as an example, how an automatic distribution system, as it is used by most delivery companies nowadays, replaced the previous paper-based system. Delivery truck drivers could receive up-to-date routing and delivery instructions on their mobile devices and were also able to scan containers on delivery, electronically capturing the customer's signature. The quality of service improved, inventory accuracy became higher, and delivery errors were eliminated, reducing claims for incorrect deliveries (Stanford 2002).

Organisations themselves and the way work is organised change along with the introduction of ubicomp. However, most ubiquitous technologies will not be implemented in the way they were envisioned but rather will be subject to a more gradual, iterative change as contingencies arise and people (and organisations, policy makers, and governments) negotiate how these technologies can or should be used. Often, ubicomp technologies will, in turn, introduce new complexities; for example, they may introduce new accountability demands (Boos et al. 2012) and may not always turn out to bring the promised advantages.

For example, radio-frequency identification (RFID) systems were seen for a long time as a way to solve traditional enterprise resource planning problems, such as warehousing, supply chain inefficiencies, or lack of transparency. RFID tags enable identification of objects into which they are embedded (or attached to) from a distance without a direct line of sight and some even store data. In this way, for example, a product's manufacturer and other attributes may be stored directly on the item itself (Fano & Gershman 2002). However, today, the largest implementations of RFID chips are public transport or other ticket systems (Konomi & Roussos 2007) and tags used to prevent theft in stores.

Concepts, developments, and important texts

A major conceptualisation of the phenomenon of computers starting to spread into everyday environments that is still influential today was Mark Weiser's 1988 vision of a future in which 'invisible' computers would make personal desktop computers obsolete. According to Weiser, traditional desktop computers forced users to approach computers rather than computers assisting users in accomplishing everyday tasks. Computers had therefore so far remained largely in a world of their own. Weiser thought that the idea of a 'personal' computer was 'misplaced' and that eventually computing should become an integral, invisible part of people's lives (Weiser 2002). In order to achieve the aim of such a vision of 'ubiquitous computing', computers needed to disappear from people's consciousness and become embedded into everyday objects: 'The most profound technologies are those that disappear. They weave themselves into the fabric of everyday life until they are indistinguishable from it' (Weiser 2002: 19).

Weiser further envisioned a mobility of devices, the use of sensors, the development of 'smart tools', and new forms of interaction. Invisible computers would communicate with each other, creating new environments with which people would interact (Weiser 2002). The term 'ubiquitous computing' in Weiser's definition referred not merely to computers leaving the office. Instead, ubiquitous computing 'is invisible, everywhere computing that does not live on a personal device of any sort, but is in the woodwork everywhere' and 'activates' the world (Abowd et al. 2002: 48).

Today, smart homes and self-driving or interconnected communicating cars are but some examples of how computers are spreading into various everyday spaces. Other examples include closed-circuit television (CCTV) cameras, GPS navigation systems, and, most of all, mobile devices; mobile phones, for example, are often already fitted with or connected to various sensors such as health-monitoring wristbands and fitness apps that have not only turned out to be very successful commercially but have in fact become an integral part of many everyday activities. An 'Internet of Things' is created, a term which highlights how both technologies and the data they produce are connected, encouraging the merging of home, work, and public spaces (Lyons et al. 2004: 28). A variety of further terms are used to describe how computing technologies are being taken out of the office and away from being mere desktop computers in order to enhance previously non-computerised everyday situations: pervasive computing, cyberphysical systems, mobile computing, and wearable computing are but some of them.

Ubicomp within organisations

Ubiquitous computing technologies and organisations are changing simultaneously and together; Orlikowski (2007) has called this process of change a 'constitutive entanglement'. The social and the material are entangled in everyday life within organisations: a considerable amount of materiality is entailed in every aspect of organising in both visible forms (e.g., desks, computers) and less visible forms (e.g., electricity, data networks).

Various authors (e.g., Suchman 1987, Latour 2005, Orlikowski 2007) have pointed out that, when looking at technology (or: the material), we must neither solely focus on technology effects (technocentric perspective) nor merely on interactions with technology (human-centric perspective). Both perspectives are limited. A technocentric perspective assumes that technology is predictable and stable, performing as intended and designed across time and space. Such a perspective produces technologically deterministic claims about the relationship of technology and organisations. Heidegger has shown how such thinking is also problematic from an ontological point of view: whenever we take note of ourselves, we find ourselves already engaged in ongoing, everyday activity in which things already and immediately show up as 'possibilities for' this or that practical intention – 'for there is no such thing as a man who, solely to himself, is only man' (Heidegger 1977: 31).

A human-centric perspective, on the other hand, focuses on how humans make sense of and interact with technology, and often the technology thus vanishes from view. In order to move beyond these conceptual difficulties, a way needs to be found to engage with the everyday materiality of organisational life without ignoring it, taking it for granted, or treating it as a special case and at the same time focussing solely on technology effects.

There have been several proposals that challenge the conventional distinctions between technology and the social that involve 'reverting to a limiting dualism that treats them as separate (even if interacting) phenomena' (Orlikowski 2007: 1438). Examples include actor-networks (Latour 2005), sociotechnical ensemble (Bijker & Law 1992), and relational materiality (Law 2004). Generally, these concepts seek to decentre the human subject and to reconfigure notions of agency. Latour argues that agency is not an essence that inheres in humans but a capacity realised through the associations of actors so that it is relational, emergent, and shifting in nature (Latour 2005). Non-human agency can thus be conceptualised. Suchman (1987) also stresses the necessity to look at capacities for action, instead of at an individual actor living in a world of separate things. Orlikowski suggests 'that we can gain considerable analytical insight if we give up on treating the social and the material as distinct and largely independent spheres of organizational life' (Orlikowski 2007: 1438). To stress how the social and the material constitute each other, we need to replace the idea of materiality as preformed substances with that of 'performed relations' that emerge in ongoing, situated practice. This notion departs from the idea of common or reciprocal interaction that assumes that distinct entities interact with each other, but also presupposes some *a priori* independence of these entities. In contrast, Orlikowski's notion of constitutive entanglement assumes that there are no such independently existing entities:

> Humans are constituted through relations of materiality – bodies, clothes, food, devices, tools, which, in turn, are produced through human practices. The distinction of humans and artifacts, in this view, is analytical only; these entities relationally entail or enact each other in practice.
>
> (Orlikowski 2007: 1438)

Rather than talking about 'social practices', we should therefore talk about 'sociomaterial practices' (Suchman 1987). Orlikowski has illustrated, with the example of a study on the use of BlackBerry devices at a UK company, how constitutive entanglements of technology and organisations can be examined (2007: 1439). BlackBerrys were some of the first mobile computing devices that allowed reading and writing email on the go. In the example, professionals' communication practices had been reconfigured through their engagement with BlackBerrys. For example, professionals explained that they experienced a strong obligation to check incoming 'pushed' messages, so as to keep in the loop with what was going on. They also expected others to be available via their BlackBerrys and anticipated that others would expect the same of them. These expectations became generalised over time. As the lines between work and nonwork blurred, professionals' communication practices acquired tensions. At the researched company, individual desires to disconnect were in conflict with the collective expectations of the sociomaterial network. It was, however, not the case that the BlackBerrys had a certain social impact or that the new affordances were making communication more efficient: rather, 'The performativity of the BlackBerrys is sociomaterial, shaped by the particular contingent way in which the BlackBerry service is designed, configured, and engaged in practice' (Orlikowski 2007: 1444).

It follows that materiality cannot be treated as separate from the social. This has implications on how we view ubiquitous computing: ubiquitous computing technologies are expected to change the way people work, for example, by saving time or increasing efficiency. However, bearing in mind the earlier critiques of techno- and human-centric perspectives, one analytical problem lies in the assumption that the technologies change something. It then seems more beneficial to assume that there are co-constitutive entanglements of technology and the social, in which neither of the two precedes the other.

Case example: ubicomp in organisations

The following case example is taken from a long-term ethnographic study of road construction and maintenance workplaces in the UK between 2005 and 2009 (see Kinder & Kortuem 2008, Kinder et al. 2008), with a particular focus on existing technologies in the area of ubiquitous computing.[2] Mainly, this concerned technologies such as GPS tracking of vehicles, swipe card door systems, or mobile and stationary CCTV cameras. One aim of the study was to inform design of new ubicomp technologies by analysing why ubicomp technologies had become interesting to the construction and maintenance industry. Why had they become necessary in order to remain competitive and successful? What was the context for the decisions to introduce the technologies? What were the changes in society that made these technologies important and how, in turn, were the technologies part of establishing these changes? The research aimed to situate individual workplaces within the wider organisation but also to link the negotiations of ubicomp within wider discourses in the news media at the time.

In our study of UK road construction and maintenance workplaces, we found the reasons, thinking processes, and discussions about why ubicomp technologies had become desirable in order to remain competitive and successful as an organisation and to be connected to improving 'transparency' and managing liability issues.

Within the road construction and maintenance sector in the UK, there was a complex arrangement of contracts and subcontracts. Several larger companies competed in bidding for highway contracts. Local councils sometimes, but not always, employed private contractors for all or parts of the road maintenance and construction work in a county. Previously, all workers had been employed by the councils themselves. In addition, outsourcing could involve multiple organisations, as contracted companies might employ subcontractors for specific types of work. Responsibilities and alliances changing over time and the emergence of public-private partnerships creating various exceptions and special circumstances for specific types of work made the situation even more complex. For example, a company might be maintaining one specific stretch of a high-speed road independently of the other contracts in the area. This complex situation had led to shared responsibilities for work being carried out, which became hard to oversee and manage. Consequently, companies were seen to be under pressure from clients, insurance companies, and regulatory bodies to become more 'transparent'. Increasingly, companies were required to provide proof of work having been carried out and to give details about when, by whom, and under what circumstances this had occurred.

Due to their capacity to record assets' locations and to collect data about work activities, ubicomp technologies promised to provide the tools for collecting data that would allow for making 'transparent' and eventually controllable exactly such complex situations. New sensors, global positioning systems (GPS), and cameras were able to collect digital data in places previously only accessible to nondigital data capture and thus met the demand for a specific kind of transparency and control over information. Many records of work carried out were still being submitted in handwritten form at the end of the day, with workers having to estimate working times including especially relevant information such as exposure times to vibration (when operating specific tools), which may pose a long-term health risk.

While the situation within the road construction sector may be specific, contemporary organisations increasingly operate under conditions of complexity with organisational boundaries being hard to define. Within our setting, an example of managing pressures imposed by transparency demands with the help of ubicomp technology was the introduction of GPS tracking systems into vehicles. At the time of our study, using GPS for tracking vehicles was a relatively new technology that not all companies employed. For example, at the company we looked at, a GPS tracking system for gritting vehicles had only recently been introduced. GPS units in the vans were transmitting location data to a central server. The company expected to be able to track stolen vehicles, to prove to customers where, when, and what type of work had been carried out, and to collect data about vehicle speed and location in the case of accidents. These capacities were driven by demands for transparency from various actors: for example, clients expected the improved service of being provided with information on the status of contracted work. In this way, demands for transparency posed to the organisation from the outside led to a considerable change in the workplaces, e.g., of the gritting vehicle drivers whose everyday activities while at work were becoming visible within a digital system and thus were being made 'transparent'. Drivers themselves then often felt that they were the target of surveillance. For example, workers saw the implementation of GPS tracking technologies as part of a trend to increase control over their work, comparable to electronic timecards or automatic door controls. Drivers, construction workers, and maintenance workers thus often interpreted the introduction of such technologies as an effort of the middle management to enhance control over their everyday work practices. Local managers, however, were exposed to demands for transparency from clients and upper management, which they in turn had to conform to. They themselves, being closer to the way actual operations were performed, would often see how seemingly efficiency-increasing technology might not actually increase efficiency, due to situational complexities, or might cause frictions between middle management and workers.

However, transparency demands could also stem from inside the organisation and the lower and middle management. Various actors hoped that ubicomp technologies might increase transparency so as to be beneficial for the tasks that they needed to perform in the line of their work. An example in the setting of road construction was the growing convergence of new data, i.e., data gathered by systems such as vehicle cameras, highway cameras, and GPS systems. All these data converged in traffic control rooms that facilitated continuous monitoring of traffic, work, and incidents. Control rooms organised the response management and the road space booking system for construction or maintenance work. Wall mounted screens would show live feeds from stationary motorway CCTV cameras, and smaller screens showed the feed of other cameras, such as supply depots' security cameras. Cameras could be controlled remotely, and control centre personnel were able to zoom in on areas of interest. Such area control rooms received calls from bigger, regional control centres and the police, and if there was an incident, they would send out response vehicles. All response vehicles were monitored by a GPS location system that reported their location to the control room where units were visualised on a map. There was also a log so that vehicles' movements could be 'replayed' on the map. Control centre employees were very enthusiastic about the benefit of data gathered by cameras and GPS systems as it allowed controlling all vehicles and work going on in an area. They aimed to obtain as much data as possible and were very keen to extend monitoring to areas as of yet inaccessible to data capture, such as motorways not yet equipped with cameras or vehicles not yet fitted with GPS tracking systems. One control centre employee said: 'You can actually see what is going on' and stated that this was 'such a benefit'.

A more indirect way in which demands for transparency were carried into the organisation was the increased importance of liability, which also made data capture in previously inaccessible domains interesting. For example, we found that the health and safety context was very strongly defined by liability risks (Kinder et al. 2008). While there were obviously genuine concerns about health and safety in the risky environments of road work sites, health and safety also needed to be seen within the context of what many managers described as a rise in a 'blame culture'. Companies were increasingly becoming a target for insurance and other legal claims. In order to remain efficient, companies had to find ways to prevent lawsuits, or at least to be able to give proof that they had fulfilled their legal obligations. Within the fieldwork, three main areas were identified in which liability played an important role in the industry. First, health and safety in workplaces could be enforced to avoid legal cases brought against employers by injured employees. Second, civil lawsuits could be fought off, which often were brought against companies by members of the public. Third, companies were enabled to prove fulfilling work promised in contracts with clients.

Several managers told us that they held a growing 'claims culture' or 'blame culture' in the UK responsible for the increase in (wrongful) suits. One manager said: 'We've very, very much got a claims culture in this country, which is frightening'. Managers would also see the increasing blame culture as a reason why health and safety was even more important, as accidents became even costlier through follow-up lawsuits. For example, one operations manager said: 'It's a culture thing, that they sue you at the drop of a hat .... If you cut corners, not only for the safety of your operatives but also for the safety of the public you tend to find that it'll be counterproductive because if you get sued for tripping somebody up on a footpath and they break their arm, you can guarantee you lose 10,000 pound'.

Workers' desire to not be the target of enhanced control and surveillance thus was in conflict with 'the collective expectations of the sociomaterial network' (Orlikowski 2007: 1444), i.e. the expectation for the industry per se to become more 'transparent' and to gain an advantage in the fight for survival within an industry that was increasingly subject to a 'blame culture'. Managers' desire to be able to deal with transparency demands and other pressures connected to company growth and the need to generate profit in turn often were in conflict with the emerging sociotechnical networks in individual workplaces, where promises for more efficiency and control would not only meet resistance by the workers but also be inefficient due to unexpected contingencies and complications. For example, certain local knowledge that experienced road inspectors had about road layout or ownership would not necessarily be easy or even possible to transfer to electronic devices (see Kinder et al. 2008).

To summarise, in the UK road construction workplace study, we saw how ubicomp technologies collected digital data in places previously only accessible to nondigital data capture and thus met the demand for specific kinds of transparency and control over information, which had become especially necessary in order to mitigate liability risks. The ways in which opportunities and restrictions through suddenly available data would play out were diverse, with both benefits and drawbacks for individual actors being possible and sometimes unforeseeable. It became clear, however, that ubicomp technologies were discussed within the context of the data they would provide and the transparency and control over complexity that this data promised. In the concrete workplace settings, we could observe the wider societal trend towards more transparency and the perceived rise of a 'claims culture' play out and provide a background for why and how technologies would be used.

Conclusion

We have presented some lessons learned from a study of ubicomp within an organisation in the expectation that today's Internet of Things and the 'Industry 4.0' initiative in Germany in particular require ethnographers to study this digitisation of everyday workplace environments. In this endeavour, researchers could build on the theoretical framings around sociomaterial practice as they were used in the studies 10 years ago. Ethnographic studies with their focus on 'tacitly known scripts and schemas that organize ordinary activities' (Ybema et al. 2009: 2) could reveal the complexities of everyday encounters with the Internet of Things in organisational settings that otherwise often tend to be overlooked.

Further resources

The *IEEE Pervasive Computing Magazine* is a good place to track the progress of both the cutting edge developments in computer science and the current discussion of the field more generally. It has been publishing contributions about Mobile Systems, Ubiquitous Computing, and the Internet of Things since its first issue, published in 2002.

Link: http://ieeexplore.ieee.org/xpl/aboutJournal.jsp?punumber=7756

The ACM International Joint Conference on Pervasive and Ubiquitous Computing (Ubicomp conference) is one of the most important conferences in the field.

Link: www.ubicomp.org

Chapter 5 of the book *Ubiquitous Computing Fundamentals* by Alex S. Taylor 'aims to focus on some particular issues raised by ethnography in ubicomp and the implications these issues have for real-world ethnographic practice' (Taylor 2010: 205).

Notes

1 See www.bmbf.de/en/the-information-society-2353.html.
2 This case example has been abbreviated and adapted from a previous publication in Kinder-Kurlanda and Boos (2015).

References

Abowd, G., Mynatt, E., and Rodden T., 2002. The human experience. *IEEE Pervasive Computing*, 1 (1), pp. 48–57.

Bijker, W. and Law, J., 1992. *Shaping technology/building society: studies in sociotechnical change.* Cambridge: MIT Press.

Boos, D., Günter, H., Grote, G., and Kinder, K., 2012. Controllable accountabilities: the Internet of Things and its challenges for organizations. *Behaviour and Information Technology*, 32 (5), pp. 449–467.

Emmett, I. and Morgan, D.H.J., 1982. Max Gluckman and the Manchester shopfloor ethnographies. In R. Frankenbger (Ed.), *Custom and conflict in British society*. Manchester: Manchester University Press, pp. 140–165.

Fano, A. and Gershman, A., 2002. The future of business services in the age of ubiquitous computing. *Communications of the ACM*, 45 (12), pp. 83–87.

Federal ministry of education and research n.d. The information society. Available at: www.bmbf.de/en/the-information-society-2353.html.

Fleisch, E. and Tellkamp, C., 2006. The business value of ubiquitous computing technologies. In G. Roussos (Ed.), *Ubiquitous and pervasive commerce: new frontiers for electronic business.* London: Springer, pp. 93–113.

Heidegger, M., 1977. *The question concerning technology and other essays.* Translated and with an introduction by William Lovitt. New York: Harper Torchbooks.

Kinder, K. and Kortuem, G., 2008. NEMO project: using technology models to explore organizational issues and using ethnography to explore cultural logics in design, deployment and use. In F. Michahelles (Ed.), *Adjunct proceedings. First international conference on The Internet of Things. 26th–28th March 2008.* Zurich, Switzerland.

Kinder, K.E., Ball, L.J., and Busby, J.S., 2008. Ubiquitous technologies, cultural logics and paternalism in industrial workplaces. *Poiesis and Praxis: International Journal of Technology Assessment and Ethics of Science*, 5 (3–4), pp. 265–290.

Kinder-Kurlanda, K.E. and Boos, D., 2015. Socio-ethical issues of ubicomp: societal trends, transparency and information control. In K.E. Kinder-Kurlanda and C. Ehrwein Nihan (Eds.), *Ubiquitous computing in the workplace: what ethical issues? An interdisciplinary perspective.* Heidelber: Springer, pp. 61–74.

Konomi, S. and Roussos, G., 2007. Ubiquitous computing in the real world: lessons learnt from large scale RFID deployments. *Personal and Ubiquitous Computing*, 11 (7), pp. 507–521.

Latour, B., 2005. *Reassembling the social: an introduction to Actor-Network-Theory.* Oxford: Oxford University Press.

Law, J., 2004. *After method: mess in social science research.* London: Routledge.

Lyon, D., 2007. *Surveillance studies: an overview.* Cambridge: Polity Press.

Lyons, M.H., Potter, J.M.M., Holm, D.A.M., Venousiou, R., and Ellis, R., 2004. The socio-economic impact of pervasive computing – intelligent spaces and the organization of business. *BT Technology Journal*, 22 (3), pp. 27–38.

Orlikowski, W., 2007. Sociomaterial practices: exploring technology at work. *Organization Studies*, 28 (9), pp. 1435–1448.

Schulz-Schaeffer, I. and Meister, M., 2015. How situational scenarios guide technology development – some insights from research on ubiquitous computing. In D. Bowman, A. Dijkstra, C. Fautz, J. Guivant, K. Konrad, H. van Lente, and

S. Woll (Eds.), *Practices of innovation and responsibility: insights from methods, governance and action*. Amsterdam: IOS Press, pp. 165–179.

Stanford, V., 2002. Pervasive computing goes to work: interfacing to the enterprise. *IEEE Pervasive Computing*, 1 (3), pp. 6–12.

Suchman, L., 1987. *Plans and situated actions: the problem of human-machine communication*. Cambridge: Cambridge University Press.

Taylor, A.S., 2010. Ethnography in ubiquitous computing. In J. Krumm (Ed.), *Ubiquitous computing fundamentals*. Boca Raton: CRC Press, pp. 203–236.

van Dijck, J., 2014. Datafication, dataism and dataveillance: big data between scientific paradigm and ideology. *Surveillance & Society*, 12 (2), pp. 197–208.

Weiser, M., 2002. The computer for the 21st century. *IEEE Pervasive Computing*, 1 (1), pp. 18–25.

Ybema, S., Yanow, D., Wels, H., and Kamsteeg, F., 2009. Studying everyday organizational life. In S. Ybema, D. Yanow, H. Wels, and F.H. Kamsteeg (Eds.), *Organizational ethnography: studying the complexity of everyday life*. London: Sage, pp. 1–20.

Zuboff, S., 2015. Big other: surveillance capitalism and the prospects of an information civilization. *Journal of Information Technology*, 2015 (30), pp: 75–89.

11 Calculating spaces
Digital encounters with maps and geodata

Ina Dietzsch and Daniel Kunzelmann

Introduction

In the context of the comprehensive digitisation of everyday life alongside a general increase of communication based on geodata, maps, and mapping have undergone massive change. This brings new challenges to the field of cultural analysis. Focusing on geographic maps and so-called geomedia – 'media for which spatial coordinates and/or physical localisation are a necessary condition to their functioning' (Döring & Thielmann 2009: 19) – the following text will inquire what concepts we possess that might be appropriate to meet these challenges.[1] The term 'cultural technique' is essential here in the sense of Bernhard Siegert (2011), whose understanding embraces the classical meaning as the 'mastery of writing, pictorial and numerical systems', but also a wider meaning that includes: (1) systems of order and representation, such as diagrams, grids, indexes; (2) 'operational techniques', such as graphical operations in the field of art; (3) analogue and digital measurement and data procedures in the field of science and humanities, as well as (4) 'topographic, architectonic and media dispositives of the political' (ibid: 19).[2] Using the term cultural technique of mapping to view maps allows us to understand the production and use of maps as a closely entangled process of spatialisation, as well as to reconstruct all social practices into which maps are embedded, by which they are configured and through which they emerge (Schüttpelz 2006), while remaining sensitive to historical implications. Bearing this in mind, the first part of this chapter introduces the most important dimensions of mapping in general by answering the question: what exactly happens when something is 'mapped'? This is followed by a discussion of three key concepts that we believe are fruitful for the examination of digital processes in the field of mapping. Finally, these concepts are applied to examples from current research projects.[3]

Basic dimensions of an ancient cultural technique: mathematics, narrative, and politics

Maps are not what they seem. In everyday life, they appear as efficacious representations of the physical world, but they are not. Rather, they are spatial and social constructions built upon mathematical (geometrical) models.

Mathematics: why maps do not represent territory

Mapping in modern Western societies has been closely related to the development of scientific cartography, within which, for many years, the power of professional experts over the knowledge of map-making was located. For a long time, from a technocratic perspective, maps seemed to project an objective and objectifiable truth of an exact scientifically quantifiable territory. It was only the professional culture of cartography that brought about the ways in which we perceive space today, and in which maps appear to us as a self-evident representation of the physical world around us (Wood 1992).

It has mainly been voices from critical cartography who have pointed out that maps do not represent a prior existing territory but serve to shape and visualise relations according to the powerful rules of cartography. The major mathematical problem for technically producing maps is 'how to represent the round earth on flat paper. The technique to do this is called "map projection", and there are hundreds of ways to do it. Each projection has certain strengths. Each has profound weaknesses' (Wood, Kaiser, & Abramms 2006: 5). Map projections are thus geometric models that solve the problem of the 'round earth on flat paper' or, to put it differently, the transformation from a solid to a plane in a variety of ways. Each of these models must be a distortion, but as each seeks to serve a different purpose, each will distort in a different way. Such a non-representational perspective, as human geographers call it (Thrift 2004, Laurier 2011), is compatible with the reflexive understanding of mapping processes in cultural theory, like those held by Stephan Günzel, for example, who consistently pleads for what he calls a 'topological turn' (Günzel 2008: 220).[4] Instead of putting the pictorial power of a 'physical space', howsoever it is imagined, into the foreground, the topological view makes 'spatiality' the analytical term, privileging attention given to arrangements, relations, and 'neighbourhoods' of any kind. Günzel demonstrates his argument using the network plans of the London Underground. Figure 11.1a shows a conventional map by the constructor Fred H. Stingemore, which he produced in 1926, and Figure 11.1b one by the designer Harry Beck, who began drawing topological maps in 1933.

Each of the maps relates to the physical space in a different way and to a different extent, with the topological map (Figure 11.1b) fundamentally challenging the claim of maps to refer to specific points on the surface of the earth. What remains is the significance of 'neighbourly relations' (ibid: 222) of single lines and stations. Formerly taken for granted, the connection of map and territory has become something that requires further explanation. In this way, it becomes an epistemological object of its own, elucidating the fact of how preconditioned the quotidian notion of maps as presentation of a territory really is.

Showing and narrating: maps understood as diagrams

Against the background we have outlined previously, maps can be understood and analysed in a more promising way as visualising and narrating

(a)

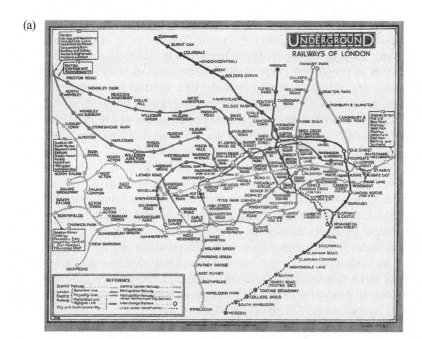

(b)

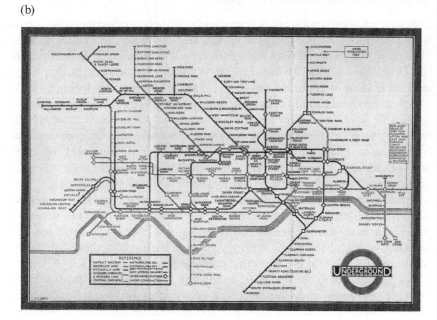

Figure 11.1 Network plans of the London Underground.[5]
Source: (a) Wikimedia Commons, map by: Fred H. Stingemore; (b) Wikipedia, map by: Harry Beck.

(mobile) diagrams that organise and communicate spatial relations, making these more accessible and imaginable. As much as any other diagram, a map consists of what Ingold (2007) calls 'guide-' or 'gridlines' and 'plotlines' (ibid: 156). *Gridlines* are those lines that form an integrative part of a plane, such as projection grids, resulting from a map projection and being visible as lines of longitude and latitude, or cadastral grids on city maps.[6] *Plotlines*, by contrast, are those shaping narratives on grids and planes (Dietzsch 2015). If plotlines were to disappear, gridlines are still there; they remain untouched. In order to avoid the misunderstanding of potential linearity in what follows, we will speak of grid and plot only. This is because plots – on diagrams in particular – are not meant to be fully formed linear narratives but *narrative elements*, a cluster of points organised according to a specific logic, ready to be *seen* but not necessarily a linear pattern ready to be *read*. Connecting the distinction of grid and plot with the cartographical principle of layers allows for an even more complex analysis.[7] Classic thematic maps (on demographics or the distribution of natural resources), as well as contemporary digital geoinformational systems (GIS), organise relations between *geographic* information and *factual* information in layers. The empirical examples in this text will show how the two principles (grid/plot and layers) can be linked in a productive way.

Maps are political: the concept of 'posting'

Maps are as highly political as the practice of their production. The classic work by Anderson (2006) helps to understand this nexus. In a chapter called 'The map' (ibid: 170–178) in his book *Imagined Communities: Reflections On the Origin and Spread of Nationalism*, Anderson demonstrates that mapmaking is far more than a form of 'geographic engineering' trying to technically reproduce already existing territory on a piece of paper (e.g., to then being able to put a country's boundary stones at the 'correct' geographical location). One key function of a map is the formative character it develops by helping to imagine the communities it seems to only represent. Together with other colonial instruments of governance, like the census or the museum, maps brought into being national communities (ibid: 163–187). This political mechanism can be observed in contemporary contexts of mapping too, although whatever social coherence the map creates need not necessarily to be a 'community', nor has its scope to be 'national'. By referring to a seemingly natural territory, it brings about a feeling of a shared sociality. Maps are still used for precisely this political power.

For Latour (1987), the modern, nondigital map serves as a paradigmatic example and indispensable part of networks of early Western modernity, networks 'built to mobilise, cumulate and recombine the world' (ibid: 228). The map allowed for remote mastery of places far away. The mutual interaction of diverse networked actors preserved local knowledge in the map as a mobile medium and, in doing so, made this knowledge transportable. These

Calculating spaces 213

'data' have been brought to 'data centres' where they were accumulated and recombined with other information. Former local knowledge of the world was thus transformed into new local knowledge and lifted onto a new scale. In Latourian terms, it is a chain of transformation. At the beginning stands the territory and at the end is the map. Part of each link within this chain of transformation is political, because mapping is to make a claim for symbolic power.

Every narrative element of a map poses a special type of ontological relation. These 'postings' bring into being what initially they purport to be 'real'. By insisting that 'something' exists, they help to provide existence to this very 'something', e.g., a border (Wood, Fels, & Krygier 2010: 53–57). 'Such *postings* add up, they construct, they perform the territory' (ibid: 61). They lend power to a map by transforming a (potentially) unstable reality into seeming ontological security (Kitchin & Dodge 2007). '[T]he power of the map is, quite literally, a function of the power of the posting which, by embedding a fundamental, ontological proposition inside a locative one, leverages the power of both into a ... performance of the real' (ibid: 52). Hence, the process of creating a map forms part of what Bourdieu (1991) depicts as 'struggles over the monopoly of the power to make people see and believe, to get them to know and recognise, to impose the legitimate definition of the divisions of the social world' (ibid: 221). If in what follows it is argued that digital maps provide context-sensitive solutions for everyday problems of orientation, this notion of postings helps focus on the political consequences that are encompassed by mapping as a cultural technique in the world; it hints at the normativity of mapping and the negotiations and the historicity of territory and spatiality contained with the practice (Kitchin, Gleeson, & Dodge 2012: 4).

Digitising mapping

What changes when the processes of producing and using maps are digitised? How can the entanglements of maps and knowledge based on geodata be described against the background of rapidly changing technological developments? Before we address these questions, it is important to acknowledge that digitisation is a meta-process that occurs at various points within the transformative chain from territory to the map stated by Latour and in different ways at different points in time, interrelating with other processes. The following argument focuses on the recombination and reordering of visibility and non-visibility, as well as calculating spaces. The argument builds on a specific contemporary state of technology that makes it nearly impossible to imagine one single part of the mapping process without digitisation.[8]

Recombination

Describing the functional logic of modern processes of mapping with the help of an analytical terminology, including terms such as data, data centre, and network, might nurture the impression that contemporary digital mapping does not differ fundamentally from classical modern forms, yet:

> [i]f inventions are made that transform numbers, images and texts from all over the world into the same binary code inside computers, then indeed the handling, the combination, the mobility, the conservation and the display of the traces will all be fantastically facilitated.
> (Latour 1987: 228)

Also, maps work increasingly within digital connections, in contexts in which every bit of information can be transformed into a 0–1 code that makes all information of a map 'programmable, alterable and subject to algorithmic manipulation' (Miller 2011: 14). In this process, mobility and invariance of signs are intensified (Schüttpelz 2009: 70). On the one hand, all praxis already included in analogical processes is made more efficient; on the other hand, mapping praxis becomes mobile in a completely new way. Maps, then, constitute a conglomerate of mobile units of information that can be recombined again and again. This increases the *variability* of connections of a new data variety that can be involved and visualised on the map. It also extends the functionality of those data. Contemporary digital maps, based on geodata, become 'interactive', 'networked', 'hypermediated', 'automated' (Miller 2011: 15–21), and are increasingly interesting as a means of fulfilling various purposes in everyday life for an extended circle of actors. Such purposes can range from the visualisation of numerical relations over the orientation in urban spaces to the handling of group management. Basically, digitisation as a meta-process extends and shifts the spectrum of practical techniques related to mapping. It is not only the manner of how this is produced and used that changes, but what is contained in the map; every single bit of the grid and the plot becomes changeable. Already existing (data-)modules can be combined in any desired way. Furthermore, these modules themselves can be transformed in how they form part of the grid or plot. The fact that each single element of a map can potentially be defined anew is of particular relevance in respect to the political dimensions of mapping, as every new attempt at a new definition can now be made the subject of political negotiation.

At this point, it is also worth reconsidering the previously outlined concepts of grids, plots, and layers, which are inherent to all contemporary geographical maps. The plot of an analogue map, once chosen, is fixed for the time being, whereas digitised maps consist of data saved on various levels of information in several databases in order to be later switched on and off in layered visualisations. In this context, new analytical questions arise, first about the specific practice of recombination and second about the related ordering systems. Which layer of information is defined to serve the grid best? What is the related plot? Or put another way: which layers contain information that is perceived as stable and which is perceived more mobile (or deliberately hold mobile) and mutable of the plot? Furthermore, if maps are seen from the perspective of *crowdmapping* and *participatory mapping*, it must be asked who is able and allowed to participate in writing on which layer by feeding in data, interlinking data, or even adding or rewriting categories? In a very practical way, this also means one must be sensitive to the question of who is entitled to set postings and/or to write and change databases and under what conditions?

Reordering visibilities and invisibilities

A second group of problems is set on the cultural analytical agenda by the omnipotent overview ostensibly promised by digital maps. While bringing together different parts of a paper map, fractures become directly visible. In contrast, digital visualisations now seem to enable the perfection of a totalised way of seeing. For example, on the display of a navigation system, a seamless move between different views becomes possible. Digital maps thereby become mutually connected 'panels of visualisation', and its usage transforms into a 'parcours' characterised by the consistent shift between panoramic and oligoptic perspectives.[9] In addition, the continuous shift between different technologies of access (smartphone, paper, tablet, desktop computer, and so forth) determines the shift of the borders of media, by further changing the relations between the part and the whole. Therefore, Johannes Passmann and Tristan Thielmann (2013) argue that not even Google Earth and its zoom options can provide the perfect panoptic view. They rather produce 'fragile panoramas' by constantly moving the horizon, yet maintaining the perception of being able to visually pace the entire space (ibid: 78). This perception of being able to seamlessly read a digital map is produced and supported by the movements of the user him or herself and databases updating themselves permanently. It is therefore an important peculiarity of digital maps that the still existing breaks are made invisible on screen. This is in sharp contrast to a geographic atlas, for example, the pages of which have to be turned one by one.

We will argue for another change of perspective with regard to digital maps. In contrast to the paper map, which contains a selection of information that had been considered relevant by producers concerning the specific purpose of the map (Thielmann 2013: 38), digital geomedia make visible the entire chain of transformation from the territory to the map: data gathering, administration, recalculation, printing, and using the map in order to make one's way and navigate territory (ibid). Such visibility has a price – the price of another invisibility on the level of calculation codes. Here is another point at which sound cultural analysis raises new questions. The question is no longer what is shown on a map and how, but also what is made to 'disappear' by design, which aims to make everything intuitive and easy (Galloway 2004, Klemmer, Hartmann, & Takayama 2006). Where and how is 'calculation' happening and to what extent does this influence the everyday production and perception of space?

Calculating spaces[10]

As we have shown, cultural analytical research can build upon a variety of insights from critical cartography (Kitchin 2008, Wood & Fels 2008, Della Dora 2009). Commentators from media studies who deal with geomedia also provide interesting starting points (Schüttpelz 2013, Thielmann 2013).[11] At the centre of contemporary research lies a shift of focus away from *maps as objects* and towards the open and ever incomplete *process of map-making* (Kitchin, Gleeson, & Dodge 2012). Rather than looking at the *ontology* of maps, such analysis investigates their *ontogenesis* (Kitchin & Dodge 2007). Mapping seems to be increasingly conceptualised as a socially networked form of *doing data*, a collective activity of capturing, measuring, and realising shared spaces (Southern 2013). Such practices can only be adequately examined empirically, by taking the materiality of databases and digital technology (hard- and software) into account. From a practice-theoretical perspective, maps and map-making might be understood as an interplay of mutually structuring resources (Lave 1988, Beck 1996). Such resources may be found and analysed in many 'socio-technological ensembles' of human and non-human actors and agencies, that is, within an interoperation of '[u]sers, preferences, infrastructures, providers, technology, laws, etc.' (Callon 1986, Latour 1991, Ilyes & Ochs 2013: 80).

What does this mean for specific research contexts? Mobile broadband connections increasingly allow actors to move in virtual spaces while they are physically in motion. The concept of 'hybridity' of spatial behaviour highlights the fact that currently produced and perceived spaces are always a mixture of digital and nondigital elements that can be separated analytically but merge in actuality (de Souza e Silva 2006). This does not exclude approaches that examine virtual spaces as spaces in their own right (Boellstorff 2008). On the contrary, commentators arguing for hybrid spaces do not deny the existence of independent cyberspaces. They simply highlight the fact that technologically determined virtual and physical spaces work according to different functional logics and often refer to one another in multifaceted ways. Locative media – smartphones equipped with mobile Internet and GPS – are part of such hybrid spaces. The analytically interesting question is, therefore: where can 'the digital' be located exactly within such hybrid spatial agglomerations? To quote Latour (1987) again, where does the translation into 'the same binary code' actually take place? (ibid: 228). How does this computational power affect the spatial orientation of today's actors?

As previously mentioned, maps have always been modelled from top to bottom by mathematics. Digitisation adds a further dimension of mathematically modelled processing to the nondigital map. Mapping now occurs not just as producing a map once but as a process of continuous actualisation and ongoing communication of digital signals between and with satellites. Activities that were previously directly related to the praxis of mapping are now increasingly outsourced into software. The praxis of map-making is strongly permeated by what Kitchin and Dodge (2011) call '*code*'. They argue that – be it the software that allows us to send and receive emails on a laptop, the ATM around the corner, or the digital map on the smartphone – code is increasingly part of quotidian activities:

> Taken together, coded objects, infrastructures, processes, and assemblages mediate, supplement, augment, monitor, regulate, facilitate, and ultimately produce collective life. They actively shape people's daily interactions ..., and mediate all manners of practices in entertainment, communication, and mobilities.
>
> (ibid: 9)

The authors further make a distinction between two ways in which programmed software exercises its power on spatial production: '*Coded Space*' and '*Code/Space*'. Coded space refers to a space in which software plays only a supporting role. 'In coded space, software matters to the production and functioning of a space, but if the code fails, the space continues to function as intended' (ibid: 18). One of the examples is a paper given by someone who has prepared a PowerPoint presentation in order to support his or her performance. In principle, the paper can also be given in the case of the PowerPoint software failing. In contrast, Code/Space describes how only calculating operations of software basically create and then continue to stabilise a social space. With the help of the example of an airline's check-in desk, the authors demonstrate that this social space only works based on the structuring power of the software. If this software fails, the Code/Space would cease to exist, becoming nothing more than a chaotic waiting room.

> Code/Space occurs when software and the spatiality of everyday life become mutually constituted, that is, produced through one another. Here, spatiality is the product of code, and the code exists primarily in order to produce a particular spatiality.
>
> (ibid: 16)

Digital mapping today is mostly still a practice within and in interplay with coded spaces, in the context of which maps are experienced as visualisation through interfaces. From this point of view, the challenge of cultural analysis in the future will be to include the logic of computers beyond such visual planes (Kunzelmann 2015). This shift of perspectives ultimately leads to the necessity of a sound and detailed examination of infrastructures that calculate in the background of collaborative and individual mapping practices. Only this outlook opens up the view to the complete picture of sociotechnological ensemble of contemporary actor-media-spaces (Schüttpelz 2013).

Maps in different sociotechnical ensembles

The following three examples have been selected to demonstrate how the discussed analytical concepts can play out in specific research contexts.

The children's map of basel: grids, plots, and layers

Taking a closer look at city maps, it is apparent that the grid on which the plot is told is not the one of longitude and latitude. More often, it is the cadastral plan, completed with a plot of streets, places, and buildings, or the network of streets itself becomes the grid on the background of which additional information such as tourist sites, organisations of social care, and so on are located. When a map is digitised, the single layers, which are essentially different databases, can be recombined; grids can be transformed into plots and *vice versa* as the children's map (shown in Figure 11.2) graphically demonstrates. Funded by several bodies, and in cooperation with other institutions, Basel's city government put this map online in 2014. The purpose of doing so was to provide a digital version of the recent paper map in order to be able to equip it with more options and to render it more efficient in terms of participation.[12]

The result of this effort is a map, the parts of which are easily to be recognised as the immutable (the grid) and the mutable (the plot). The contribution of map projection to the grid is undisputed – it is clearly a topographic map – but kept hidden. Only visible are the street and area map, where living areas and public places are drawn in different colours and therefore are distinguishable into 'private' and 'public'.[13] This grid gives a schematic and stable order to the space it produces. By moving the cursor appropriately, an additional layer can be engaged: the one showing the borderlines of the urban districts. The map also inheres clear instructions as to how to use it. A limited number of immutable layers can be exclusively combined with a limited number of mutable ones. Projection, street map, and the borders of urban districts as grids remain unaffected during the entire process of digitisation. This also applies to the borders of the chosen section for the entire city map. Very similar to the paper version, this section shows the core area of the city of Basel. Potential plot data outside of this section cannot be included.

All narrative elements making the plot (see Figure 11.3a and b) are very strongly mobilised by digitisation. To a limited extent, potential users determine displayed information while switching additional databases by switching mutable layers on and off. In doing so, they are enabled, to a certain extent, in the creation of new connections. They make their choices, find their own tracks on the map, or even plan their way through the town, transcending the previously disputed borders of urban districts.

People create stories and counterstories with, on, and through the map. Therefore, the map allows users to collectively narrate and negotiate stories. The amount of information that can be used, however, remains preselected

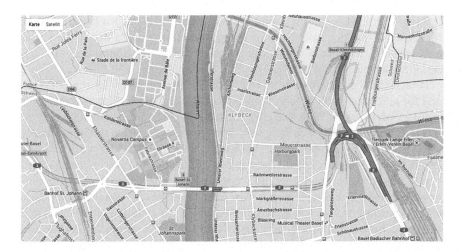

Figure 11.2 The 'clotted' grid of the children's map of Basel.[14]
Source: Christoph-Merian-Stiftung. Map data: © GeoBasis-DE/BKG / Google.

(a)
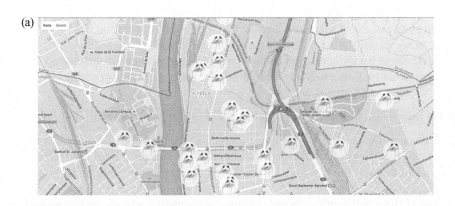

(b)
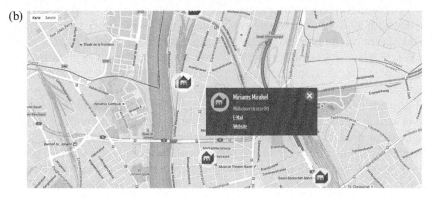

Figure 11.3 Narrative elements of the plot as visualised data layers.[15]
Source: Christoph-Merian-Stiftung. Map data: © GeoBasis-DE/BKG / Google.

by producers who, in this sense, also remain the authors of the map. Moments of participation within the production process are to be found only where a designer worked collaboratively with children as potential users, for example.[16]

Points and icons being connected to different plots open up the opportunity for *parcours*, implying changing panoramic and oligoptic perspectives (as outlined previously). With one click, another plane can be opened up and inspected; hints to other websites, new data levels, or additional diagrammatic visualisations. In this sense, digital maps are 'hypermedia' (Miller 2011: 17–19). Points and postings are indeed the smallest elements of the plot. The integrated hyperlinks, however, turn them into connecting hubs generating further narrative contexts. The digital map is *transplan,* insofar as its visualisation planes make connections to other linked databases. In the example of the children's city map, this particular functionality of the map is also supported on the level of geodata, because the points of the map are also linked with the more 'authentic'-seeming images of Google Street View. This leads to an interesting mixture of forms. On the one hand, the diagrammatic elements of the map are clearly recognisable. Postings on the digital map render a 'Web of Trails', which cannot be read linearly any longer and in which they become connection points in a net of references and conjunctions of information (ibid: 18). On the other hand, the schematic presentation is related to images that nurture the fiction of an authentic representation of the physical urban space with the intention of making exactly the schematic characteristics disappear.

The map of the pirate party: mapping as doing everyday logistics

In the previous example, the main focus was directed on the map as a result of the mapping process. The following example of an application programmed by the German Pirate Party emphasises the dynamics of collaboratively creating plots. The 'pirates' map' was made to provide an overview of the existence, location, and state of election posters in urban spaces during regional election campaigns. The application can be used on smartphones with a GPS function and enables members of the party to access the updated plot on the map, wherever they are. In this sense, the digital map of the Pirate Party can be understood as a supporting tool to solve an everyday logistic problem: the coordination of putting up, maintaining, and removing election posters in a larger public area.

The grid, or base map, of the pirates' map is an OpenStreetMap, on the top of which a variety of other layers can be placed.[17] The precise process can be described as follows: a potential user of the application accidentally discovers a damaged poster in a quite prominent, frequently viewed, position. Instantly, the user makes a post from this position on a mobile device, adding a commentary such as: 'poster damaged'. A pirate who is in charge of the maintenance of the election ads in that town or urban district takes note of the new posting on a map shown on a home office computer. The

Calculating spaces 221

pirates' map has changed and prompted for action: in this case, the pirate moves to the physical site in question with a toolbox (knife, cable tie, etc.) and a new poster to replace the damaged one. After it is replaced, the pirate updates the map with a comment like 'poster hanging'.

What makes digital maps fundamentally different to their paper predecessors is the collective mode of plot creation. By solving a specific 'set of relational problems' (Kitchin, Gleeson, & Dodge 2012: 3), the pirates collectively constitute a shared urban space of campaigning, which might best be understood as what Dodge and Kitchin conceptualise as *coded space*.

In their case, not only geodata of one individual is recorded on the base map as a trace of subjectivity (Thielmann 2013: 45). Rather, the short episode demonstrates traces of collaboration organised and coordinated by the fact that data relate to each other on the map. Local knowledge about current developments at different places is produced collectively. Cloud technology makes it possible for this knowledge to be produced, updated, and used simultaneously by all users. It also ensures that all participants get the latest, identical version of the map. In the moment users link geodata with information in a meaningful way (e.g., 'damaged posters', 'efficient location for election poster'), the map is not just a material tool of geographic orientation, but a social tool for coordinated collective action. On the spot, the digital map does not only work as the technological connection between various hypermediatised panels of visualisation. In the context of the pirates' map described earlier, it becomes a sustaining element within a collectively shared space of (inter-)action.

This practice of crowdmapping takes, in a continuous mutual interplay, the digital to the streets and, all in one breath, back to the map. By mapping actors in such a connected and interactive data practice, prospective actions of others are influenced, although the pirates' map does not produce algorithmically generated propositions for placing posters yet.[18] However, by purposefully switching data layers on and off, representations of the space of electoral campaigning are produced that suggest particular decisions: 'here no more posters (there are plenty!)' or 'put one here (there is no poster!)'. The digital map is the result of a participatory mapping practice as well as an actant within an actor-media-network. In the following example, the implied 'secondary agency' (Mackenzie 2006: 8), which unfolds within this process, will be elaborated on more thoroughly.

Politics (within) calculating spaces: Waze as an example[19]

Tel Aviv. On the display of the navigation system everything seems normal: streets, place names, arrows showing directions and a moving spot: the cab in which I am sitting. Suddenly, an icon in the shape of two colliding cars emerges. 'An accident in about one kilometre', explains Aaron, the driver. I had already noticed the iPhone clamped on the lower left side of the front window. Aaron continuously clicked on the touch screen there. 'This is Waze' he answered to my question about the navigation system in use. It was quite trendy software, he said, almost every one of his colleagues used it, because it worked much better than anything else. The particular trait of this system was that every single driver helps to keep the data up to date: 'We are the map. We are Waze.' The danger icon flashes again on the display. 'We're very close to the accident. 200 meters. Look!' I cannot see anything. Surprised, I tell him that I can't see an accident. 'Exactly!' he says. 'This is how the map works. When you come to the site of, let's say, an accident, Waze will ask you to confirm or deny the existence of the event at this particular location.' The display flashes again. Three windows have been popped up. Aaron presses the one in the middle. 'Since there is no accident, I have just pushed 'not there'. ... Waze will only delete the accident from the map if a certain number of drivers selects 'not there' on their own navigation device.' But it also works the other way round. Every user also can add a new event, such as an announcement of altered status, a traffic control or an accident. 'You just report it. It's really easy and convenient, just like turning on your radio', Aaron hints to the easy handling of the application. And if a defined number of users confirm the event in the same position, it turns up on every map and every GPS-running device using Waze: 'As I told you, we are the map.'[20]

Many characteristics of Waze are very similar to what was outlined with the previous example. Our intention is to place more emphasis on the inscribed political dimension of the very practice of digital mapping. To what extent can it be said that the users *are* the map? How much freedom do they have within the process of mapping? How closely are they and their actions entangled with technology and to what degree do the politics of calculating spaces structure their action? Waze continuously contributes to an alteration of traffic flows, because human beings involved in traffic make decisions, which would not have been shaped the same way without the map on the display. Databases of already classified events (see Figure 11.4a, b, and c) serve an algorithm as the basis to make specific suggestions for action in a particular situation. Those suggestions are offered to users of the software in ways like this: 'Route is recalculated. Please turn right in 500 meters'. By doing so, the software is an active part (actant) of modifying movements in space and the social dynamics of spaces.[21] As soon as one user links an action with the Coded Space of Waze, decisions made are also determined algorithmically in the sense that software calculates, interconnects, and builds networks and interactions.

The metaphor of calculating spaces not only embraces the fact that human beings translocate particular facilities into technology, which then sink into unreflected routines and seem to have lost significance for everyday action. Physical geography is increasingly a technologically mediated experience. Take the example of taking a train in Germany to get from one city to another. Although you might have chosen the 'shortest' route, it may simply be the fastest one that is, in fact, longer. Thus, for everyday logistics,

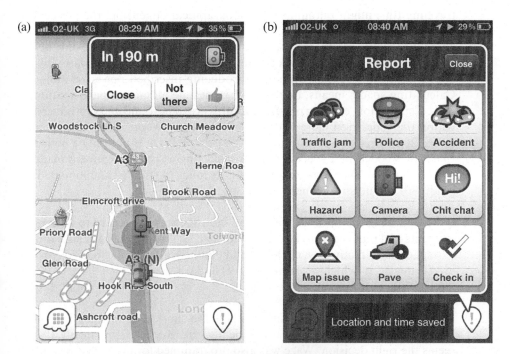
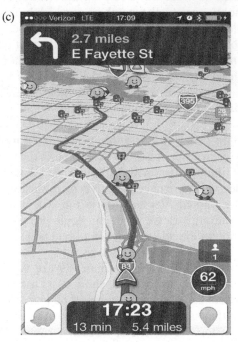

Figure 11.4 Screenshots of the geodata-based navigational software Waze.[22]
Source: Flickr.

traveling time might outnumber the geographical dimension, the physical length of the trip. Nevertheless, the train system would not be able to work without geographical mapping in the first place. Geographic space only becomes the infrastructural background, transferred into the software, disappearing from attention but still reproducing inscribed power relations. Calculating spaces means, therefore, that, despite all factual participation and collaboration, existing power structures are to be found and reproduced within the code, even more unobserved than ever (Wood, Fels, & Krygier 2010: 120). The space produced with the help of Waze is not only a coded space, it is coded in a specific way. Transferring a spatial argument made by Lefebvre (2009) to a coded space, it is as much 'fashioned and moulded' (ibid: 170) and penetrated normatively as any other space.

> If space has an air of neutrality and indifference with regard to its contents and thus seems to be 'purely' formal, ... it is precisely because this space has already been occupied and planned, already the focus of past strategies, of which we cannot always find traces.
>
> (ibid: 171)

Users of navigation systems usually follow the shortest or fastest route on the map, not the most beautiful or politically loaded one. Pragmatically, this seems the right decision. Waze was also programmed for precisely this purpose. In a hypermobile society, the members of which build on highly efficient flows of people and goods, lack of information about traffic jams can present a real problem. The narrative elements in the plot of navigation systems like Waze symbolise this ideology of efficiency. Without a critical mass of committed, interacting users making their decisions according to this logic of efficiency, the world of Waze would not contain enough categorised information for this kind of efficient mobility and simply would be useless.

Moreover, the code penetrating calculating spaces exists within specific power relations. This becomes even more visible when moving with Waze in Israeli territory or in Palestinian territory. There, the range of functions that can be used is limited, first for the simple reason that there are often no zip codes available, which are necessary to type in the desired destination. Another reason for limited functionality is the lack of accessible mobile broadband. Such administrative or technologic-infrastructural requirements, which are not apolitical at all, touch upon two essential functioning conditions for collaborative mapping processes.[23]

It is also productive to look at the base map of Waze. As long as the digital map successfully serves its purpose – to show to its user the most convenient possible route from A to B – the map will be perceived as a mirror of geographic reality; the map is the world and the world is the map. Postings work magically here, too. Nevertheless, Waze also enables the change of the world. It contains a feature with the help of which the alleged representation can be challenged. The software allows not only for collaboratively narrating the plot (by mapping 'accidents' or 'information about traffic jams'). By being allowed to edit place names on the *base map* itself, users also gain access to the 'backstage'. They can draw and name new routes on the grid or 'correct' existing ones (Thielmann 2013). Google, when they bought Waze in 2013, limited this option in all areas with contested political borders. The feature can neither be used in Spain nor China with full functionality. It also reduced its options in Israel after fierce debates about the base map. Is the line on Waze's base map a 'security fence' or 'separation wall' (*The Times of Israel* 2014)?

In places where territory is politically debated, the functionality of the technology of digital maps has been cut back, because it is more than just the solution of problems of everyday logistics. If one is simply concerned about efficient traffic logistics, it does not matter whether a line on the map is named 'security fence' or 'separation wall'. For the political lifeworld of navigating users, however, there is a significant difference concerning the symbolic and factual meaning and in the context of the ontological claim related to the mapping process, as stated by critical cartography for maps in general: 'this is not only there, but there is this' (Wood, Fels, & Krygier 2010: 56). It has been demonstrated very clearly that digital maps allow for new fields of political negotiation that go as deep as the very basic grid of a map and they also indicate at what point political participation ends.

Conclusion

The digitisation of maps opens up a wide and promising field for cultural analysis. At first glance, one might think that maps have to be understood first and foremost as tools producing locality. The shift of perspective in the direction of mapping as cultural techniques, however, brings questions to the fore about the 'how' of production and use, as well as another whole range of questions and perspectives. The reflexive perspective of cultural analysis, which argues topologically, primarily challenges the seemingly self-evident connection between the map and the territory it purports to represent and makes this very connection the subject of analysis. We have taken a diagrammatical perspective that understands maps as diagrams and (co-)acting part of a sociotechnical ensemble. The place of maps within the particular ensemble is therefore also always a subject of investigation itself. This allows for a wide range of definitions as to what a map can be: a stabilised result of spatial and visualising practices; material products being used for quotidian purposes or actants in the context of locative, socially networked data practice.

In the contexts of maps and mapping, digitisation can be observed at various points of the chain of transformations between the measurement of a territory and the data visualisation in the shape of a map, according to each particular research question and each particular view about what kind of use of binary codes is of interest. All of our examples suggested a second agency of 'coded space' involved in the context of digital technology (Kitchin & Dodge 2011). Yet, today, everyday spatial practices are still closely entangled with offline activities. The extent to which mapping practices will develop in the direction of 'Code/Space' (ibid) will be a question of interest for future research.

Though considering maps and mapping as a kaleidoscopic object of research, in all cases mapping must be understood as a spatializing cultural technique with a long history. Even the mathematical model of order inheres several dimensions in the shape of historical sedimentations. They must not be lost from view, particularly not if political practices are the focus of research. To put it another way: the former power of cartography and the cartographer is now replaced by transnational operating companies and the figures of programmers and administrators who do not negotiate spatial dimensions of political influence but the price of data 'hoarded' in order to transform it into economic capital.[24]

In this context, one of the prospective questions will be: who will actually own the posting on a digital map that has been put on and connected with geodata once? One lone posting will not be of any economic value in the future of 'Big Data' accumulated in a database; however, it might indeed render a value. As cultural anthropologists, we answer this question with a plea not to stay with the cultural aspects of visualisations but rather to take a very sound look at the databases behind them; not to try to circumvent the materiality of data and first and foremost, not to let the respect vis-a-vis mathematics take over by ignoring the *logics and agencies of calculating spaces*.

Notes

1 All quotes from German were translated by the authors.
2 For an English summary of the debate about 'cultural technique' as a term with an explicit German history, see the special issue of *Theory, Culture & Society* (2013).
3 The empirical examples are drawn from the project 'Media Worlds and Everyday Urbanism' at the Institute of Cultural Anthropology and European Ethnology (University of Basel), which is funded by the Swiss Research Foundation, and the dissertation project of Daniel Kunzelmann (LMU Munich and University of Basel) 'Digitally changing democracies? Techno-cultural transformations of local politics in Munich, Murcia and Tel Aviv'.
4 Topology is the field of mathematics dealing with relations.
5 The map by Stingemore (Figure 11.1) is licensed under Creative Commons' 'public domain' license, the one by Beck (Figure 11.2) under 'fair use' license. Figure 11.1: Wikimedia Commons. Tube map 1926. Available at: https://commons.wikimedia.org/wiki/File:Tube_map_1926.jpg (accessed 11 October 2015). Figure 11.2: Wikipedia. Beck map 1933. Available at: https://en.wikipedia.org/wiki/File:Beck_Map_1933.jpg (accessed 11 October 2015).
6 In the context of (digital) geomedia, *base maps* are the 'basic grid' of a map.
7 For an overview see de Lange (2013).
8 For an overview on contemporary key elements of digitisation see Miller (2011: 15–21).
9 We draw on Latour's concept of 'oligopticons' (Hermant & Latour 1998), which has been discussed and criticised by Johannes Passmann and Tristan Thielmann (2013).
10 For the concept of 'calculating spaces' (in German: 'Rechnende Räume') see Koch (2014). For an attempt to analyse the logics and agencies within such

spaces drawing on the example of Facebook's News Feed algorithm see Kunzelmann (2015).
11 For articles by Thielmann in English see Thielmann (2007, 2012).
12 The appointed content management system 'Magnolia CMS' was developed by Magnolia, a locally rooted but transnational operating company.
13 There is only one choice to be made: between topographic or satellite view.
14 Figure 11.3 used with permission of the Christoph-Merian-Stiftung with map data from GeoBasis-DE/BKG / Google. Christoph-Merian-Stiftung. Kinderstadtplan Basel. Available at: www.kinderstadtplan-basel.ch/de.html (accessed 11 October 2015).
15 Figures 11.4 and 11.5 used with permission of the Christoph-Merian-Stiftung with map data from GeoBasis-DE/BKG / Google. Christoph-Merian-Stiftung. Kinderstadtplan Basel. Available at: www.kinderstadtplan-basel.ch/de.html (accessed 11 October 2015).
16 The project 'Kinderstadtplan Basel' is available at: https://kubusmedia.com/kinderstadtplan-basel-internet-auftritt/ (accessed 11 October 2015).
17 The collaborative work on the base map and all its connected aspects are left out here, because the next empirical example will deal with it in detail.
18 Regarding the reciprocation between 'digital' and 'physical' spaces see Bastos, Recuero, and Zago (2014).
19 In 2008, Waze was developed by a small start-up company in Israel. It is a navigation software for smartphones and other mobile devices with a touchscreen and GPS.
20 Section from fieldnotes of Daniel Kunzelmann.
21 Visualisations produced by Waze demonstrate the social dynamics of the city's traffic space very vividly. Available at: http://youtube/LOHIWsK-Mf0 (accessed 11 October 2015).
22 All images are licensed under Creative Commons' license 'BY-SA 2.0' and accessed 11 October 2015. Figure 11.6: Available at: https://flic.kr/p/aExSow, Figure 11.7: Available at: https://flic.kr/p/aEu2Hz, Figure 11.8: Available at: https://flic.kr/p/ndccbf.
23 Seemingly only technical devices also have a political dimension. The Israeli territories give another example (see Saunders 2015).
24 Waze is connected to such an attempt of 'capitalising' data. Google, as Waze's new owner, now uses the socially generated data of Waze's users in order to feed them into their software Google Maps (Google 2013).

References

Anderson, B., 2006. *Imagined communities: reflections on the origin and spread of nationalism.* London: Verso.
Bastos, M., Recuero, R., and Zago, G., 2014. Taking tweets to the streets: a spatial analysis of the Vinegar Protests in Brazil. *First Monday*, 19 (3). Available at: http://firstmonday.org/ojs/index.php/fm/article/view/5227/3843 (accessed 1 April 2015).
Beck, S., 1996. *Umgang mit Technik. Kulturelle Praxen und kulturwissenschaftliche Forschungskonzepte.* Berlin: Akademie Verlag.
Boellstorff, T., 2008. *Coming of age in second life: an anthropologist explores the virtually human.* Princeton: Princeton University Press.
Bourdieu, P., 1991. *Language and symbolic power.* Cambridge: Polity Press.
Callon, M., 1986. The sociology of an Actor-Network: the case of the electric vehicle. In M. Callon, J. Law, and A. Rip (Eds.), *Mapping the dynamics of science and technology.* London: The Macmillian Press, pp. 19–34.
De Lange, N., 2013. *Geoinformatik in Theorie und Praxis.* Berlin: Springer.

De Souza e Silva, A., 2006. From cyber to hybrid: mobile technologies as interfaces of hybrid spaces. *Space & Culture*, 9 (3), pp. 261–278.

Della Dora, V., 2009. Performative atlases: memory, materiality and (co-)authorship. *Cartographia*, 44 (4), pp. 240–255.

Dietzsch, I., 2015. Erzählen mit Zahlen – Diagramme als Orte des [Er]Zählens. *Zeitschrift für Volkskunde*, 2015 (1), pp. 31–53.

Döring, J. and Thielmann, T., 2009. Mediengeographie. Für eine Geomedienwissenschaft. Introduction. In J. Döring & T. Thielmann (Eds.), *Mediengeographie. Theorie-Analyse-Diskussion*. Bielefeld: Transcript, pp. 9–64.

Galloway, A., 2004. Intimations of everyday life: ubiquitous computing and the city. *Cultural Studies*, 18 (2/3), pp. 384–408.

Google, 2013. *New features ahead: Google Maps and Waze apps better than ever*. Available at: http://google-latlong.blogspot.de/2013/08/new-features-ahead-google-maps-and-waze.html (accessed 11 October 2015).

Günzel, S., 2008. Spatial Turn – Topological Turn. Über die Unterschiede zwischen Raumparadigmen. In J. Döring and T. Thielmann (Eds.), *Spatial Turn. Das Raumparadigma in den Kultur- und Sozialwissenschaften*. Bielefeld: Transcript, pp. 219–237.

Hermant, E. and Latour, B., 1998. *Paris: ville invisible*. Paris: La Découverte. Available at: www.bruno-latour.fr/sites/default/files/downloads/PARIS-INVISIBLE-GB.pdf (accessed 13 January 2016).

Ilyes, P. and Ochs, C., 2013. Sociotechnical privacy. *Tecnoscienza*, 4 (2), pp. 73–91.

Ingold, T., 2007. *Lines: a brief history*. London: Routledge.

Kitchin, R., 2008. The practices of mapping. *Cartographica*, 43 (3), pp. 211–215.

Kitchin, R. and Dodge, M., 2007. Rethinking maps. *Progress in Human Geography*, 31 (3), pp. 331–344.

Kitchin, R. and Dodge, M., 2011. *Code/Space: software and everyday life*. Cambridge: MIT Press.

Kitchin, R., Gleeson, J., and Dodge, M., 2012. Unfolding mapping practices: a new epistemology for cartography. *Transactions of the Institute of British Geographers*, 38 (3), pp. 480–496.

Klemmer, S., Hartmann, B., and Takayama, L., 2006. How bodies matter: five themes for interaction design. In *Proceedings of the 6th Conference on Designing Interactive Systems*, ACM, New York, pp. 140–139. Available at: https://hci.stanford.edu/publications/2006/HowBodiesMatter-DIS2006.pdf (accessed 12 February 2016).

Koch, G., 2014. *Digitale Texturen urbaner Räume. Überlegungen zum Ortsbezug von Öffentlichkeit und Privatheit*. Available at: www.forum-privatheit.de/forum-privatheit-de/aktuelles/veranstaltungen/veranstaltungsdokumente/20140704_workshop_rechnende-raeume/Digitale-Texturen_Vortrag-Gertraud-Koch_WS-Rechnende-Raeume.pdf (accessed 11 October 2015).

Kunzelmann, D., 2015. Die stille Politik der Algorithmen. Das Beispiel Facebook. *Kuckuck. Notizen zur Alltagskultur*, 2015 (2), pp. 30–35.

Latour, B., 1987. *Science in action: how to follow scientists and engineers through society*. Cambridge: Harvard University Press.

Latour, B., 1991. Technology is society made durable. In J. Law (Ed.), *A sociology of monsters: essays on power, technology and domination*. London: Routledge, pp. 103–132.

Laurier, E., 2011. Driving: pre-cognition and driving. In T. Creswell and P. Merriman (Eds.), *Geographies of mobilities: practices, spaces, subjects*. Farnham: Ashgate, pp. 69–81.

Lave, J., 1988. *Cognition in practice: mind, mathematics and culture in everyday life*. Cambridge: Cambridge University Press.
Lefebvre, H., 2009. Reflections on the politics of space. In N. Brenner and S. Elden (Eds.), *State, space, world: selected essays*. Minneapolis: University of Minnesota Press, pp. 167–184.
Mackenzie, A., 2006. *Cutting code: software and sociality*. New York: Peter Lang.
Miller, V., 2011. *Understanding digital culture*. Los Angeles: Sage.
Passmann, J. and Thielmann, T., 2013. Beinahe Medien. Die medialen Grenzen der Geomedien. In R. Buschauer and K. Willis (Eds.), *Locative Media. Medialität und Räumlichkeit*. Bielefeld: Transcript, pp. 71–104.
Saunders, V., 2015. The challenges of being an entrepreneur in Palestine. Available at: http://vickisaunders.com/2015/02/19/the-challenges-of-being-an-entrepreneur-in-palestine/ (accessed 11 October 2015).
Schüttpelz, E., 2006. Die medienanthropologische Kehre der Kulturtechniken. *Archiv für Mediengeschichte*, 6, pp. 87–110.
Schüttpelz, E., 2009. Die technische Überlegenheit des Westens. In J. Döring and T. Thielmann (Eds.), *Mediengeographie*. Bielefeld: Transcript, pp. 67–110.
Schüttpelz, E., 2013. Elemente einer Akteur-Medien-Theorie. In E. Schüttpelz and T. Thielmann (Eds.), *Akteur-Medien-Theorie*. Bielefeld: Transcript, pp. 9–70.
Siegert, B., 2011. Kulturtechnik. In H. Maye and L. Scholz (Eds.), *Einführung in die Kulturwissenschaft*. München: Wilhelm Fink, pp. 95–118. Available at: www.uni-weimar.de/medien/kulturtechniken/kultek.html (accessed 11 September 2015).
Southern, J., 2013. Comobile perspectives. In R. Buschauer and K. Willis (Eds.), *Locative Media. Medialität und Räumlichkeit*. Bielefeld: Transcript, pp. 221–242.
The Times of Israel, 2014. Waze traffic app becomes Israeli-Palestinian battleground. Available at: www.timesofisrael.com/politics-drives-waze-users-to-edit-apps-israeli-map/ (accessed 11 October 2015).
Theory, Culture & Society, 2013. Special issue: cultural techniques, 30 (6), November 2013.
Thielmann, T., 2007. 'You have reached your destination!' position, positioning and superpositioning of space through car navigation systems. *Social Geography*, 2 (1), pp. 63–75.
Thielmann, T., 2012. Geobrowsing behaviour in google earth: a semantic video content analysis of on-screen navigation. In T. Jekel, A. Car, J. Strobl, and G. Griesebner (Eds.), *GI_Forum 2012: geovisualisation, society and learning*. Berlin: Wichmann, pp. 2–13.
Thielmann, T., 2013. Auf den Punkt gebracht: Das Un- und Mittelbare von Karte und Territorium. In I. Gryl, T. Nehrdich, and R. Vogler (Eds.), *geo@web. Medium, Räumlichkeit und geographische Bildung*. Wiesbaden: Springer, pp. 35–59.
Thrift, N., 2004. Driving in the city. *Theory, Culture, & Society*, 21 (4/5), pp. 41–59.
Wood, D., 1992. *The power of maps*. New York: The Guilford Press.
Wood, D. and Fels, J., 2008. *The natures of maps: constructions of the natural world*. Chicago: The University of Chicago Press.
Wood, D., Fels, J., and Krygier, J., 2010. *Rethinking the power of maps*. New York: The Guilford Press.
Wood, D., Kaiser, W., and Abramms, B., 2006. *Seeing through maps: many ways to see the world*. Amherst: ODT.

12 Augmented realities
Gertraud Koch[1]

Relevance for cultural analysis

Through technologies that can create augmented reality (AR), it is becoming possible to blend the borders between the virtual and the physical world or to almost let them disappear altogether, and thus to award a presence and reality to the media-produced illusions, imaginations, or realities which are hardly different from physical, materially tangible entities.[2] Even more than before, perception is thus organised via media. In AR, this occurs in a manner that often makes it imperceptible that media have entered into the perception of the world because they themselves are no longer visible as media but are inscribed into the spaces and their functions themselves. If the emerging technical possibilities are pursued in the manner that is currently apparent in medicine, in the culture and education sectors, in entertainment, and particularly in areas of the economy, then in the mid to long term, new ways of perception will be created via the medialisation of spaces. The possibilities for the design of social and cultural spaces through their material, symbolic, and practical appropriation are fundamentally changing with AR. Because media technologies enter between people and their perception of the world in a significant manner, that is, media technologies are significant for how the world can be perceived, their constructive contribution in the creation of reality must be taken up and considered in cultural analysis. In this context, US technology philosopher Don Ihde suggests a post-phenomenology that discusses this media-guided creation of the world as a starting point for the social construction of reality. With these changed principles of the creation of the world through media, the epistemological foundations of current approaches in cultural analysis, which mainly argue from a phenomenological perspective, are in need of revision.

Yet, it is not entirely new ideas that AR builds upon. Rather, it recurs to existing practices and techniques that it recombines into new ones, extends them technologically, and thus creates more than the sum of existing parts. The informational enhancement of spaces through monuments, memorials, and many other symbolic objects or signs is a form of appropriation of space, which is probably as old as humanity itself, which has constantly been developed further along with media and technology, and which articulates itself in many forms. Furthermore, the overlaying of spaces by means of symbolic interpretations is a cultural form that AR connects to. Urban anthropological cultural analysis has shown how potent the symbolic interpretation of spaces via stories, images, myths, and tales is across all media formats, and how much these textures and these imaginations can guide our experience and our perception but also the material design of spaces and can become inscribed into the habitus of cities and places (Lindner 2006, 2008).

As a concept and terminology, AR has thus originated in computer science; however, it indicates clear social and cultural analytical references to theory via the term reality. Aside from social constructivism (Foerster & Glasersfeld 2007), the lifeworlds (Berger & Luckmann 1966, 1969/2000) and phenomenology, respectively post-phenomenology, are also theoretical points of reference here that invoke further subject-related perspectives in these traditions such as, for instance, the communicative construction of spaces (Knoblauch 1995, Christmann 2015). In contrast to social and cultural theoretical approaches, AR is strongly driven by the exploration of technological possibilities and thus by a constructive moment. The social and cultural positings in the construction of reality that are set with these AR technologies justify the requirement for a cultural analysis of the cultural and social implications of AR, to use the insights gained for the development of cultural theoretical concepts, and thus also to question or contradict the power of interpretation of computer science in the positing of constructions of reality. From this cultural, analytical perspective, it is thus more advisable to speak of augmented realities (ARs) in the plural, in order to clarify that, while the respective technologically created construction of reality does trace back to the implementation of standardised processes, these are taken up and respectively translated into highly different social and cultural contexts of reference and that the resulting constructs of reality originate from complex contexts of origin and are more than the application of standardised technological processes. A cultural analytical theorisation of these digital forms of informational enhancement via AR is still pending. It will probably contribute to a more concrete empirical grasp on the postmodern perspective of the liquefaction of spaces (Bürkner 2007) and help explain how the AR as interface and information flow can put spaces into new, specific contexts of reference.[3]

Furthermore, in cultural analysis, AR can also be a means and a possibility for drawing on epistemologically in research and also for illustrating the mediation of research results, such as AR applications in the museum sector.

Definition of augmented reality

In computer science, augmented realities are conceived of as a form of interface, that is, a point of intersection through which computer applications become accessible to the user. In the case of AR, digitally created physical and virtual objects are aligned with each other, that is, oriented towards each other in a manner that lets them interact in real time (Haller et al. 2006). This moment of interaction between virtual and physical objects is seen as critical, because it distinguishes AR from simple informational feeds into a space, such as the playing of a film (Bimber & Raskar 2005). Augmented reality, therefore, inextricably also presupposes, aside from the ability of a person to interact with the computer system via a screen or other interface to control it, the interaction of virtual and physical objects; that is, it requires an interface between things through which these are aligned with each other. Because of the desired aim of the best possible amalgamation of both realities, such interfaces between things are not usually present in human perception but operate in the background (Billinghurst et al. 2015).

AR is seen in this as a specific state of entanglement of medially created virtual and physical realities, which reside on a continuum of an environment existing entirely without media and an entirely, medially created environment, that is, a virtual reality. All states between these poles are termed as Mixed Reality, in which virtual, that is, medially created, and physical objects are linked interactively with each other and thereby constitute a specific reality perceptible to all (Figure 12.1).

Aside from AR, augmented virtualities are therefore also of interest. They enhance virtual environments, such as a digital learning environment, in real time with images of the participating learners and teachers, that is, in a certain sense, organised complementarily to AR, which enhances physical environments virtually. Correspondingly, virtual reality, which as such is more well known to the public, is also part of this informational development.

AR as a concept is closely related to developments in computer science. A concept that sets itself explicitly apart from this more technologically led perspective and argues from a spatial sociological perspective is that of 'hybrid reality' (de Souza e Silva 2006, de Souza e Silva & Hjorth 2009), which also attempts to conceptually grasp the interleaving of digital and physical spaces. Hybrid reality is conceived of in a much broader manner as a multidimensional concept that shares the basic assumption of AR: that the

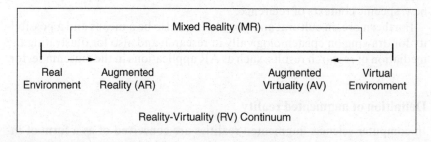

Figure 12.1 Reality-Virtuality Continuum according to Milgram et al.
Source: Milgram et al. 1995: 283.

borders between physical and digital spaces are increasingly dissolving and that social networks thus reach far into physical spaces. Yet, from a spatial perspective, the conceptual reformulation takes a far more open approach than AR: 'For this purpose, hybrid spaces are conceptualised according to three distinct but overlapping trends: hybrid spaces as connected spaces, as mobile spaces, and as social spaces' (de Souza e Silva 2006: 261). This openness facilitates highly different phenomena for paying heed to the virtual hybridisation of actual spaces, regardless of the technological conditions these were generated by. However, as a consequence, a definition of what can count as 'hybrid reality' must remain relatively unspecific and therefore difficult to narrow down. For the analysis of what is at the core of this volume – that is, theories and concepts that make phenomena of digitalisation accessible for empirical cultural analysis – the concept therefore remains fuzzy, particularly for the phenomena of digitalisation to be analysed. Yet, it is precisely these that must be clearly tangible if, as is to be discussed in this contribution, it is to be discussed how the appropriation of spaces and perceptions of reality via digitalisation by means of AR can be conceptually rethought and therefore also require new epistemological approaches.

US digitalisation and media researcher Lev Manovich also argues with an alternative term to AR when he speaks of 'the poetics of augmented spaces'. With this alternative term, he emphasises the spatial dimension (Manovich 2006) coupled with an opening of understanding how information and space relate to and are connected with each other. While AR as specifically defined in computer science is indeed integrated into Manovich's understanding, for him, loose couplings are also associative links such as those created in 'hybrid spaces' (de Souza e Silva 2006) or in the user's mind and are thought of in urban anthropological research as texture or as an imaginaire (Lindner 2008). Ideationally, Manovich draws on cultural analytical research, though without evaluating this systematically and relating it to the different digital approaches of creating such augmented spaces. The understanding of augmented spaces thus remains similarly unspecific and somewhat incompatible with empirical cultural analytical research as the hybrid spaces. It remains open in Manovich's work what the human characteristic of symbolically appropriating spaces and loading them with specific meanings means for the development of culture, or how this has already been taken up in cultural analysis, just like the question wherein – from a cultural anthropological perspective – the new qualities lie which – with digitalisation – have been carried into this practice of symbolically, materially, and practically appropriating spaces (Figure 12.2).

Another concept that argues 'transversely' to the technical understanding of AR, yet links it to the definition formulated in computer science as guidelines and parameters, is a social and cultural analytical determination of the informational enhancement of spaces as *augmented realities* in the plural. This plural points to the contribution of social and cultural dimensions to the creation of the phenomenon, and that in doing so, the seemingly fixed

Figure 12.2 WiFi box, London, Summer 2015.
Photo: Gertraud Koch.

technological infrastructures integrate specific contexts and also improvise, change, or reconstitute them, so that partly different sociotechnical solutions also originate that, similar to AR, lead to an interactive and dynamic alignment of virtual and physical objects, respectively spaces, in human perception. This can occur via the overlaying of medial images, sounds, and so forth with factual spaces, in the sense of the already mentioned textures and the imaginaire; the interactive, dynamic alignment of virtual, fictional, and of physical objects is generated imaginatively in the user's mind. User generated, not professionally created augmented realities, today are often created by means of the interleaving of physical locations with social media that have a geographical reference. Even if these overlaps are not as immediately perceptible in the cityscape as would be the case for the AR created by computer science, the dynamics and interactions between virtual and physical constructs in their effects on the perception of a city space are nevertheless evident. In the absence of technical possibilities for generating this augmentation directly on location in the specific space, they are based in a highly significant manner on the imaginations of humans who, in doing so, bridge a gap that otherwise, according to the definition of AR, stipulates that there must be an immediate interaction between physical and virtual objects. In the imagination, in the admittedly somewhat staggered links that are recreated time and again in social media, something like 'small' augmented realities originates beyond the technical means. These ARs are, due to the missing technical possibilities, to be conceived of as fragmented and as 'seamful spaces' in which different available digital media platforms and applications are brought into overlap with the social space. By means of imagination, the user creates the 'missing link', the interactive alignment of virtual and physical objects, in the mind. ARs, which are conceived of as seamful spaces and as sociotechnical systems, in such a way also point to the fact that it is cultural ideas that inspire and guide such media development. The AR created by computer science is such a motivating factor for the ordinary user in its potentiality that users immediately realises them with the means at their disposal in a reduced form. 'Where media development itself is not advanced enough yet to redeem the principle of immediacy, he takes advantage of the hypermediality, that is, the variety and diversity of the media repertoire' (Bolter & Grusin 2000). Hypermediality is thus, in a certain way, 'compensatory' until the specific AR technology has become suitable for mass use of a large bandwidth of applications for a large strata of consumers (Koch 2016).

State of research and phenomenological areas of augmentation

Although AR development has been undertaken in computer science for several decades already, it has only taken on a stronger dynamic with the beginning of the millennium and, at present, still has many technical difficulties to contend with (Carmigniani et al. 2011). The link between physical and virtual objects and even more so the interaction between these in real time are technically ambitious. As a technology, AR has been developing since the 1960s, even if, at that time, only stationary computers and 'head mounted displays' were available as technical devices (Sutherland 1968) through which these were implemented. Thus, mobile applications were unthinkable, except if they were built into fighter jets, as was done for military use and that, for a long time, remained their primary application (Billinghurst et al. 2015). Since the 1990s, the link of AR with geo-referenced data has been trialled. As a first mobile AR application, the so-called

'Touring Machine' (Feiner et al. 1997) was developed and used as a campus guide, in which virtual annotations for physical buildings could be made (Schmalstieg et al. 2011: 16).

Since then, further developments in computer technology have led to AR being available for use on ever-smaller devices up to the mobile phone. The current paradigm made possible in this way is augmented reality 2.0, in which the infrastructural conditions for user-generated content are to be generated also for mobile AR applications. This focus of the already-established Web 2.0 applications that is undertaken here is an attempt at a market-oriented extension of the use of AR technology for mass consumption, that is, accessibility for mobile phone users who are not particularly technology adept (Schmalstieg et al. 2011: 14).[4] This development paradigm allows us to expect that AR applications will, in the not too distant future, become far more relevant in everyday life than they have been so far.[5]

Aside from this orientation of AR towards mobile applications via the smartphone, a further development paradigm in the area of 'head mounted displays' is becoming apparent, in which screens enhance a public space informationally. Here, AR development distinguishes between the implementation in inner and outer spaces, because these create very different framing conditions for technological solutions.

On the basis of existing applications, a broad range of AR applications is becoming apparent in the industry and medicine sectors, as well as the areas of culture, entertainment, and education. The spread of these technologies into different areas of everyday life, as well as among a broad public – which, in addition to infrastructures that are suitable for mass use as well as efficient computer science applications, presupposes an availability of the respective affordable devices – is a significant factor for AR to be relevant in a cultural manner as computer science developments and thus, in the sense sketched earlier, also increasingly relevant as augmented realities for cultural analysis.

Augmented reality on public screens

Visual projections on walls in public spaces have a long tradition to which screen-based augmented realities can connect.[6] Scandinavian media researcher Erkki Huhtamo reports in his archaeology of mobile media how the Laterna Magica technology was already used to undertake mobile projections onto urban house walls with the aid of trucks (Huhtamo 2009). In addition to projection technologies, screens have increasingly been integrated directly into buildings as a sort of ornament (Caspary 2009), that is, as a fixed component of urban construction: 'contemporary urban architecture – in particular, many proposals of the last decade that incorporate large projection screens into architecture and project the activity inside onto these screens' (Manovich 2006: 231). This mode of augmentation is sometimes also described as space related or 'spatial augmented reality' (Bimber & Raskar 2005) (Figure 12.3).

Augmented realities 237

Figure 12.3 Piccadilly Circus advertising screens.
Photo: Gertraud Koch.

Starting with the broadcasting of TV features, a much broader field of applications has developed where, for a long time, it was almost exclusively commercial advertisement that was visible on public screens (Struppek 2006). In connection with the event culture around sport and entertainment, public screens have taken on a far more important meaning. The use of these screens for social and cultural interests has thus, however, not been exhausted by any means. Partly, their potential for the rediscovery of the public sphere is emphasised and also trialled, yet it is particularly their digital character that renders them into visual fields of experimentation at the interface of virtual and physical urban spaces. Mirjam Struppek, a freelance urbanist and curator of art festivals and conferences, illustrates with a variety of examples of artistic, participation-oriented projects the bandwidth of ideas and uses beyond commercial purposes that has already been achieved – even if there are always issues regarding financing. She deducts from this a potential meaning of public screens for the social development of cities:

> In considering the social sustainability of our cities, we need to look closer at the 'liveability' and environmental conditions of public space; if people are to be encouraged to appropriate public space, new supportive strategies are needed in which they can take on the role of pro-active citizens, not just law-abiding consumers. Several recent media installations in public spaces have explored various possibilities of reactivating urban space and its public sphere.
>
> (Struppek 2006: 178)

238 *Gertraud Koch*

Therefore, it is precisely the option of linking the public screens with the Internet that facilitates the involvement of social media and triggers participation and interaction in the public sphere (Yue 2009), respectively, with the development of specific platforms and making them available. The screens in the public sphere themselves have become diverse and different in their character. In addition to the digital projection planes, one can now also find large format displays in the public sphere, which have been equipped with touch functions similar to the screens of mobile phones and can thus be operated directly and immediately via the graphic user interface. One example of this and of the broad link between highly different dimensions of information and service functions is public screens such as the one in Oulu, Finland (Ylipulli et al. 2014).[7] Similar to this example, such public information can be used for touristic, cultural historical, or everyday purposes – such as the bus timetable or gaming functions on a public screen, with touch function positioned in the inner city, which can serve as a public contact point for all who are on the move in the urban space, even though access to mobile internet via WiFi can constitute a certain competition to such public screens (Ylipulli et al. 2014).

Another example of how public touch displays are implemented for cultural purposes is the Museum of Soho, the London suburb around Piccadilly Circus, which does away with its own space and works via temporary exhibitions in different locations, mainly however via a public interactive screen in the centre of Soho near Piccadilly Circus (Figure 12.4).

An important principle of museum work in this is the use of the interactive potential of digital media for drawing the public into the activities of the museum. This is also true for AR applications on public screens and

Figure 12.4 MoSoho, interactive screen at the Museum of Soho, London.[8]
Photo: Gertraud Koch.

projection planes that are used in participatory planning processes, such as via AR 3D models of cities, which can be modified by the inhabitants over the Internet and can enter into negotiations between urban planners and decision makers or as catalysts for interactions between different people and groups among themselves (Struppek 2006: 184–186).

Mobile AR applications in the museum, education, and entertainment sector

Numerous cultural institutions have begun to take AR as an approach to communication and participation into their work, and many different forms of an augmentation of spaces between analogous and digital have resulted from this. Particularly, applications for mobile devices are now frequently used. A comparative study by Lena Wulf demonstrates that different approaches are followed here. She points out that while some applications merely refer to opening hours and entrance fees for museums, others supersede conventional communicative media, such as accompanying booklets or audio guides. These, she writes, are extended both in terms of content and in terms of function with exhibition guides or a combination between museum and city guide, which facilitates access to the museum contents irrespective of location or time (Wulf 2016: 1). The large majority of museum applications are set up as information systems, such as we know them, in connection with the Internet; up to now, only a few of them work with AR approaches such as, for instance, the Museum of London App, which makes the overlaying of urban scenes, such as the London Bridge, with images and other objects of the museum holdings available directly on location in the city space.[9]

On the one hand, AR applications are interesting for museums insofar as they facilitate a more intuitive and discrete way for the interaction with the objects on display, while on the other hand, it is interesting for AR developers to cooperate with museums as these context-intensive inner spaces are attractive for experimenting with AR application; in contrast to other public spaces, they are accessible to a broad public with very different prior knowledge and skills in dealing with information technologies (Damala et al. 2008: 122, Wulf 2016). The current state of AR technology, as well as the perspectives for development that were sketched at the beginning of this section, can let us expect that this will change fundamentally in the not-too-distant future, as is already beginning to become apparent in the area of museum communication guides, playful approaches to communication, and virtual presentation of objects. In these applications, it is important that AR can be implemented in museums without the infrastructures interfering too much with the museum environment. The navigation in the museum space, orientation, and interaction between physical and virtual objects currently remain critical challenges for implementation. The development that is aimed for – of museums becoming multimedial spaces enhanced with AR that facilitates a rich communication of cultural heritage – depends significantly on the speed of further developments of IT technology (Damala et al. 2008). Furthermore, the current efforts to link museum work with digital media are largely driven by practical projects and pragmatic considerations, while there is still very little reflection on how memory work and memorial culture change in the face of an already-developing, increasingly digital, cultural heritage (Koch 2015).

The spectrum on which AR applications for exhibition guides are experimented with is relatively broad and includes, in the area of exhibition guides, both inner spaces as well as public spaces (Miyashita et al. 2008). Different implementations of AR in the museum sector are fixed installations in historical places in Portugal and Belgium, which visualise for the visitors how these locations looked in earlier times. Other museums experiment with such installations in the indoor area. Further options for the use of AR are trialled in the area of virtual replicates of exhibited or non-exhibited objects. 'Museum wearables' have also been used, which play video clips in front of one eye, while the location identification takes place via infrared sensors.

> An interesting 'virtual' AR scenario was provided in 2003 by the DinoHunter project of the Senckenberg paleontological museum in Germany, where young visitors visiting the museum's web site could start a (virtual) mystery tour manipulating a (virtual) PDA that augmented the dinosaurs' skeletons, reconstituting how they would have been like.
> (Damala et al. 2008: 121)

Moreover, playful approaches that are specifically tailored to certain user groups such as children and are designed as interaction with digitally enhanced spaces are being intensively trialled (Sintoris et al. 2010).[10] Often, a better communication of knowledge is cited as a motive for the use of AR (Yoon et al. 2012). Partly, learning approaches from so-called Edutainment are at the centre here, in which knowledge is to be acquired in a playful manner. In this, it is assumed that the link between digital and nondigital objects is particularly interesting for younger people because of their high affinity with mobile devices. Principles of gamification also play a significant role here (Sintoris et al. 2010). From the beginning of a somewhat broader AR game development around the year 2002 (Flintham et al. 2003), this area has seen a dynamic development and has differentiated into many distinct subgenres, as is evident from a list of the AR games that operated at the state of research in 2010 (Tan & Soh 2011), which incidentally also play a role particularly in the museum sector (Carmigniani et al. 2011). Overall, personalisation is seen as an important option, which can be thought about more intensively with digitalisation. For instance, personalised stories enhanced with AR are an approach used at the Akropolis Museum, by initially creating a visitor profile that dynamically adapts the narration, including the

AR activities. Here, the forms of informational enhancement via AR that are provided are (a) virtual reconstructions of original aspects, (b) the positioning in original locations, (c) the visual emphasis of interesting details and annotations, and (d) the re-enactment of mythical appearances (Keil et al. 2013). The extension into the outer spaces and, respectively, the linking of inner and outer spaces via AR have thus also become common in AR applications (Piekarski & Thomas 2002).

Future developments and difficulties

AR development, which was initially strongly focused on visual elements, has long since been developing towards audio augmentation (Bederson 1995, Eckel 2000). This, however, has so far not been at the focus of cultural analytical attention or its aspects are being only partially discussed under other terminologies, such as 'auditory augmentation' (Bovermann et al. 2012), 'sonification', or 'auditive display', respectively, 'auditive interface' (Kramer 1994, Hermann 2008, Vazquez-Alvarez et al. 2012). A vivid example of how auditory and physical objects can be linked is sound stairs, usually in the shape of piano keys in subway stations, shopping centres, and other public spaces, which enhance stairs and floors in an auditory manner so that piano music can be generated by the users' feet.[11] This rather more amusing and astounding application points to the fact that virtually any physical object can be overlaid with a new digital texture (interface), thus becoming informationally enhanced and, in this, can potentially address very different sensual, visual, auditory, tactile, olfactory, or taste-related perceptions. Even if this has so far occurred only very selectively and is being explored mostly in areas of the arts, the synaesthetic addressing of many senses will guide AR development in the mid to long term, as has already been formulated in the constitutive phase of AR as information as vision. Such multimodal interfaces, as these new interfaces of digital applications towards the user are being termed in computer science, mark the transition towards the Internet of Things (cf. the contribution by Shah in this volume) and an embodiment of the operation of computer applications in the everyday dealings with things, so that the technical character of action or the construction of reality remains largely unnoticed (Carmigniani et al. 2011: 352–353). If the T-shirt that was just purchased communicates from the shopping bag with the large advertising screen via a RFID tag and presents further articles of the same brand to the customer, the underlying construction of reality can hardly be traced. Partly, however, applications are also drawn on in the medical sector or for assisting with disabilities (van Scoy et al. 2005).

In a correspondingly unnoticed manner, the surveillance and evaluation of the behaviour of AR users can occur because the interactive alignment of physical and real world – generally referred to as tracking – thus becomes the norm in many lifeworld connections (Hohl 2009, Hong 2013). Respectively, many questions about social acceptance are being raised such as, among others, the recall of the Google Glasses for the European market has demonstrated. However, the connected group of themes are not receiving much research at present (Billinghurst et al. 2015, S. 231 f.). Tools that are already on the market, which allow 'simple' user-generated AR, which are, among others, being trialled in the academic context (Klopfer & Sheldon 2011), will continue to further these tendencies if there are no other precautions taken by means of legal regulation. Due to the new possibilities that they will bring for the average user or media amateurs (Reichert 2008), a quick dissemination will most likely nevertheless occur.

Perspectives for analysis and considerations about possible operationalisations

In a post-phenomenological perspective that researches augmented realities as a sociotechnical construct and that is being suggested here as an analytical perspective in contrast to a technical understanding of augmented reality, as well as those suggested by Lev Manovich and Adriana de Souza e Silva, the question of how media enter between the world and perception and how this relationality creates new conditions for perception in the world (Ihde 1993) is at the centre. The question is how, in the course of this remediation (Bolter & Grusin 2000), through the informational enhancement via AR, new forms of ARs and thus the perception, the being in, and the relationship to the world through medial overlays are also being reconfigured.[12] In cultural analytical perspectives on the ARs, it will therefore initially be about reconstructing and deconstructing the specific relationships between man and technology, which are in the process of constituting themselves. How are these technologies linked with people's practices so that reality is seamlessly enhanced, as has been formulated as a perspective to be aimed at? In what way do the respective implementations play a role for the constitution of reality and in this form specific ways of perception – for instance, QR codes, which lead to websites and create overlays over certain physical settings or measuring stations in parks, which communicate with a chip in the running shoes of registered users and invisibly register data on a website, which then makes the running in this setting reconstructable, measurable, comparable, and analysable? Which possibilities exist to make this seamlessness in cooperation visible after all, for instance through Software Studies or an Ethnography of Infrastructure, and thus to reconstruct the constitutiveness of AR as such in its conditions in the sense of an STS approach? What knowledge and what prior knowledge must exist in order to apply ARs and use them meaningfully in one's everyday life or to reconstruct them for oneself via user tools that gradually enter the market for mobile devices? In which referential context do virtual and physical spaces beyond the mere technological implementation interrelate at a level of content and perception, and in what way, in doing so, do they constitute reality? This also includes the question of embodiment, that is, what the user needs to learn in order to apply this technology, which concrete knowledge and abilities must the user have or develop in dealing with the technologies, and which mental models of their way of working and the conditions of origin must the user develop or will be needed in order to work with them? How does the user appropriate the augmented technologies? Which social and cultural structuring of this appropriation process exist? How do they then become relevant in everyday lives?

In contrast to the ethnography of infrastructure, which takes an approach from the social practices that become possible in interleavings with digital infrastructures or only originate because of these, a research perspective on augmented realities accentuates the ways of perception and also the changed appropriations and ways of construction of social spaces that are created with digital media. An explorative study by Willis (Willis 2008) shows, for instance, that the perception of spaces is still largely structured by the physical space and that the WiFi networks that are superimposed on this are only partly being noticed in their physical reach as an element that structures spaces.

Summary and further perspectives

The relevance of augmented realities as cultural and social developments is only beginning to emerge. They are in the process of originating as a cultural analytical research perspective on digitalisation; correspondingly, there are, to date, few empirical or theoretical approaches that could be called on in order to lend contours to the field. This contribution sketches the relevance of these technological developments for cultural analytical research, particularly in the area of the construction of reality and the perception of the world, as well as in the appropriation and construction of space, that is, dimensions that can be considered fundamental for the being in the world of humans, without wishing to portray these areas as already exhaustive. Rather, a further differentiation of cultural analytical perspectives according to the subject areas in which AR technologies are currently being explored as areas of application is to be assumed: in medicine; in the cultural, education, and entertainment sectors; and in industry. The contribution, therefore, differentiates distinct understandings of the informational enhancement of spaces in computer science, the social sciences, and the humanities in order to gain an understanding of the phenomenon as a sociotechnologically guided practice that can, depending on the context, have many different faces. An excerpt of the bandwidth of what is apparent in the applications of ARs is sketched with the aid of public screens and mobile applications, that is, the areas through which informational AR primarily organises its approaches to research. The developments sketched here are driven 'hands-on', beyond cultural, theoretical considerations, initially by pragmatic points of view while taking into account the experts in the subject areas, yet nevertheless mainly in a technologically oriented scientific manner. It has not been worked out yet with which approaches cultural analysis can tackle these topics; however, this is sketched in a post-phenomenologically oriented argumentation insofar as perspectives for questions and thus analytical perspectives are named and can be taken up for empirical research.

Notes

1 Translated from the German by Dr. Stefanie Everke Buchanan.
2 This contribution uses the term augmented reality (in short: AR) to denote the technological developments as distinct from augmented realities in the plural (in short: ARs), which is used to highlight the embeddedness of the technological developments into social and cultural contexts and to indicate that a cultural-theoretical understanding must include those. A detailed perspective on this relation will be developed in the course of this contribution.
3 The original expression in German is 'Verflüssigung von Räumen'.
4 On the flip side, there is a tendency in social media platforms to increasingly integrate functions that facilitate a mobile geographical reference and thus to also create the preconditions for the interleaving in the mode of the AR, cf. Wilken 2014.
5 See www.nytimes.com/2016/07/12/technology/pokemon-go-brings-augmented-reality-to-a-mass-audience.html?_r=0 (accessed 14 July 2016).
6 From a technical point of view, one differentiates between screen-based, projector-based, and head-mounted AR.
7 See www.ubioulu.fi/en/UBI-hotspots (accessed 14 July 2016).
8 See also www.mosoho.org.uk/interactive-screen (accessed 3 July 2016).
9 See www.museumoflondon.org.uk/Resources/app/you-are-here-app/home.html (accessed 3 July 2016).
10 See also www.itpro.co.uk/apps/26886/snapchat-hires-vr-designer-to-work-on-augmented-reality.
11 Bonn: www.youtube.com/watch?v=rdBGx_JBLtw. Odenplan, Stockholm, Sweden: www.youtube.com/watch?v=ipMib6ejGuo. Auckland, New Zealand: www.youtube.com/watch?v=gKuyhfLlXzA. Monster Piano, Place unknown: www.youtube.com/watch?v=qlZYGZEHMgU (accessed 1 July 2016).
12 The term remediation as coined by Bolter and Grusin refers to the fact that existing media need to 'reinvent' themselves if new media technologies arise and thus reconfigure the existing media landscape. The readjustment of newspapers in the age of the Internet is an illustrative example for this.

References

Bederson, B.B., 1995. Audio augmented reality: a prototype automated tour guide. In I. Katz, R. Mack, and L. Marks (Eds.), *Proceedings CHI'95 conference companion on human factors in computing systems*. New York: ACM, pp. 210–211.
Berger, P.L. and Luckmann, T., 1966. *The social construction of reality: a treatise in the sociology of knowledge*. Garden City: Anchor.
Berger, P.L. and Luckmann, T., 1969/2000. *Die gesellschaftliche Konstruktion von Wirklichkeit. Eine Theorie der Wissenssoziologie*. 17th ed. Frankfurt am Main: Fischer.
Billinghurst, M., Clark, A., and Lee, G., 2015. A survey of augmented reality. *Foundations and Trends in Human-Computer Interaction*, 8 (23), pp. 73–272.

Bimber, O. and Raskar, R., 2005. *Spatial augmented reality. Merging real and virtual worlds.* Wellesley: CRC Press.
Bolter, D.J. and Grusin, R., 2000. *Remediation: understanding new media.* Cambridge: MIT Press.
Bovermann, T., Tünnermann, R., and Hermann, T., 2012. Auditory augmentation. In K. Curran (Ed.), *Innovative applications of ambient intelligence: advances in smart systems: advances in smart systems.* Hershey: Business science reference, pp. 98–112.
Bürkner, H-J., 2007. Stadtentwicklung in einer sich zur Wissensgesellschaft verändernden Industriegesellschaft – Herausforderungen für die Stadtplanung. In D. Baum (Ed.), *Die Stadt in der Sozialen Arbeit.* Wiesbaden: Springer, pp. 288–304.
Carmigniani, J., Furht, B., Anisetti, M., Ceravolo, P., Damiani, E., and Ivkovic, M., 2011. Augmented reality technologies, systems and applications. *Multimedia Tools and Applications*, 51 (1), pp. 341–377.
Caspary, U., 2009. Digital media as ornament in contemporary architectural facades: its historical dimension. In S. McQuire, M. Martin, and S. Niederer (Eds.), *Urban screens reader.* Amsterdam: Institute of Network Cultures, pp. 65–74.
Christmann, G.B. (Ed.), 2015. *Zur kommunikativen Konstruktion von Räumen. Theoretische Konzepte und empirische Analysen.* Wiesbaden: Springer.
Damala, A., Cubaud, P., Bationo, A., Houlier, P., and Marchal, I., 2008. Bridging the gap between the digital and the physical: design and evaluation of a mobile augmented reality guide for the museum visit. In S. Tsekeridou, A.D. Cheok, K. Giannakis, and J. Karigiannis (Eds.), *Proceedings of the 3rd international conference on digitial interactive media in entertainment and arts.* New York: ACM, pp. 120–127.
de Souza e Silva, A., 2006. From cyber to hybrid mobile technologies as interfaces of hybrid spaces. *Space and Culture*, 9 (3), pp. 261–278.
de Souza e Silva, A. and Hjorth, L., 2009. Playful urban spaces: a historical approach to mobile games. *Simulation & Gaming*, 40 (5), pp. 602–625.
Eckel, G., 2000. Immersive audio-augmented environments. In E. Banissi, F. Khosrowshahi, M. Sarfraz, and A. Ursyn (Eds.), *Proceedings fifth [IEEE] international conference on information visualisation: 25–27 July 2001, London, England, Bd. 128.* Los Alamitos: IEEE, pp. 571–573.
Feiner, S., MacIntyre, B., Höllerer, T., and Webster, A., 1997. A touring machine: prototyping 3D mobile augmented reality systems for exploring the urban environment. *Personal Technologies*, 1 (4), pp. 208–217.
Flintham, M., Benford, S., Anastasi, R., Hemmings, T., Crabtree, A., Greenhalgh, C., Rodden, T., Tandavanitj, N., Adams, M., and Row-Farr, J., 2003. Where on-line meets on the streets: experiences with mobile mixed reality games. In *Proceedings of the SIGCHI conference on human factors in computing systems.* New York: ACM, pp. 569–576.
Foerster, H.v. and Glasersfeld, E.v., 2007. *Wie wir uns erfinden. Eine Autobiographie des radikalen Konstruktivismus.* Heidelberg: Carl-Auer-Systeme Verlag.
Haller, M., Billinghurst, M., and Thomas, B. (Eds.), 2006. *Emerging technologies of augmented reality: interfaces and design.* Hershey: IGI Global.
Hermann, T., 2008. Taxonomy and definitions for sonification and auditory display. In P. Susini and O. Warusfel (Eds.), *Proceedings of the 14th annual international conference on auditory display (ICAD).* Paris: ICAD.
Hohl, M., 2009. Beyond the screen: visualizing visits to a website as an experience in physical space. *Visual Communication*, 8 (3), pp. 273–284. doi:10.1177/1470357209106469.

Hong, J., 2013. Considering privacy issues in the context of Google glass. *Communications of the ACM*, 56 (11), pp. 10–11. doi:10.1145/2524713.2524717.

Huhtamo, E., 2009. Messages on the wall: an archaeology of public media displays. In S. McQuire, M. Martin, and S. Niederer (Eds.), *Urban screens reader*. Amsterdam: Institute of Network Cultures, pp. 15–28.

Ihde, D., 1993. *Postphenomenology: essays in the postmodern context*. Evanston: Northwestern University Press.

Keil, J., Pujol, L., Roussou, M., Engelke, T., Schmitt, M., Bockholt, U., and Eleftheratou, S., 2013. A digital look at physical museum exhibits: designing personalized stories with handheld augmented reality in museums. In A.C. Addison, G. Guidi, L. De Luca, and S. Pescarin (Eds.), *Proceedings of the 2013 digital heritage international congress*. Volume 2. Marseille: IEEE, pp. 685–688.

Klopfer, E. and Sheldon, J., 2011. Augmenting your own reality: student authoring of science-based augmented reality games. In M. Umaschi Bers, G.G. Noam (Eds.), *New media and technology: youth as content creators*. Hoboken: John Wiley & Sons, pp. 85–94.

Knoblauch, H., 1995. *Kommunikationskultur. Die kommunikative Konstruktion kultureller Kontexte*. Berlin: De Gruyter.

Koch, G., 2015. Kultur digital. Tradieren und produzieren unter neuen Vorzeichen. In E. Bolenz, L. Franken, and D. Hänel (Eds.), *Wenn das Erbe in die Wolke kommt. Digitalisierung und kulturelles Erbe*. Essen: Klartext.

Koch, G., 2016. Augmented spaces of protest. Imaginative Medienpraktiken und Hypermedialität als Ressourcen des zivilgesellschaftlichen Widerstands. *Hamburger Journal für Kulturanthropologie*, (5), pp. 89–110 [online]. Available at: https://journals.sub.uni-hamburg.de/hjk/article/view/1041.

Kramer, G. (Ed.), 1994. Auditory display: sonification, audification, and auditory interfaces. In *Proceedings Volume 18, Santa Fe Institute Studies in the Sciences of Complexity*. Reading: Addison-Wesley Publishing Company.

Lindner, R., 2006. The cultural texture of the city. In J. Fornäs (Ed.), *The ESF-LiU conference cities and media: cultural perspectives on urban identities in a mediatized world. Linköping electronic conference proceedings*. Linköping: Linköping University Electronic Press, pp. 53–58.

Lindner, R., 2008. Textur, imaginaire, Habitus: Schlüsselbegriffe der kulturanalytischen Stadtforschung. In H. Berking and M. Löw (Ed.), *Die Eigenlogik der Städte. Neue Wege für die Stadtforschung*. Frankfurt am Main: Campus, pp. 83–94.

Manovich, L., 2006. The poetics of augmented space. *Visual Communication*, 5 (2), pp. 219–240.

Milgram, P., Takemura, H., Utsumi, A., and Fumio, K., 1995. Augmented reality: a class of displays on the reality-virtuality continuum. *Proceedings of SPIE, Telemanipulator and Telepresence Technologies*, 2351, pp. 282–292.

Miyashita, T., Meier, P., Tachikawa, T., Orlic, S., Eble, T., Scholz, V., Gapel, A., Gerl, O., Arnaudov, S., and Lieberknecht, S., 2008. An augmented reality museum guide. In: M.A. Livingston, O. Bimber, and H. Saito (Eds.), *7th IEEE international symposium on mixed and augmented reality 2008*. Piscataway: IEEE, pp. 103–106.

Piekarski, W. and Thomas, B., 2002. ARQuake: the outdoor augmented reality gaming system. *Communications of the ACM*, 45 (1), pp. 36–38.

Reichert, R., 2008. *Amateure im Netz. Selbstmanagement und Wissenstechnik im Web 2.0*. Bielefeld: Transcript.

Schmalstieg, D., Langlotz, T., and Billinghurst, M., 2011. Augmented reality 2.0. In S. Coquillart, G. Brunnett, and G. Welch (Eds.), *Virtual realities*. Vienna: Springer, pp. 13–37.

Sintoris, C., Stoica, A., Papadimitriou, I., Yiannoutsou, N., Komis, V., and Avouris, N., 2010. MuseumScrabble: design of a mobile game for children's interaction with a digitally augmented cultural space. *International Journal of Mobile Human Computer Interaction*, 2 (2), pp. 53–71.

Struppek, M., 2006. The social potential of urban screens. *Visual Communication*, 5 (2), pp. 173–188. doi:10.1177/1470357206065333.

Sutherland, I., 1968. A head-mounted three-dimensional display. In *AFIPS '68 Conference Proceedings*. New York: ACM, pp. 757–764.

Tan, C.T. and Soh, D., 2011. Augmented reality games: a review. In C.H. Tan (Ed.), *Proceedings of Gameon-Asia 2011: 3rd Asian conference on simulation and AI in computer games*. EUROSIS, pp. 17–24.

van Scoy, F., McLaughlin, D., and Fullmer, A., 2005. Auditory augmentation of haptic graphs: developing a graphic tool for teaching precalculus skill to blind students. In *Proceedings of ICAD 05–11th Meeting of the International Conference on Auditory Display. International Community for Auditory Display*, pp. 1–5.

Vazquez-Alvarez, Y., Oakley, I., and Brewster, S.A., 2012. Auditory display design for exploration in mobile audio-augmented reality. *Personal and Ubiquitous Computing*, 16 (8), pp. 987–999.

Wilken, R., 2014. Places nearby: Facebook as a location-based social media platform. *New Media & Society*, 16 (7), pp. 1087–1103. doi:10.1177/1461444814543997.

Willis, K.S., 2008. Places, situations and connections. In A. Aurig and F. de Cindio (Eds.), *Augmented urban spaces: articulating the physical and electronic city*. Aldershot: Ashgate, pp. 9–26.

Wulf, L., 2016. Von der Vitrine in die Hand. Smartphone-Apps in der Museumsvermittlung. *Hamburger Journal für Kulturanthropologie*, (5), (Forthcoming).

Ylipulli, J., Suopajärvi, T., Ojala, T., Kostakos, V., and Kukka, H., 2014. Municipal WiFi and interactive displays: appropriation of new technologies in public urban spaces. *Technological Forecasting and Social Change*, 89, pp. 145–160.

Yoon, S.A., Elinich, K., Wang, J., Steinmeier, C., and Tucker, S., 2012. Using augmented reality and knowledge-building scaffolds to improve learning in a science museum. *International Journal of Computer-Supported Collaborative Learning*, 7 (4), pp. 519–541.

Yue, A., 2009. Urban screens, spatial regeneration and cultural citizenship: the embodied interaction of cultural participation. In S. McQuire, M. Martin, and S. Niederer (Eds.), *Urban screens reader*. Amsterdam: Institute of Network Cultures, pp. 261–278.

Part IV
Concepts of culture revisited

13 The political economy of digital technologies

Outlining an emerging field of research

Andreas Wittel

Introduction

The political economy of digital technologies is a relatively young field of research that has rapidly grown over the last two decades. I am not aware of an attempt to outline this field in a more systematic way. In fact, it might be debatable whether such an attempt makes much sense in view of the rapid growth of this field and its dynamic transformations and expansions. For this reason, the following reflections are an attempt to provide a provisional appraisal of this field.

What is also debatable is the newness of this field of research. While I am claiming that the political economy of digital technologies (PEDT) is an emerging field, it does not come out of nowhere. It has a predecessor; it is emerging from another field of research, the political economy of media and communication (PEMC). I propose two things: First, that PEDT has its roots in PEMC and that it has developed from PEMC; second, that PEDT has developed with a number of significant transformations and expansions. These expansions from PEMC are so profound that it is indeed legitimate to conceptualise PEDT as a new field of study, which focuses on different themes and questions, which uses and produces different theoretical concepts, and which finally embarks on a different understanding of what it means to create scholarly work.

An exploration of these expansions will be at the centre of this chapter. In particular, I will outline three expansions. However, in order to discuss these expansions, it is necessary to outline first the field of PEMC, PEDT's predecessor. The structure of this chapter is simple. It is an argument in four steps. First, I will briefly outline the research field of PEMC. In steps 2, 3, and 4, I will discuss the three expansions.

The first expansion is a *thematic* expansion *from media to technologies*. Digital technologies are not just media technologies but technologies that are at the heart of all industrial sectors. In the digital age, a differentiation between media technologies (or technologies for the media industries) and other technologies (or technologies for other industrial sectors) is becoming increasingly problematic. The second expansion is a *theoretical expansion*.

More specifically, this is an expansion of Marxian concepts. PEMC did not – or better, could not – use many concepts central to Marx's political economy such as labour, value, commodity, property, and alienation for an analysis of mass media. In the age of digital and distributed media, however, questions relating to labour, value, and property are at the very heart of PEDT's concern.

The third expansion is that of a *scholarly ethos*, from mere critique to *activism*. PEDT is not just interested in a critique of digital capitalism, but it is also concerned with a search for alternatives to neo-liberal capitalism and with ways out of the current crisis. This is a perspective, to paraphrase Marx, that does not aim to interpret digital capitalism, but to change the political-economic system altogether. This activist turn is as much about imagination as it is about understanding the contemporary predicament.

The political economy of media and communication (PEMC)

The political economy of media has been constituted as an academic field in the age of mass media, which are characterised by linear forms and one-way flows of communication, where content is being distributed from a small number of producers to a large number of recipients (Poster 1995).

Outlining the key issues, questions, debates, and findings of an academic field in a few paragraphs is always a difficult undertaking that leads to oversimplifications, questionable generalisations, and the privileging of a coherent narrative at the expense of a more nuanced perspective. This is also true for the field of political economy of media and communication. It is quite surprising, however, that a rather broad consensus of what this field is about does exist. Comparing a number of introductions to this field (Mosco 1996, Devereux 2003, McQuail 2005, Durham & Kellner 2006, Laughey 2009, Burton 2010), it becomes rather obvious that there is not much disagreement about key issues, questions, and findings that have been produced in the political economy of media and communication.

It starts with the observation that media institutions have increasingly become privatised and turned into businesses. This is seen as problematic, as media industries are seen as not just any industry. To understand the unusual character of the media industries, one has to examine the dual nature of the content being produced, which is simultaneously a commodity and a public good. It is a private good – a commodity – as media industries are using their products for the accumulation of profit. At the same time, this content is a public good as it constitutes, to some degree, the public sphere. So, on the one hand, media institutions have a social, cultural, and political function; on the other hand, they are driven by economic interests. It is this dual nature of media content that makes the assumption that media are an independent force, naturally safeguarding democracy and the public interest, rather questionable. Equally doubtful is the assumption that mass media just mirror public opinion.

The political economy of media is based on the premise that media are powerful, that they are able to influence public opinion, and that they can shape public discourse. Therefore, it is crucial to focus on the production of media content within a wider political and economic context. It is this focus on materiality and the political, economic, and technological conditions in which media content is being produced that distinguishes the political economy of media from other academic fields, such as the more affirmative strands within cultural studies and audience studies, which generally locate power and control not with media institutions but with an active audience as the true producer of meaning. The political economy of media is as much social analysis as media and communication analysis.

This field is mainly concerned with the following issues: First, with an understanding of the media market. How do media companies produce income and generate profits? Second, with an inspection of questions of ownership of media organisations (public, commercial, and private, non-profit organisations) and an analysis of the implications of ownership structures with respect to media products (obviously, this is especially relevant for the production of news). Third, the field is concerned with changing dynamics of the media sector, in particular with developments such as internationalisation of media industries, concentration and conglomeration of media organisations, and diversification of media products. This leads into debates on cultural imperialism and media imperialism. The fourth issue is about media regulation, media policy, and media governance, originally on a national level but increasingly with a global perspective. It is important to note that these areas of inquiry are closely connected; in fact, they overlap considerably.

Raymond Williams (1961), who is usually not portrayed as someone who is part of the inner circle of political economy of media, was in fact among the first to develop such an approach. In an essay on the growth of the newspaper industry in England, he starts with the observation that 'there is still a quite widespread failure to co-ordinate the history of the press with the economic and social history within which it must necessarily be interpreted' (194). He sets out to develop such a perspective, studying empirically a period of 170 years. His findings are highly sceptical:

> These figures do not support the idea of a steady if slow development of a better press. The market is being steadily specialised, in direct relation to advertising income, and the popular magazine for all kinds of reader is being steadily driven. This does not even begin to look like the developing press of an educated democracy. Instead it looks like an increasingly organised market in communications, with the 'masses' formula as the dominant social principle and with the varied functions of the press increasingly limited to finding a 'selling point'.
> (Williams 1961: 234)

Clearly, Williams anticipated many of the themes and results that will be debated within this field over the next five decades.

PEDT and expansion 1: from media to technologies

As I have mentioned already, PEMC was a political economy of mass media and mass communication. In the age of mass media, a distinction between media industries and other industries was not a difficult thing at all. Media industries produced media objects such as newspapers, magazines, books, films, songs, and television series. Media industries did not produce cars, fridges, or agricultural objects. This clear distinction between media industries and other industries was based on objects or commodities that were produced, but also on the technologies with which these objects were created. In order to produce newspapers, the newspaper industry used the printing press. Car manufacturers did not use the printing press; they used other technologies. In the age of mass media, the distinction between media technologies and other technologies (or technologies in other industrial sectors) was simple and straightforward.

It is exactly this distinction between media technologies and other technologies and between media products and other products that has become increasingly blurred in the digital age. Now, almost all technologies are at the same time digital technologies. Let's introduce some examples to support this claim. Only in the digital age is it possible that a (former) computer company such as Apple can move into the music marketplace and dominate this industry within a few years. Kodak has built its reputation for many decades not as a media company but as a photo company. Instagram, its logical successor, is clearly part of the media industries or the creative industries. Google has started as a software company, but has soon turned into a media conglomerate. But this is still not an accurate description, as Google has steadily broadened its business operations and is now the market leader in areas such as robotics and artificial intelligence. Let's take Google's driverless car. Has Google now become a car manufacturer? Will Google soon dominate the automobile industry? Is the Google car really a car, or is it a computer on four wheels? Is Bitcoin a financial product or a media product?

These examples should demonstrate how clouded indeed the distinction between media industries and other industries has become over the last decade. Manuel Castells (1996) was probably among the first theorists to analyse the role of digital technologies in the information age.

> What characterises the current technological revolution is not the centrality of knowledge of information, but the application of such knowledge and information to knowledge generation and information processing/communication devices in a cumulative feedback loop between innovation and the uses of innovation.
>
> (1996: 31)

It is this relation between knowledge and information, on the one hand, and digital technologies and their feedback loops, on the other hand, which has produced technological change at such an accelerated pace. Castells outlines five features of the information technology paradigm. The first two of the features he mentions are:

> The first characteristic of the new paradigm is that information is its raw material: these are technologies to act on information, not just information to act on technology, as was the case in previous technological revolutions. The second feature refers to the pervasiveness of effects of new technologies. Because information is an integral part of all human activity, all processes of our individual and collective existence are directly shaped (although certainly not determined) by the new technological medium.
>
> (1996: 70)

He argues that, in the digital age, information has become a product. So, the fundamental difference between industrialism and informationalism is that now all industrial sectors (agriculture, manufacturing, the service industry, finance) operate with digital technologies.

The implications of this development are profound. While PEMC stayed clearly focused on the media industries, PEDT is broadening the focus. PEDT is interested in 'digital capitalism' (Schiller 1999). PEDT is about very diverse fields of research. To name a few, it is about Castells's (1996, 2009, 2012) work on the information age, about David Graeber's (2002, 2011, 2015) economic anthropology, about David Harvey's (2007, 2010, 2012) political economy, about Yochai Benkler's (2006) work on non-market production, and about the work of the Italian post-operaismo school (Lazzarato 1998, Hardt & Negri 2000, 2004, 2009, Virno 2004, Berardi 2009). In digital capitalism, PEDT is about a political economy that is as much interested in the financial sector as it is interested in media. To use again the example of Google, it started as an innovative search engine. However, since Google morphed into Alphabet, it has essentially become a holding company that buys and sells technology companies (Rushkoff 2016).

PEDT and expansion 2: from a limited to an unlimited Marxian political economy

The first expansion was about an opening up of the research topic, an opening up from a focus on media to a focus on digital technologies. The second expansion is about another opening up, not a thematic but a theoretical one. The argument I want to make here goes like this: In the age of mass media PEMC has engaged with Marxist concepts in a rather limited way. In the age of digital media, Marxist theory can be applied in a much broader sense. In order to support this claim, I will make an argument in two steps. The first step is to produce evidence for the claim that political economy of mass media engages with Marxist theory in a rather limited way and also to explain the logic behind this limited engagement. The second step is an exploration how PEDT can use other key concepts of Marx's political economy for a critical analysis of digital capitalism. In particular, I will focus here on two key concepts: labour and property.

PEMC and the base-superstructure model

The theoretical roots of PEMC – at least, their critical tradition (which is all I am concerned with) – are usually located in Marxism. So, how much engagement with Marx do we get in this academic field? The short answer: there is some engagement but it is fairly limited. In order to support this claim with some evidence, I will check a number of texts that are generally considered to be important contributions.

The first and rather surprising insight is that a considerable number of books (Herman & Chomsky 1988, Schiller 1989, Curran 1990, Curran & Seaton 1997, Herman & McChesney 1997, Grossberg et al. 1998, Curran 2000, Nicols & McChesney 2006) have either no reference or less than a handful of references to Marx or Marxism. In the latter case, these references function usually as signposts (such as to distinguish Marxists from liberal traditions of political economy). They do not engage with Marxist theory in a more profound manner.

Nevertheless, they are all rooted in Marxist theory, or to be more precise, in one particular part of Marxist theory. They are all directly linked to the base and superstructure model. According to Marx, human society consists of two parts, a base and a superstructure. The material base consists of the forces and relations of production, while the superstructure refers to the non-material realm, culture, religion, ideas, values, and norms. The relationship between base and superstructure is reciprocal; however, in the last instance, the base determines the superstructure.

> The ideas of the ruling class are in every epoch the ruling ideas, i.e. the class which is the ruling material force of society, is at the same time its ruling intellectual force. The class which has the means of material production at its disposal, has control at the same time over the means of mental production, so that thereby, generally speaking, the ideas of those who lack the means of mental production are subject to it. The ruling ideas are nothing more than the ideal expression of the dominant material relationships, the dominant material relationships grasped as ideas; hence of the relationships which make the one class the ruling one, therefore, the ideas of its dominance ... Insofar, therefore, as they rule as a class and determine the extent and compass of an epoch, it is self-evident that they do this in its whole range, hence among other things rule also as thinkers, as producers of ideas, and regulate the production and distribution of the ideas of their age: thus their ideas are the ruling ideas of the epoch.
> (Marx & Engels 1974: 64–65)

The texts mentioned above directly or indirectly apply the base and superstructure model to the media industry, which, like no other industrial sector, contributes to the production of the superstructure. However, they apply this model in various ways and there is considerable disagreement about what some see as a deterministic model with a linear, non-dialectical, and reductionist perspective.

There are, however, texts that engage with Marx and in particular with his base and superstructure model in a more profound way. Mosco (1996), who provides perhaps the most detailed analysis of the literature in this field, starts his books with an introduction to Marxist political economy. Murdock (1982) focuses in particular on the base and superstructure model and compares it with a more praxis-oriented perspective. Williams (1958: 265–284) engages in great detail with this model and argues that it is more complex than usually acknowledged (e.g., that this relation is reciprocal, rather than a one-way street).

> The basic question, as it has normally been put, is whether the economic element is in fact determining. I have followed the controversies on this, but it seems to me that it is, ultimately, an unanswerable question.
> (Williams 1958: 280)

Like Williams, Nicholas Garnham (1990) also counters charges of economic reductionism. He insists that Marx's model offers an adequate foundation for an understanding of the political economy of mass media. He moves away from a deterministic view of the relation between base and superstructure towards a model that is more anchored in reciprocity and a dialectic relation.

To conclude, apart from some rare exceptions – most notably, Dallas Smythe, who will be discussed later – PEMC incorporates Marxist theory in a rather limited way. This academic field refers predominantly to Marx's concept of base and superstructure (either directly or indirectly) to make claims about the relationship between ownership of means of production (and concentration of ownership, media conglomerates, and so forth) and questions of media content, ideology, manipulation, power, and democracy.

To avoid any misunderstandings: this is not meant at all as a critique of political economists of mass media. I do not see this limited appropriation of Marxists concepts as a failure of this academic field. My point is very different. I want to argue that this limited appropriation made complete sense in the age of mass media. It has a logic to it that lies very much in mass media technologies. There was not much need for a more systematic application of other concepts within Marx's political economy. This has profoundly changed with PEDT. In the age of digital and distributed media, new problems and questions emerge that are linked to other key concepts of Marx's political economy, such as labour, property, commodity, value, alienation, class, and struggle. In the following sections, I will explain how PEDT explores the concepts of labour and property for an analysis of digital capitalism.

PEDT and labour

Throughout the last century, labour has been analysed in the western hemisphere as wage labour only. Apart from very few exceptions, alternatives to wage labour have hardly entered public discourse. It was a common perception that there was just no alternative to wage labour. Obviously, this theoretical orientation was a reflection of an economic reality characterised largely by wage labour as the dominant form of production. One of the few exceptions is Karl Polanyi, who has introduced the concept of fictitious commodities to point out that labour (like land and money) is not a real or discrete commodity, but artificial; as it cannot be regulated by market only, it depends on the state and on law for it to function as a commodity. Another exception is André Gorz (1989, 1999), whose reflections on alternatives to wage labour had at least some impact on public debates outside the academic discourse. But Gorz was perceived as a utopian theorist. It is certainly correct that Gorz has developed a humanist perspective and that his critique of wage labour is most of all a critique of alienation. He has rather neglected the relation of wage labour and value.

This is the merit of the Italian post-operaismo school. Theorists such as Antonio Negri, Michael Hardt, Maurizio Lazzarato, and Paolo Virno have pointed out in the late 1990s that wage labour is not the only source for the generation of profit. The post-operaismo school puts much emphasis on the growing significance of immaterial labour. Immaterial labour, which is both intellectual labour and affective labour, involves a number of activities that would not be considered work in Fordist work environments.

> It is not simply that intellectual labor has become subjected to the norms of capitalist production. What has happened is that a new 'mass intellectuality' has come into being, created out of a combination of the demands of capitalist production and the forms of 'self-valorization' that the struggle against work has produced.
> (Lazzarato 1998: para. 3)

The concept of immaterial labour is inspired by a few pages in *Grundrisse*, where Marx writes about wealth creation and the production of value, which is increasingly independent of labour.

Let's go back one step and introduce two concepts of labour in Marx's political economy and two concepts of creation of exchange value and profit. Let's start with labour and Marx's distinction between labour and labour process.

Labour is not merely an economic but a human activity. It is a universal category of human existence and it is independent of any specific economic or social forms. Labour is what keeps us alive and what makes us develop.

This is a rather broad concept. Labour can be equated with action or with praxis. Labour is what we do.

> Labour is, in the first place, a process in which both man and Nature participate, and in which man of his own accord starts, regulates, and controls the material re-actions between himself and Nature. He opposes himself to Nature as one of her own forces, setting in motion arms and legs, head and hands, the natural forces of his body, in order to appropriate Nature's productions in a form adapted to his own wants. By thus acting on the external world and changing it, he at the same time changes his own nature.
> (Marx & Engels 1996: 177)

In stark contrast to labour, his concept of labour process refers to specific historic modes of production and to specific historic societies and economies. With this historical approach, he wants to demonstrate that the labour process, the specific organisation of work, is not inevitable. Existing labour processes can always be overcome. Marx is particularly interested in the difference between a feudal and a capitalist labour process. In capitalism, the labour process is based on wage labour, on the fact that the worker sells his labour power as a commodity to the capitalist. Comparing the feudal labour process with the capitalist labour process, Marx highlights two things:

> First, the labourer works under the control of the capitalist to whom his labour belongs; the capitalist taking good care that the work is done in a proper manner, and that the means of production are used with intelligence, so that there is no unnecessary waste of raw material, and no wear and tear of the implements beyond what is necessarily caused by the work. Secondly, the product is the property of the capitalist and not that of the labourer, its immediate producer. Suppose that a capitalist pays for a day's labour-power at its value; then the right to use that power for a day belongs to him, just as much as the right to use any other commodity, such as a horse that he has hired for the day ... The labour-process is a process between things that the capitalist has purchased, things that have become his property.
> (Marx & Engels 1996: 184–185)

Here, Marx has identified two forms of alienation that did not exist in feudalism or in any other mode of production before capitalism. The first form of alienation refers to the product of the worker's own work and the inability to use the product of this own work for his or her living. The second form of alienation refers to the inability to organise the process of work, which lies exclusively in the hands of the capitalist who owns the means of production.

With respect to the generation of surplus value, Marx offers two rather contradictory explanations. In *Capital Vol. 1*, he argues that surplus value can only be created in the conditions of wage labour and that it can only derive from the exploitation of wage labour. Important for his argument is the distinction between productive and unproductive labour. Productive labour is labour that is productive for capital. It produces commodities, exchange value, and profit (surplus value). Unproductive labour does not produce surplus value. To give an example, a person employed in a private household to perform tasks such as cooking and cleaning does not produce a commodity. While his or her labour power is sold as a commodity, the product of this labour power is not. Therefore, this is unproductive labour. A cook working as an employee in a restaurant, however, produces commodities, as he or she produces meals that are sold to customers. Therefore, this is productive labour. So, productive and unproductive labour are not distinguished with respect to what people do (in both cases, they cook), but with respect to their relation to capital and the commodity form. It is not surprising that this concept has received a lot of criticism, particularly from Marxist feminists (Haug 1977), who have pointed out that the exploitation of wage labour is not independent from reproductive labour.

However, Marx comes to rather different conclusions in *Grundrisse*. Marx writes about wealth creation and the production of value which is increasingly independent of labour.

> (T)he creation of wealth comes to depend less on labour time and on the amount of labour employed ... but depends rather on the general state of science and on the progress of technology ... Labour no longer appears so much to be included within the production process; rather the human being comes to relate more as watchman and regulator to the production process itself ... He steps to the side of the production process instead of being its chief actor. In this transformation, it is neither the direct human labour he himself performs, nor the time during which he works, but rather the appropriation of his own general productive power, his understanding of nature and his mastery over it by virtue of his presence as a social body – it is, in a word, the development of the social individual which appears as the great foundation-stone of production and of wealth.
>
> (Marx 1973: 704–705)

It is especially this line of thought that inspired post-operaismo theorists to develop their concept of immaterial labour in cognitive capitalism. Immaterial labour is not like labour in Fordism. It produces more than commodities; it produces subjectivity. In cognitive capitalism, labour becomes synonymous with life itself.

I have earlier argued that PEMC has not much used the concept of labour for their analysis of the political economy of mass media. While this is largely correct, there is one very important exception to this claim – Dallas Smythe. In the 1970s, he developed the concept of audience commodity. Smythe argues that media audiences are a commodity. They are turned into a commodity by media producers. The activity of watching television connects media audiences to advertisers. Thus, media audiences perform labour. For Smythe, this is a tragedy with three players: the two bad guys are media producers and advertisers, while the victim is audiences. Media producers construct audiences. They also sell time to advertisers. Therefore, they deliver audiences for advertisers. His argument for why audiences perform labour is developed as follows. In modern capitalism, there is no time left that is not work time. Capitalism makes 'a mockery of free time and leisure' (Smythe 1977: 47). He explains how this observation relates to Marx's theory of labour power.

> Under capitalism your labor power becomes a personal possession. It seems that you can do what you want with it. If you work at a job where you are paid, you sell it. Away from the job, it seems that your work is something you do not sell. But there is a common misunderstanding at this point. At the job you are not paid for all the labor time you do sell (otherwise interest, profits, and management salaries could not be paid). And away from the job your labor time *is* sold (through the audience commodity), although you do not sell it. What is produced at the job where you are paid are commodities...What is produced by you away from the job is your labor power for tomorrow and for the next generation: ability to work and to live.
>
> (Smythe 1977: 48)

With his concept of audience commodity, Smythe is the godfather of the free labour debate – even though he has never used this term. More recently, the debate on free labour was initiated by Tiziana Terranova (2004). In an essay first published in 2000, before Wikipedia, Facebook, and all other social web platforms, she conceptualises free labour as the 'excessive activity that makes the Internet a thriving and hyperactive medium' (2004: 73). This includes 'the activity of building web sites, modifying software packages, reading and participating in mailing lists and building virtual spaces' (2004: 74). Consistent with the operaismo discourse on immaterial labour, she situates the emergence of free labour with post-Fordism.

> Free labour is the moment where this knowledgeable consumption of culture is translated into excess productive activities that are pleasurably embraced and at the same time often shamefully exploited.
>
> (2004: 78)

For Terranova, free labour is, first, unpaid labour. It is free in the sense of free beer; it is voluntarily given. Second, it is free in the sense of freedom.

It is more autonomous and less alienating than wage labour. It is not a factory but a playground. Thus, it can be enjoyed. Third, it is exploited by capital. She does not explain, however, how free labour can be exploited. It is not possible here to introduce the debate on free labour in more detail. For this, I would refer to contributions by Andrejevic (2008, 2009, 2011), Fuchs (2014, 2015), Pasquinelli (2008), Scholz (2013), and special issues by Burston et al. (2010) and Beverungen et al. (2013). For a critique of the free labour concept, I refer to Hesmondhalgh (2010) and Wittel (2012, 2013).

I will conclude this section with a brief reference on three books that are particularly insightful for an analysis of labour in digital capitalism. Both Christian Fuchs (2015) and Nick Dyer-Witheford (2015) provide an excellent analysis of all forms of digital labour, from the miners in Congo who extract coltan (which is needed for mobile phones), manufacturers of semiconductors in the Silicon Valley, workers in call centres in India, hardware producers in Southeast Asia, software programmers, and free labourers for companies such as Facebook and Twitter. David Graeber (2015) makes a compelling argument about the connection between the rise of digital technologies and the rise of bureaucratic procedures in capitalism. Bureaucracy, according to Graeber, produces an increasing amount of bullshit jobs and bullshit talks. These are jobs and tasks that generate admin or paperwork and which make life and work hugely ineffective. Writing about his own area, the rise of managerialism in academic cultures, he makes a rather interesting argument. He states that the last academic giants have been scholars such as Deleuze, Foucault, and Bourdieu and that they have all risen in prominence at a time when bureaucracy was still somehow contained. In the age of total bureaucratisation, however, creative thinking is not fostered any more but largely suffocated.

PEDT and (intellectual) property

In the age of mass media, property has always been significant with respect to the ownership of the means of production. However, an interest on property in terms of media content was rather limited. Ronald Bettig (1996) is perhaps overly careful to say that the area of intellectual property and copyright in particular has been relatively unexplored. He is one of very few political economists who examined the property of media content. Interestingly, this is a study just at the beginning of the digital turn.

Bettig is interested in the difference between the normative principles of intellectual property and the actually existing system. The central normative justification for intellectual property is built on the assumption that the creators of intellectual and artistic work need an incentive to be creative. The copyright is meant to give the creator exclusive rights to exploit their work, which in turn will provide an income for the creator and provide motivation to produce new work. However, the actual copyright system does not operate according to this ideal. Most artistic and intellectual work relies on a process of production, reproduction, and distribution that involves many people and expensive technology. According to Bettig, 'ownership of copyright increasingly rests with the capitalists who have the machinery and capital to manufacture and distribute' (1996: 8) the works.

> Precisely because the capitalist class owns the means of communication, it is able to extract the artistic and intellectual labor of actual creators of media messages. For to get 'published', in the broad sense, actual creators must transfer their rights to ownership in their work to those who have the means of disseminating it.
>
> (1996: 35)

This is a very correct analysis for the age of mass media that does not leave much room for hope. Still, he states with astonishing foresight that 'the enclosure of the intellectual and artistic commons is not inevitable or necessary, even though the emphasis on the logic of capital makes it seem as if it is' (1996: 5). Bettig must have felt that times are changing. In the mid-1990s, when his book was published, sharing cultures and the digital commons were largely restricted to the open source movement. There was no file-sharing software such as Napster, no legal experiments with copyright such as the Creative Commons, and there was no social web. In the age of mass media, the expansionary logic of capital has not left much room for intellectual and artistic commons. An overwhelming part of media content was not common property but captured by capital. In this respect, Bettig's statement has some prophetic qualities. Furthermore, it is important to mention the studies of Celia Lury (1993) and Scott Lash and Celia Lury (2007). In a number of empirical studies on trademarks, they show how commons goods have been transformed into intellectual property.

Since the emergence of the social web and digital commons, it has become very clear that the enclosure of the intellectual and artistic commons is not inevitable at all. Various forms of non-market production (Benkler 2006) are now competing with commercial content. Kleiner (2010: 7) writes about a battle between artistic and intellectual labour and those who want to rescue the digital commons, on one side of the battlefield, and those who aim for enclosure, on the other side. He describes it as a battle to the death:

> What is possible in the information age is in direct conflict with what is permissible ... The non-hierarchical relations made possible by a peer network such as the internet are contradictory with capitalism's need for enclosure and control. It is a battle to the death; either the internet as we know it must go, or capitalism as we know it must go.
>
> (2010: 7)

While there are other ways to conceptualise the relationship between non-market production and commercial production, there is no doubt that copyright and intellectual property have become a key issue. Copyright and intellectual property issues are a complex problem now with immense social, cultural, and economic implications. Two things have changed fundamentally. First, the technologies that are needed for the production of cultural content have become affordable to many people. No more printing press, TV channel, or recording studio are quintessential any more – all that is needed is a laptop or a mobile phone. Second, the costs for the reproduction of this (digital) content are close to zero now. As all digital objects can be reproduced endlessly and distributed with minimum additional costs, they count as non-rival goods. In fact, most intellectual property is non-rival, meaning they can be used by one person without preventing other people from using the same goods. Digital objects, however, are not only non-rival, but they are also abundant by nature. Therefore, all attempts to rescue the idea of copyright via digital rights are absurd in the sense that they create artificial scarcity. They turn objects that are abundant into legally scarce goods. In the digital age, only the creation of artificial scarcity can feed capitalist accumulation. It is exactly because digital things are not just non-rival but also abundant that the issue of intellectual property has moved from a sideshow to centre stage. Bettig has developed a convincing argument with much empirical backup as to why the copyright arrangements – as legitimate as they are in an ideal normative sense – have not really supported the creators of intellectual and artistic work, but those who control the communication flows. With the digital turn, this rather problematic arrangement is becoming even worse.

It is impossible to summarise the free culture debate in a few lines. I still want to make a few remarks, only to situate the key positions with respect to Marx. The first thing to note is that there is a relatively straightforward line between critical political economists and liberal political economists such as Yochai Benkler (2006) and Lawrence Lessig (2004). The latter celebrate free culture without giving up on the legitimacy of intellectual property. They merely suggest modifications to copyright law. They also applaud the digital commons as a progressive development without being overly concerned about the free labour that goes into the building of the digital commons. For Benkler (2006), commons-based peer production enhances individual freedom and autonomy. This is where critical political economists take a different position. For them, free labour is a problem that needs to be addressed.

The debates within the camp of critical political economists of digital media are not so clear cut. While both positions exist, a passionate defence of free culture (e.g., Cory Doctorow 2008 or Kevin Carson 2011) and a passionate concern about free labour and the exploitation of this free labour by capital (Pasquinelli 2008, Kleiner 2010), in most accounts, we find a general acknowledgement of this dilemma, a dilemma that is hard to crack, with many commentators sitting on the fence. One way out of the free culture dilemma resulted in the search for new models to guarantee the creators of artistic or intellectual work some income (e.g., Peter Sunde's 'Flattr' or Dmytri Kleiner's 'copyfarleft' and 'venture communism' suggestions).

Apart from some rare exceptions (notably Wark 2004 and Kleiner 2010), these debates circumvent, however, a discussion on property itself. Even those who passionately defend free culture support their position with rather pragmatic arguments – for example, with the claim that free culture ultimately stimulates creative production and innovation, whereas copyright brings about a reduction of creative and innovative work. While these are important arguments, I do find it astonishing that a fundamental critique of intellectual property itself has so far not been put on the table. Badiou asks a good rhetorical question: Why do we 'keep tight controls on all forms of property in order to ensure the survival of the powerful?' (2010: 5).

This is where Marx could come in rather handy. The first thing we can learn from Marx is that property is not a natural right. It is a historic product. Property relations are subject to specific historic conditions.

> The French Revolution, for example, abolished feudal property in favour of bourgeois property. The distinguishing feature of Communism is not the abolition of property generally, but the abolition of bourgeois property. But modern bourgeois private property is the final and most complete expression of the system of producing and appropriating products, that is based on class antagonism, on the exploitation of the many by the few. In this sense, the theory of the Communists may be summed up in the single sentence: Abolition of private property.
> (Marx 1988: 68)

The second thing to note is that Marx's perspective on property is innovative and very distinct from liberal political theorists, as he does not focus on the relationship between a person and an object. Instead, Marx conceptualises property as a relation that one person establishes to other people with respect to commodities. So, fundamentally, property relations are an expression of social relations. In capitalism, property is based on the antagonism between capital and wage labour. It is based on the accumulation of profit on the side of those who own the means of production.

> Self-earned private property, that is based, so to say, on the fusing together of the isolated, independent laboring-individual with the conditions of his labor, is supplanted by capitalistic private property, which rests on exploitation of the nominally free labor of others, i.e., on wage-labor. The capitalist mode of appropriation, the result of the capitalist mode of production, produces capitalist private property.
> (Marx & Engels 1996: 762–763)

As such, capitalist private property is not so much about the ownership of things, but about the right to exclude others from using them. Dismantling the widespread myth that private property is justly earned by those who are intelligent and willing to work hard while the rest are 'lazy rascals', Marx comes up with an alternative explanation on the origin of property:

> Such insipid childishness is every day preached to us in defence of property ... In actual history it is notorious that conquest, enslavement, robbery, murder, briefly force, play the greater part.
>
> (Marx & Engels 1996: 713–714)

The third – and, for our purposes, more important – observation is Marx's distinction between *private and personal property*. In capitalism, *private* property is bad; it is not only the result of alienated labour (wage labour) but worse, it is also the means that makes alienated labour possible in the first place and the means to maintain this unjust relation between capital and labour. Private property is productive property. It is property that is crucial for capitalist production. It is property that can be used for the creation of surplus value. It might be a bit simplistic, but, in general, Marx equates private property with privately owned means of production. This is very different from *personal* property or property for consumption (for reproduction, for subsistence), which should not be socialised, as there is no need for doing so. Unproductive property or property based on needs is rather harmless after all.

> When, therefore, capital is converted into common property, into the property of all members of society, personal property is not thereby transformed into social property. It is only the social character of the property that is changed. It loses its class character ... The average price of wage-labour is the minimum wage, i.e. that quantum of the means of subsistence, which is absolutely requisite to keep the labourer in bare existence as a labourer ... We by no means intend to abolish this personal appropriation of the products of labour, an appropriation that is made for the maintenance and reproduction of human life, and that leaves no surplus wherewith to command the labour of others.
>
> (Marx 1988: 68–69)

No doubt intellectual property is not personal but *private* property. No doubt these are productive commodities. They produce surplus value and also lay the foundation for future commodities that produce even more surplus value. Information produces more information, news produce more news, knowledge produces more knowledge, and art produces more art. Therefore, intellectual property is an invention that, in capitalism, does not protect the creators of these immaterial objects. Instead, it helps capitalist accumulation. Bettig has supported this claim in great detail with rich empirical evidence.

In my view, the debate between those who support free culture and those who are concerned about the exploitative nature of free labour got stuck.

Both positions should be supported from a Marxist point of view. They contradict each other, but they do so in perfect harmony with what Marx sees as internal contradictions of capitalism. Furthermore, the development of new business models for intellectual and artistic workers does not look promising, neither theoretically nor practically. It all boils down to the simple fact that capitalists are not willing to support free labour for altruistic reasons and those who are exploited earn just enough to maintain their own subsistence.

The only way out of this dilemma is a debate on the legitimacy of private property itself. This has to include not just intellectual property, but material property, too. Property relations reflect social relations. Our property relations in digital capitalism are the result of a class-based society with ever fewer people owning ever more.

PEDT and expansion 3: from critique to activism

In the age of mass media, PEMC has critically analysed and commented on developments in media and communication. PEMC did not, however, engage in a search for alternatives. This is hardly surprising. In the age of mass media, there were no significant public debates on alternatives to capitalist political economies. Neither was it possible to see alternatives to the two dominant forms of media institutions, which were either public media (such as the British Broadcasting Corporation or the German ARD and ZDF) and publicly funded media, or commercial media, which operated as firms and did not received public funding.

In the age of digital and distributed media, however, it is possible to imagine and to develop such alternatives. PEDT is part of this activism. PEDT is interested in academic activism. It is based on a scholarly ethos that consists of interestedness, engagement, partiality, and a partisan nature of knowledge production. To put it simply and paraphrase Marx, it is a scholarly practice that is more interested in changing the world than merely interpreting it. Obviously, the difference between change and interpretation is rather flawed conceptually. A critique – which is always a form of interpretation – only makes sense as an attempt to induce change. Why write a critique of, say neo-liberalism, if there is no desire to influence public opinion in order to overcome neo-liberalism eventually. Still, the comparison of change with interpretation (or critique) holds a great deal of value, as interpretation or critique does not have to be concerned with a vision of how to initiate change. Academic activism cannot be content with critique only. It has to offer more than that; it has to be concerned with the development of visions, ideas, and proposals that can enable change.

Let's demonstrate this with an example. In *Cyber-Proletariat* (2015), Nick Dyer-Witheford turns his focus on a rather old-fashioned topic – class analysis. According to his account, class analysis has become a major taboo. It would only be used in order to point out that the concept of class does not make sense anymore and that classes do not exist. Those very few theorists who do use the concept of class in a critical way would be condemned 'as actively hostile towards social harmony, if not inciting civil war' (Dyer-Witheford 2015: 8). This is how he continues:

> And it is indeed in such a spirit, let us confess, that we insist on class analysis, as that instrument required to recognise the inhuman, abstract and unearthly reductions forced onto people and planet by an economic system founded on a constitutive state of civil war, even if, today, this is a class war waged effectively only from above.
> (Dyer-Witheford 2015: 8)

I want to outline two developments that are particularly important to understand the rise of an activist PEDT. The first development refers to the ever-intensifying crisis of capitalism; the second refers to developments in the social web.

The crisis of capitalism is, in fact, multiple crises. It is not only an economic, but also a social, a political, and an ecological crisis. The social crisis manifests itself in a growing, unequal distribution of wealth. It is, as Warren Buffett, one of the richest people on the planet, correctly observed in 2011, played out as a class war. 'There is class war, all right, but it's my class, the rich class, that's making war, and we are winning' (Stein 2006). It is a political crisis as the failure of the political system to reverse or even stop the polarisation of wealth distribution is becoming increasingly evident. While, in theory, the political system is supposed to regulate capital, the exact opposite is closer to the truth: capital is controlling the political system. This development is often described as a crisis of representative democracy. The ecological crisis is not just about the life-threatening consequences of climate change but also about air pollution, shortages of water supply, and the impacts of deforestation.

There is much agreement that the multiple crises of capitalism will not just fade away. Wolfgang Streeck (2014) has recently identified three long-term trends in the trajectory of all highly developed capitalist states. The first is a persistent decline in the rate of economic growth, the second is an equally persistent rise in overall indebtedness (including states, firms, and private households), and the third is a persistent rise in economic inequality. As all three trends are long-term developments, one does not have to be a prophet to realise there is not much hope for economic recovery. In fact, the various moments of this crisis will deepen and reinforce each other. The ecological crisis, for example, is likely to increase migration and the number of refugees, thus intensifying the social crisis with new levels of militarisation and protection of the rich against the poor. The current debates about fortress Europe mark only the beginning of this development.

Strictly speaking, the search for alternative ways to organise life in the 21st century is not an academic project. It is not something that can be theorised, neither can it be researched empirically. The search for alternatives is not merely based on an analysis of reality. Instead, it requires imagination, desire, and a certain readiness for utopian thinking. While all these attributes seem rather unacademic, scholarly activism draws its legitimacy from the notion of relevance. In times of crisis, the most pressing question is how to overcome the crisis. It is for this reason alone that academic activism it not just legitimate, but an obligation. It is the order of the day.

The second reason for PEDT's activist turn are developments in the social web. In the age of mass media, there was little room for experimentation, as alternatives to mass media were simply not available (Hands 2011). The social web has expanded the ownership of media beyond the public/private dichotomy. Wikipedia, WikiLeaks, Democracy Now!, open source software, and many other platforms and initiatives are communication commons; they are neither public nor commercial media.

With the emergence of the social web and its platforms of user-generated content, the need for gatekeepers and mediators has vanished. The social web fosters a mutual and non-hierarchical form of communication between peers. The social web has turned the Internet into a field of social experimentation. On local, regional, national, and international levels, we see a rise of initiatives, institutions, and projects that explore alternatives to the status quo. These alternatives emerge in a number of social fields. I want to briefly introduce six fields of academic activism.

The first area is about the rebuilding of the web or – more modestly – some platforms. Perhaps the first initiative was organised by Douglas Rushkoff: his *ContactCon* in 2011 brought together theorists, activists, and practitioners to discuss possibilities to create an alternative web – open source, distributed, and non-commercial. Astra Taylor's *People's Platform* (2014) is another call to arms for such a project. The 'Unlike Us' mailing list, initiated by Geert Lovink, has led to a number of publications – most notably the *Unlike Us Reader* (Lovink & Rasch 2013) – which are not only a critique of corporate social media – Facebook in particular – but also an exploration of open-source projects that develop non-commodified social media platforms such as 'Diaspora'. The most recent example in this area is *Platform Cooperativism*, a conference organised by Trebor Scholz in 2015. It aims at a collaboration between academic activists, trade unions, and software programmers to develop alternatives to the so-called sharing economy and to challenge companies such as Airbnb and Uber with the building of firms that are rooted in cooperative models of ownership.

The second area of academic activism refers to copyright and other forms of intellectual property. Initiatives in this area include the *free culture movement* and *A2K* (Access to Knowledge). While the free culture movement focuses its energies on free access to cultural products, A2K campaigns for a lift of intellectual property for all knowledge, be it pharmaceutical, scientific, or technical. *Open education resources* (OER) is an initiative to make available educational resources for everyone who has access to the Internet.

Perhaps the most influential initiative is the *Creative Commons*, a non-profit organisation that enables the use of knowledge and cultural products through free and legal tools. It provides a variety of copyright licenses, which are all less restrictive than copyright law. The Creative Commons was founded in 2001 by Lawrence Lessig. *Copyfarleft* – a radical alternative to copyleft – was originated by Dmytri Kleiner in 2010 and has since been used as a licensing model by the *P2P Foundation*.

The third area involves new publishing initiatives. Most prominently, these are open access initiatives that make academic writing available for free. There is a wide range of open access initiatives, so I refer only to those initiatives that do not seek profit. There are too many journals to be included in this list, however *ephemera, tripleC,* and the *Journal of Peer Production* stand out in that they have all received considerable acclaim. More recently, open-access initiatives extended to book publishing. The flagship project here is perhaps the *Open Library of Humanities*. While open-source publications clearly dominate the agenda of publishing initiatives, there is more to be discovered. *The Torist* is the first encrypted academic journal. The purpose of this project is beyond my understanding, but it is clearly part of this activist list. Another new initiative is activism against academic social media sites such as *academia.edu* or *ResearchGate.net,* which, on a positive side, provide links to and downloads of free academic publications, but operate with corporate investment and are based on a highly problematic quantification of academic content. *Why are we not boycotting academia.edu?* is the first conference (organised 2015 by the Centre of Disruptive Media, Coventry University) dedicated to this project. Finally, the list of new publishing initiatives should include mailing lists, which have provided early global forms of communication and interaction, long before the birth of social media. There are hundreds of such mailing lists. The first and probably still most important one for academic activists is *nettime.org*, a mailing list for networked cultures, politics, and tactics, founded in 1995 by Geert Lovink and Pit Schultz and now moderated by Ted Byfield and Felix Stalder. The mailing list had and still has a pivotal role in stimulating a critical theory of the Internet. It had around 4,500 subscribers in 2015.

The fourth area of activity refers to the building of the (digital) commons. During the last decade, talk of the common, the commons, the digital commons, the commonwealth, the commoner, and communing has profoundly shaped the debate in the political Left. Commons-related activism is both ubiquitous and heterogeneous, coming in all shapes and colours. It is impossible to present this area in a few lines. However, two institutions particularly deserve to be mentioned. One is the German Heinrich Böll Foundation, the political foundation of the Green Party, which is a think tank for policy reform and a catalyst for green visions. The Heinrich Böll Foundation has given the commons much attention over the last decade and has made it one of their priorities for the initiation of change. Conferences and other activities at the foundation have led to two significant publications, *The Wealth* of the Commons: A World Beyond Market and State (edited by Bollier & Helfrich in 2012) and *Patterns of Commoning* (edited by Bollier & Helfrich in 2015). The other institution is the P2P Foundation, founded and hosted by Michel Bauwens. The P2P Foundation provides a well-organised and updated repository of commons-related initiatives. Beyond that, it is also actively involved in some of these initiatives. *Network Society and Future Scenarios for a Collaborative Economy* (Kostakis & Bauwens 2014) is a critical reflection network on how to build a commons-oriented economy. It draws theoretically and empirically on a project of the P2P Foundation in Ecuador.

The fifth area of academic activism refers to money, currencies, finance, and debt. Perhaps the most prominent initiative here is David Graeber's (2011) call for a global *debt amnesty*. While the chances for such a call to be put into practice is probably close to zero, Graeber has immensely contributed to a global awareness of a debt spiral that is out of control. Andrew Ross was involved in the *Occupy Student Debt Campaign*. He is also one of the originators of *Strike Debt*, a coalition that was formed in 2012 to build a debtors movement and was involved in writing the *Debt Resistors Operations Manual*. There is also a great interest in virtual currencies such as *Bitcoin*. *MoneyLab* is part of a global movement for a democratisation of finance. The first *MoneyLab Reader* (edited by Lovink et al. 2015) was published at the Institute of Network Cultures. Publications on Bitcoin are rising exponentially. Brett Scott, the author of *The Heretic's Guide to Global Finance: Hacking the Future of Money* (2013), has collected a very impressive database of several hundred publications on Bitcoin. He is also the founder of the *London School of Financial Activism*.

Last but not least, the sixth area of academic activism is about higher education and a fight back against a global transformation of the public university into a corporation. This transformation is particularly visible in Anglo-Saxon countries but is by no means restricted to them. As a reaction to the privatisation of higher education, many academics have decided to create free and autonomous universities. They have engaged in the building of a higher education commons. One of the most prominent initiatives in the UK is the *Social Science Centre* in Lincoln; others include the *Free University of Liverpool*, the *Free University Brighton,* and the *People's Political Economy Group* in Oxford. An *Alternative Education Counter Cartography* (posted on the Social Science Centre website) lists worldwide 123 sites of autonomous higher education initiatives, most of them situated in the US and the UK.

Conclusion

This chapter is an attempt to outline an emerging area of research, the political economy of digital technologies (PEDT). It is an area of research that has its roots in the political economy of media and communication (PEMC). However, it has also developed in a number of ways that make PEDT very distinct from its predecessor. I have outlined three of these expansions. The first one is an expansion from media to technologies, the second one is a theoretical expansion, an expansion of Marxist concepts, and the third one is an expansion from academic critique to academic activism. Due to this emphasis on differences between PEMC and PEDT, I have ignored the similarities between both fields. Therefore, it is important to note that all debates that were central for PEMC are still relevant and ongoing in the age of digital and distributed media and technologies.

References

Andrejevic, M., 2008. Watching television without pity. *Television & New Media*, 9 (1), pp. 24–46.
Andrejevic, M., 2009. Exploiting YouTube: contradictions of user-generated labour. In P. Snickars and P. Vonderau (Eds.), *The YouTube reader*. Stockholm: National Library of Sweden, pp. 406–423.
Andrejevic, M., 2011. Facebook als neue Produktionsweise. In O. Leister and T. Röhle (Eds.), *Generation Facebook. Über das Leben im Social Net*. Bielefeld: Transcript, pp. 31–50.
Badiou, A., 2010. *The communist hypothesis*. London: Verso.
Benkler, Y., 2006. *The wealth of networks: how social production transforms markets and freedom*. New Haven: Yale University Press.
Berardi, F.B., 2009. *The soul at work: from alienation to autonomy*. Los Angeles: Semiotext(e).
Bettig, R.V., 1996. *Copyrighting culture: the political economy of intellectual property*. Oxford: Westview Press.
Beverungen, A., Otto, B., Spoelstra, S., and Kenny, K., 2013. Free work. *Ephemera*, 13 (1), pp. 1–9.
Bollier, D. and Helfrich, S. (Eds.), 2012. *The wealth of the commons: a world beyond market and state*. Amherst: Levellers Press.
Bollier, D. and Helfrich, S. (Eds.), 2015. *Patterns of commoning*. Amherst: The Commons Strategies Group.
Burston, J., Dyer-Witheford, N., and Hearn, A., 2010. Digital labour: workers, authors, citizens. *Ephemera*, 10 (3–4), pp. 214–221.
Burton, G., 2010. *Media and society: critical perspectives*. Maidenhead: Open University Press.
Carson, K., 2011. How "intellectual property" impedes competition. In G. Chartier and C.W. Johnson (Eds.), *Markets not capitalism: individualist anarchism against bosses, inequality, corporate power, and structural poverty*. London: Minor Compositions, pp. 325–334.
Castells, M., 1996. *The rise of the network society*. Maldan: Blackwell.
Castells, M., 2009. *Communication power*. New York. Oxford University Press.
Castells, M., 2012. *Networks of outrage and hope: social movements in the Internet age*. Cambridge: Polity Press.
Curran, J., 1990. The new revisionism in mass communication research: A reappraisal. *European Journal of Communication*, 5 (2), pp. 135–164.
Curran, J., 2000. *Media organisations in society*. London: Arnold.

Curran, J. and Seaton, J., 1997. *Power without responsibility: the press and broadcasting in Britain*. London: Routledge.
Devereux, E., 2003. *Understanding the media*. London: Sage.
Doctorow, C., 2008. *Content*. San Francisco: Tachyon Books.
Durham, M.G. and Kellner, D., 2006. *Media and cultural studies: keyworks*. Oxford: Blackwell.
Dyer-Witheford, N., 2015. *Cyber-proletariat: global labor in the digital vortex*. London: Pluto Press.
Fuchs, C., 2014. *Social media: a critical introduction*. London: Sage.
Fuchs, C., 2015. *Digital labour and Karl Marx*. New York: Routledge.
Garnham, N., 1990. *Capitalism and communication: global culture and the economics of information*. London: Sage.
Gorz, A., 1989. *Critique of economic reason*. London: Verso.
Gorz, A., 1999. *Reclaiming work: beyond the wage-based society*. Cambridge: Polity Press.
Graeber, D., 2002. *Towards an anthropological theory of value: the false coin of our own dreams*. New York: Palgrave Macmillan.
Graeber, D., 2011. *Debt: the first 5,000 years*. London: Melville House.
Graeber, D., 2015. *The utopia of rules: on technology, stupidity, and the secret joys of bureaucracy*. London: Melville House.
Grossberg, L., Wartella, E., and Whitney, D.C., 1998. *Mediamaking: mass media in a popular culture*. Thousand Oaks: Sage.
Hands, J., 2011. *@ is for activism: dissent, resistance and rebellion in a digital culture*. London: Pluto Press.
Hardt, M. and Negri, A., 2000. *Empire*. Cambridge: Harvard University Press.
Hardt, M. and Negri, A., 2004. *Multitude: war and democracy in the age of empire*. London: Penguin Books.
Hardt, M. and Negri, A., 2009. *Commonwealth*. Cambridge: Harvard University Press.
Harvey, D., 2007. *A brief history of neoliberalism*. Oxford: Oxford University Press.
Harvey, D., 2010. *The enigma of capital and the crises of capitalism*. London: Profile Books.
Harvey, D., 2012. *Rebel cities: from the right to the city to the urban revolution*. London: Verso.
Haug, F., 1977. *Erziehung und Gesellschaftliche Produktion. Kritik des Rollenspiels*. Frankfurt am Main: Campus.
Herman, E.S. and Chomsky, N., 1988. *Manufacturing consent: the political economy of the mass media*. London: Vintage.
Herman, E.S. and McChesney, R.W., 1997. *The global media: the new missionaries of corporate capitalism*. London: Cassell.
Hesmondhalgh, D., 2010. User-generated content, free labour and the cultural industries. *Ephemera*, 10 (3–4), pp. 267–284.
Kleiner, D., 2010. *The telekommunist manifesto*. Amsterdam: Institute of Network Cultures.
Kostakis, V. and Bauwens, M., 2014. *Network society and future scenarios for a collaborative economy*. London: Palgrave Macmillan.
Lash, S. and Lury, C., 2007. *Global culture industry: the mediation of things*. Cambridge: Cambridge University Press.
Laughey, D., 2009. *Media studies: theories and approaches*. Harpenden: Kamera.

Lazzarato, M., 1998. Immaterial labor. *Generation online*. Available at: www.generation-online.org/c/fcimmateriallabour3.htm (accessed 26 February 2012).

Lessig, L., 2004. *Free culture: the nature and future of creativity*. New York: Penguin Books.

Lovink, G. and Rasch, M. (Eds.), 2013. *Unlike us reader: social media monopolies and their alternatives*. Amsterdam: Institute of Network Cultures.

Lovink, G., Tkacz, N., and de Vries, P. (Eds.), 2015. *MoneyLab reader: an intervention in digital economy*. Amsterdam: Institute of Network Cultures.

Lury, C., 1993. *Cultural rights: technology, legality and personality*. London: Routledge.

Marx, K., 1973. *Grundrisse: foundations of the critique of political economy*. London: Penguin Books.

Marx, K., 1988. *The communist manifesto*. New York: W. W. Norton.

Marx, K. and Engels, F., 1974. *The German ideology*. London: Lawrence and Wishart.

Marx, K. and Engels, F., 1996. *Karl Marx Frederick Engels: collected works, Vol. 35: Karl Marx; Capital, Vol. 1*. London: Lawrence and Wishart.

McQuail, D., 2005. *McQuail's mass communication theory*. London: Sage Publications.

Mosco, V., 1996. *The political economy of communication: rethinking and renewal*. London: Sage Publications.

Murdock, G., 1982. Large corporations and the control of the communications industries. In M. Gurevitch, T. Bennett, J. Curran, and J. Woollacott (Eds.), *Culture, society and the media*. London: Methuen, pp. 114–147.

Nichols, J. and McChesney, R.W., 2006. *Tragedy and farce: how the American media sell wars, spin elections and destroy democracy*. New York: New Press.

Pasquinelli, M., 2008. *Animal spirits: a bestiary of the commons*. Amsterdam: NAi Publishers and Institute of Network Cultures.

Polanyi, K., 2001. *The great transformation: the political and economic origins of our time*. Boston: Beacon Press.

Poster, M., 1995. *The second media age*. Cambridge: Polity Press.

Rushkoff, D., 2016. *Throwing rocks at the Google bus: how growth became the enemy of prosperity*. New York: Portfolio.

Schiller, H.I., 1989. *Culture, Inc: the corporate takeover of public expression*. New York: Oxford University Press.

Schiller, D., 1999. *Digital capitalism: networking the global market system*. London: MIT Press.

Scholz, T., 2013. *Digital labor: the Internet as playground and factory*. New York: Routledge.

Scott, B., 2013. *The heretic's guide to global finance: hacking the future of money*. London: Pluto Press.

Smythe, D.W., 1977. *Dependency road: communications, capitalism, consciousness, and Canada*. Norwood: Ablex Publishing.

Stein, B., 2006. In class warfare, guess which class is winning. *New York Times*, Business, 26 November 2006.

Streeck, W., 2014. How will capitalism end? *New Left Review*, 87, pp. 35–64.

Taylor, A., 2014. *The people's platform: taking back power and culture in the digital age*. New York: Picador.

Terranova, T., 2004. *Network culture: politics for the information age*. London: Pluto Press.

Virno, P., 2004. *A grammar of the multitude: for an analysis of contemporary forms of life*. New York: Semiotext(e).
Wark, M., 2004. *A hacker manifesto*. Cambridge: Harvard University Press.
Williams, R., 1958. *Culture and society 1780–1950*. London: Chatto & Windus.
Williams, R., 1961. *The long revolution*. London: Chatto & Windus.
Wittel, A., 2012. Digital Marx: toward a political economy of distributed media. *tripleC*, 10 (2), pp. 313–333.
Wittel, A., 2013. Counter-commodification: the economy of contribution in the digital commons. *Culture and Organization*, 19 (4), 314–331.

14 Ludification of culture
The significance of play and games in everyday practices of the digital era

Anne Dippel and Sonia Fizek

Ludus and the empirical cultural analysis

Play and games open up new dimensions and fields for the analysis of digitisation processes and phenomena within the framework of theoretical and empirical cultural research. Play as such, as well as its current digital manifestations, may be axiomatically positioned as fundamental constituents of human behaviour, which keep unfolding their potential within and amongst us.

In recent years, the permeation of various life domains with the logics of play in particular and the *ludification of culture* in general have been gaining an ever-greater significance. A variety of play forms have been proliferating in the digitised everyday. This abundance of playfulness is reflected in the symptomatic questions posed by ludologists and media scholars: 'What if our whole life were turned into a game? What sounds like the premise of a science fiction novel is today becoming reality' (Deterding & Walz 2015). The growing presence, significance, and recognition of play has also lead to the proclamation of the 21st century as the ludic century, with games becoming the dominant sociocultural organisation form (Zimmerman 2009).

From an anthropological perspective, the above hypotheses may seem a bit too far-fetched. In its long and lively tradition of play research, anthropology has witnessed the omnipresence of playfulness independently from digitisation processes (Malaby 2009). Furthermore, the 18th century was already proclaimed the century of play. In 1756, Daniel Bernoulli, a Swiss mathematician and physicist, noticed: 'The century that we live in could be subsumed in the history books as: Free Spirits' Journal and the Century of Play' (Bernoulli 1769: 387, Bauer 2006: 377, Fuchs 2014: 131).[1]

However, it seems that, along with digitisation, new playful dimensions and fields have begun to emerge. Play and games are observed, designed, and theorised in new contexts, ranging from pastime and idleness to productivity and work. On the following pages, we are discussing the significance of the concept of the *ludification of culture*, demonstrating how the playful phenomenon has spread in digital times and how it has been influenced by the digital calculating machine – particularly, its capacity to process large amounts of data and the ability to afford communication in large networks. As will become evident in the presented examples, the qualities of the digital medium facilitate human creativity to build new forms and sorts of play – material, symbolic, and imaginary alike.

The aim of the current chapter is to lay at the reader's hands an analytical and conceptual tool that would characterise the omnipresence of play in the digital sphere of the everyday. We are first introducing the concept of the *ludification of culture* from multidisciplinary perspectives, with an emphasis on cultural anthropology, philosophy, media theory, and games studies. In our argumentation, we are connecting recent academic approaches with the already established scholarly research tradition on play, games, and the digital sphere. To clarify the concept further, we are drawing upon examples from empirical research, which illustrate how our digital everyday is permeated with the logics and metaphors of play and, what follows, how the *ludification of culture* manifests itself in specific social practices, such as work. In the concluding paragraphs, we are sketching tendencies and further possibilities for the implementation and development of the concept in combination with empirical research on ludified social phenomena.

Ludification of the everyday

The omnipresence of digital and analogue games

In her anthropological analysis of the digitised world, Gertraud Koch illustrates the growing importance of digital media through numbers. More and more people spend their time in digital spaces, which, as a consequence, alters their everyday experiences, she emphasises (2015: 180). A similar quantitative method may be applied to pronounce the ever-growing importance of digital games and playful applications in other spheres of our lives. Currently, there are 1.5 billion mobile gamers around the globe of various ages, genders, and social backgrounds (Global Mobile Games Market Report 2013).

Games have been gaining an ever-greater presence and significance in the digital sphere. They are no longer played only for the game's sake. They may be serious, educate, express purpose, or contribute to a social change and advancement in science, such as in the case of the so-called serious games, games for change, or games with purpose. Digital games are not only played in the living rooms (PC and console games), but also in offices (gamified applications for business), public means of transport (on mobile phones and portable consoles), urban spaces (augmented reality mobile gaming), medical institutions (games for health), at schools and universities as part of curricula (educational and pedagogical games), or in museums as interactive installations. More recent phenomena, such as live-streaming of online competitions and play sessions, watched by millions on the Twitch.tv platform, point to the fact that the pleasure derived from observing others play may be as amusing as playing itself.

Such diversity and indeterminateness of play, although in its analogue form, already emerged as the subject of scholarly examination a few decades ago. The cultural anthropologist Brian Sutton-Smith in his *magnum opus The Ambiguity of Play* (1997) extensively studied a variety of play forms that escape clear definitions and categorisations. At its core, play is ambivalent and vague in all its aspects – its references, intentions, sense, contradictions, and meaning (Sutton-Smith 1997: 2), also its liminality, expressing the transition between various states (Turner 1969: viii, 177). Almost anything may be included within the sphere of play: playing with metaphors, watching television, being sexually intimate, joking, celebrating birthdays, and gossiping, amongst many other activities (Sutton-Smith 1997: 5). This diversity, as Sutton-Smith further discusses, applies not only to play forms and experiences but also to players (e.g., infants, children, adolescents, adults, male and female players, gamblers, elite sports players, playwrights, performers, comedians). How do we make sense of games and play in such a diverse ecosystem? How do we understand a phenomenon that has so many variations that it almost seems not what it is? Sutton-Smith proposes to solve this impossibility by analysing seven popular ideological rhetorics of play: play as progress, play as fate, play as power, play as identity, play as the imaginary, play as the rhetoric of the self, and play as frivolous (Sutton-Smith 1997: 9–11).

Ludification of culture

In the past years, the diversity and ubiquity of play have gained an intensified visibility in game and media studies. Researchers no longer focus solely on human actors and new perspectives on their playful activities or the appearance of new playful life domains, but also on the digital media and the role of non-human agents in this playful constellation. Numerous media theorists, game scholars, and designers have discussed the increasing presence and significance of games and play in the surrounding culture, in particular within the context of digitisation processes (Adamowsky 2000, Raessens 2006, 2010, 2014, Zimmerman 2009, 2013, Deterding et al. 2011, McGonigal 2011, Fuchs et al. 2014, Sicart 2014, Rautzenberg 2015, De Mul et al. 2015). This omnipresence and permeation of games and ludic logics in our everyday contexts has been referred to as *gamification* or *ludification*. At this point, it is crucial to differentiate between the two concepts, as they are not synonymous.

Gamification may be perceived as a tool, which emphasises the usage of game design elements in non-game contexts (Detering et al. 2011). It focuses on the mechanical and iterative capacity of ludic systems. Gamification relies on the adaptation of game mechanics to daily activities in order to influence the individual's behaviour and drive engagement (Gartner Inc. 2011,

Radoff 2011, Zichermann & Cunningham 2011, Zichermann & Linder 2013). The latter are believed to be brought about mainly by implementing the elements of challenge and competition. These require the winning condition, which, in most cases, translates to a system based on the allocation of points through creating possibilities of success, leader boards, badges, and social networking elements, which in turn lead to the achievement of status. All of the above form the essence of gamification in its narrow sense of a competition-focused system, which turns otherwise tedious processes into attractive gamelike activities. Also, the rhetoric around gamification seems to be predominantly structured around accumulation and pointsification, whether in neutral, positive, or critical terms (Robertson 2010, Bogost 2011a,b, Dragona 2014).

Ludification, on the other hand, is a broader concept, which analyses the surrounding culture and its daily manifestations through the lens of playfulness and games. In his studies on the *ludification of culture,* media and games theorist Joost Raessens discusses the playful nature of the digital medium itself, which encourages the emergence of new play forms (Raessens 2006, 2010, 2014). For instance, as he notices, mobile devices and mobile social platforms afford playful impulses, which manifest themselves in the experimental usage of written language, including the so-called 'texting' and 'twittering' (2010: 6). A similar ludic tendency may be observed with reference to the medium of television and its new possibility to participate playfully in watching the visual content through second screen applications. These allow the viewers to access additional content on their mobile devices and interact with others in real time. One of such applications mentioned by Raessens, *Heineken Star Player*, connects the Champions League fans and viewers via social platforms and encourages them to gamble on the outcome of the matches (2010: 8).

On the micro level, the previous example may be treated as a concrete instance of gamification with the usage of badges, achievement scores, and other competitive game mechanics, implemented in order to make the TV programme more attractive to the audience. On the macro scale, however, the gradual change of everyday interaction patterns in the medium points to a much broader process of playful participation in cultural practices.

Also, a playful usage of graphical software (such as Photoshop, amongst many others), its contribution to the modification of visual content, and, resulting from it, spreading of 'mash-ups', 'memes', and 'GIFs' are pointing to new forms of expression born in the digital sphere. *Ludification of culture* therefore epitomises a sociocultural phenomenon, at the base of which lies play.[2]

Ludification as a universal cultural phenomenon

Already, Johan Huizinga emphasised that all the great archetypal activities of human society – language, myths, and rituals – are permeated with play (Huizinga 1938/1992: 13), to which we may add harvesting, hunting, making war, and love. Games govern our lives, infiltrating all their aspects. A walk through the British Museum in London seems to further confirm Huizinga's hypothesis. There, we may find 4,500-year-old board games, discovered by archaeologists during the excavations of the Royal Tombs of Ur in Mesopotamia (Becker 2008).[3] The fundamentality of playful behaviour (Tomasello 1999: 91) is further strengthened by the fact that it resides not only in the realm of humankind. A human being is most probably the only animal that has *logos*, but certainly – following Aristotle – not the only one that plays. Huizinga asserts boldly that playfulness is one of the fundaments of civilisation, and within it, the boundaries between humans and animals, or culture and nature, are interweaving (1938/1992: 11).

There is no culture known to anthropologists, historians, or archaeologists in which games have no presence (Bally 1966: 61, Mäyrä 2008: 37). An etymological excursion through the word 'game' itself seems to support the above statement. The word 'game', of Proto-Germanic roots, means nothing more than 'together' (prefix *ga-*) and 'men' (stem *–mann*). Its second connotation points towards one of the oldest social practices, that of hunting. Here, 'game' indicates wild animals caught in a collaborative pursuit. Taking into account this double sense of the origin of the English term leads us onto the trail of Old Stone Age cave paintings, which depict hunting scenes and game, going beyond its literal sense. In the illustration of the animals and hunting scenery, game and games are portrayed in the *ilinx* of the wild chase and the *mimicry* of nature, its play of light and shadow, forms and lines painted on the rock canvas, in which both performance and playfulness unfold in this fundamental human practice. What game signifies is a collective form of promoting togetherness and securing the survival of the group. Following the etymological trail, the archaeological findings, Hans Blumenberg's reflections on the caves (Blumenberg 1979), and, finally, the words of the editors of *Understanding Video Games,* we are contributing to the presumption that 'even our ancient cave-dwelling ancestors had rule-based systems of play' (Egenfeldt-Nielsen et al. 2013: 3).

Play as part of the workspace – case study

Ludification of workspace in the digital era

Ludification of culture, as discussed earlier, is based upon the fundamental phenomenon of humanity. As a concept, it has been gaining an increasing significance and presence in the past few years due to industrialisation and digitalisation. Nevertheless, the following question remains open – what has changed in the digital era, other than the increase in the quantity and scale of playful (virtual) worlds, or the multiplication of interconnectedness amongst human and non-human actors? Following Huizinga, Sutton-Smith, and many other scholars, we have come to the conclusion that play has been always permeating numerous domains of our lives. They all, however, belong to the sphere of pastime – language, myths, and rituals (Huizinga 1938/1992: 13), or theatre, sexual intimacy, joking, and gossiping, amongst others (Sutton-Smith 1997: 12). In the digital times, playfulness spreads thanks to quantified, networked, and interactive digital media into other spheres, which, until recently, have been considered play-free zones, such as the workspace.

Today, the distinction between work and play gradually dissolves. This process may be observed on the example of *Attent*, an application tackling the problem of post management and information overload in corporate email exchange. Its users have an imagined currency (Serios points) at their disposal that enables them to prioritise their outgoing and incoming emails by attaching virtual value to them. The design of *Attent* has been inspired by a discussion led at the Business Innovation Factory-7 Summit, during which researchers and business practitioners were wondering how to combine gaming with work, so that a boring email box interface would incite a similar level of excitement to a *World of Warcraft* (2004) session.

The superposition of work and play is particularly visible when the workday reaches a complexity level at which it cannot be anything but playful, and a corporate work office transforms into a labyrinth of play areas (Stewart 2013). Ludification, it seems, is not a one-way road. For as much as playful elements enter the domains of work, work-related aspects permeate playgrounds. The relationship between work and play can neither be fully embraced by the concept of ludification, nor by its contrasting term, that of labourisation (the process of permeation of play with work elements; Dippel & Fizek 2015, 2016). Instead, it could be theorised as practice, in which work and play overlap and pervade each other.

In order to encompass this overlay, we are introducing the concept of *interference*, borrowing a term that originally was used in physics to denote the superposition of waves. Interference encompasses the impurity of play and its 'corrupted' character, which manifests itself when the line dividing games from daily life is blurred (Caillois 1958/2001: 43). It allows us to describe the transformative dimension of otherwise contrasting phenomena. The metaphor of interference challenges strictly dualistic models, in which mechanical figures or anthropocentric interpretations describe diverse and impure social processes. In the situation of empirical fieldwork, the concept unfolds the horizons of understanding discursive complexities and the sociocultural multidimensionality of the everyday. Within the model of interference, work and play appear as polar modalities of human interaction. On the one hand, they may be described separately from each other. On the other hand, they influence each other reciprocally and within the moment of hermeneutical analysis and empirical research may be observed in their overlaying condition (Dippel & Fizek 2015, 2016).

The differentiation between work and play appears already in Aristotle's *Nicomachean Ethics* (1971). Both aforementioned qualities, according to Aristotle (1971), are required in order to achieve happiness and freedom.

Gregory Bateson (1972), on the other hand, differentiates between play and combat, drawing from the animal kingdom. Here, playing is opposed to serious activities required for life sustainment or defence against danger. The very process of blurring the work-play and seriousness-playfulness lines is brought to attention in the last chapter of *Homo Ludens* (1938/1992: 200), where Huizinga discusses the loss of the purity of a frivolous playful experience and emphasises the confusion of where play ends and non-play begins. To support his claim, he uses the example of professional sports, which systematises pure play and fills it with the principles of paid work. Huizingian distinction between play and work, and the portrayal of the latter as a productive and paid activity, partially relates to a Marxist understanding of work ethic. For Karl Marx, work is defined as a useful and productive activity that may be translated into the value of commodities produced (Marx 2015/1887). At the same time, Marx departs from the Abrahamic definition of work as toil, which seems to have been placed on the human shoulders in the moment of the ancestral sin (The Bible, KJV, Genesis 3:19), and perceives work as a chance for the 'individual's self-realization', an *a priori* act of utmost freedom, which encompasses happiness, even if, throughout history, mostly corrupted, self-alienated forms of work or 'external forms of labour' have become visible (1973/1858: 611).

The interference between work and play is rather based on the contrasting understanding of work as self-alienating and play as self-fulfilling. As such, it treats work as a term associated with drudgery and toil, pointing towards exertion of the body and possibly originating from the 14th-century notion of 'tottering under a burden' (from Latin *labere*). Etymologically, labour seems to be connected with productivity, effort, hardship, and suffering, qualities through which it fundamentally differs from play.

In such dualisms, Brian Sutton-Smith (1997) notices the rhetoric of frivolity, which carries in itself implicit work ethics, moving play into the domain of fun, non-seriousness, or nonsense. As a rule, playing is often described as an activity, which happens out of joy and functional pleasure, combined with delight stemming from its objective character and outcome (Brockhaus 1957: 102). Games and play seem to be determined by their self-sufficiency and closely defined 'magic circle', which is creating a temporary world within the ordinary one (Huizinga 1938/1992: 10). They remain on the opposite end of self-alienating work, understood as long as they are non-serious (Huizinga 1938/1992: 10), unproductive (Caillois 1958/2001: 10), joyous (Scheuerl 1979: 69), and utterly absorbing (Huizinga 1938/1992: 10), making the players lose themselves in the constellation of playful time and space. The magic element within the play experience points towards the very suspension of time, as if past and future did not exist. The time within play is defined and perceived as pastime, for the players need to be entirely captured by the game in order to play it. Pastime seems to synchronise permanence and simultaneity and enclose them within what the German pedagogue Hans Scheuerl (1979: 69) defines as presence and inner endlessness.

The experience of being suspended in time and lost within the game has also been theorised from the perspectives of flow (Csikszentmihalyi 1990/2008) and immersion (Tekinbaş Salen & Zimmerman 2003, Calleja 2011).

The digital machine itself, being a work tool and a toy at the same time, unites those two seemingly mutually exclusive qualities. On the one hand, a computer is a digital calculator based on mathematical game theory (von Neumann 1928, von Neumann & Morgenstern 1944), performing work-related tasks; on the other hand, it is an entertainment centre used in free time. From its early years, the computer has found itself entangled at the intersection between work and leisure-related playful activities. It served as a computing and simulating aid at governmental departments, universities, and research and cultural institutions. At the same time, that very same assemblage of hardware and software was used to programme the first games. In 1961, a group of researchers at the Massachusetts Institute of Technology developed *Spacewar* (1961), a space combat simulation, in order to demonstrate the capacities of the computing machines to the public in a compelling way. Today, in the developed parts of the world, the most popular digital machines (personal computers, smartphones, or tablets) are an indispensable part of work and leisure.

Developer's dilemma – case study

The complexities of work and play and their mutual interdependencies and superpositions are the subject of a recent anthropological study, describing and analysing the collaborative work practices amongst videogame developers (O'Donnell 2014). The investigation is a result of ethnographic fieldwork among developers working in 'AAA' studios in the United States and India. The material was collected in the years 2004–2008. The author performed participant observation, ran structured interviews, and had additional afterwork conversations with every game developer who was willing to share their perspectives. Observing this particular vocational group, Casey O'Donnell makes an attempt to understand what work has become in the current historical and cultural moment and:

> how the creative collaborative practice of game developers and game development work sheds new conceptual light on our understanding of work, the organization of work, and the market forces that shape and are shaped by media industries in the new economy.
>
> (2014: 4)

The primary quality, which forms the basis of the author's fieldwork and which is crucial to the thesis formulated in this chapter, is the significance of play in workspace. Building upon T.L. Taylor (2006: 72–73), he refers to this playful labour or laborious play dimension as *work/play interplay* and observes the overlaps on numerous levels, from the collaborative teamwork and the playful work conduct to the very arrangement of space in companies, where employees can climb, play volleyball, or lift weights. He refers to the latter as the *Googlefication* of the workplace.

What is also crucial in the understanding of this superposition of work and play is the fact that most developers studied by O'Donnell belong to the Nintendo Generation. As he further emphasises, '[t]his sense of shared history and experience provides foundations for how videogame developers talk about their occupations' (2014: 26). They share a specific vernacular, which becomes their insider's language through which they guard access to the metaphorical game of game development, as the author states. 'When you think and talk through/with games, they become aspects of the workplace' (2014: 42).

As idyllic as the previous vision of labour may seem, O'Donnell emphasises that the new modes of work practice, based on the blurred distinction between what is work and what is play, may as well dissolve into 'destructive work practices' (2014: 31). For as much as such playful work scheme encourages people to think creatively, it also pushes them to invest more time into work, giving the video game producers and publishers the possibility to extend the developer's work week even up to 80 hours.

This is possible due to the so-called instrumental work/play, which lies at the heart of the culture of gamers. This group draws particular importance to the 'act of working through the complex problems found in videogames. Any circumvention of this labour is often seen as a circumvention of the rules' (O'Donnell 2014: 61). In other words, the developers are imposing their underlying gamer's attitude upon their work practices, which become a riddle to be solved and a playful system to be cracked and understood. For O'Donnell, this deep exploration of the systems one works within lies at the core of instrumental work/play. The capability and the need to play seem to lie at the centre of a creative collaborative work practice (O'Donnell 2014: 5, 31).

However, the long and inhumane work hours of the developers lead to the collapse of desire altogether. At this point, the work/play as O'Donnell (2014: 137) observes turns into AutoPlay – a concept describing the point where the aspects of work/play that fostered involvement and enjoyment (fun) in work practice lead to disengagement, and workers/players cease to be desiring objects.

Ludification – new tendencies and further developments

The concept in (trans)formation

The meaning of play, games, and playing has been observed and reflected upon for centuries by many scholars with reference to numerous spheres of our lives.[4]

However, as we have presented earlier, the concept of ludification has entered the academic discourse relatively recently. It highlights the significance of renegotiation processes that have resulted from the rapid development of the digital play landscape and influence the digital practices. Ludification itself is not able to fully embrace the permeation of play in once play-free domains of life and portray their reciprocal influence. In our understanding, this complex relationship may be more accurately approached with the concept of the *interference of work and play* (Dippel & Fizek 2015, 2016) or through the notion of *work[/play]* (Taylor 2006) and *work/play interplay* (O'Donnel 2014). This interrelation has been also pronounced and discussed in the recent collected volume *The Gameful World* (2015), devoted to ludification of various domains of life. Its editors propose to complement the concept of *ludification of culture* with that of the *cultivation of ludus*, which expresses the nature of changes games undergo while migrating to new, also non-leisure, territories. According to Deterding and Walz, not only games and play move towards the centre of our cultural, social, and economic existence, but also other realms of life impress their forms onto play (2015: 7). In order to avoid a strictly dualistic mode of thinking, the authors propose to unite the two concepts within the metaphor of the *gameful world*.

New fieldworks in cultural anthropology

New tendencies and developments of *ludification of culture* may be observed not only in the theoretical reflections about the concept itself but also in the empirical, ethnographic studies of digital social spaces. Densely populated virtual worlds transform into perfect research fields and open up themes focused on players and gaming culture.[5]

For instance, Mark Chen in his ethnography *Leet Noobs* (2011) takes under examination expert players and their 'raiding' practices in *World of Warcraft*.[6] Various aspects of digital games and social practices have been also scrutinised by ethnographer T.L. Taylor in her numerous works on the multiplayer gaming life (Taylor 2006), gender and sexuality in games (Taylor 2008), or the LAN party scene (Taylor & Witkowski 2010).[7] Kiri Miller, on the other hand, uses ethnographic methods to observe and theorise the experience of *Grand Theft Auto* series players (2008).

The cultural anthropologist Tom Boellstorff has performed online fieldwork and dedicated his work in *Coming of Age in Second Life: An Anthropologist Explores the Virtually Human* (2008) to online residents of the virtual world Second Life and the way they approach gender, race, sex, money, conflict, and the interplay of self and group, amongst others, within the inhabited virtual space. In the collected volume *Ethnography and Virtual Worlds* (Boellstorff et al. 2012), together with other authors, he focuses on the methodological approaches to the ethnographic study of the virtual. The possibilities of an ethnographic method have been also reflected with reference to the concepts of individualisation in text-based online worlds of Multi-User Dungeons (MUDs) (Isabella 2007).

The *ludification of culture* has also influenced cultural empirical research and hermeneutic interpretation, which no longer focus solely on the residents of virtual worlds and the players' culture, but also on other spaces influenced by digitality and playfulness. The discussed ethnographic case study of Casey O'Donnell is one of such examples. Another representative study looks at how mobile locative interfaces influence our everyday interactions, trigger previously unknown forms of sociability, and change our experience of open public spaces. Adriana de Souza e Silva and Jordan Frith (2012, 2015) scrutinise location-based social networks and mobile applications, such as *Foursquare* (2009), that use game elements to encourage people to compete with one another by checking into urban spaces marked on the map.

In his ethnographic study on the production of the game *America's Army* (2002–2009), Robertson Allen discusses the links between work and play, as well as war and entertainment in the United States (2014). He analyses the correlations between military interests in the market-based requirement for the amusement of masses and the significance of war and war games as means of recruitment and training in the American Army.

Other forms of work and play interference are discussed within the context of the so-called 'gold farming' in Massively Multiplayer Online Role-Playing games (MMORPGs), such as *World of Warcraft* (Nakamura 2014), or professionalisation of digital gaming on the e-sport stage (Taylor 2012).[8]

Elements and logics, as well as mechanisms and dynamics of play, may also be observed in the scientific everyday. In High-Energy Physics, where the fundamental phenomena of nature are studied in large collaborative teams, playful aspects of the research practice are particularly visible. During her ongoing fieldwork at CERN (Centre Européen de la Recherche Nucléaire), Anne Dippel has been 'praxeographically' (Knecht 2012, 2013, Niewöhner et al. 2012) collecting a plethora of data, which illustrate the ludification processes of the everyday work (Dippel 2014, 2015, Dippel & Fizek 2015). As she concludes, playful elements and ludified practices may be observed and detected in all work domains of the collaboration and in every experimental system. This assumption seems to be mirrored in an anecdotal statement by the physicist Harald Lesch, who summarised the success of the largest international research institution as follows: 'How come CERN functions so well? Simply because they all play there' (Sternstunde Philosophie 2014). The play is taken extremely seriously.

Empirical fieldwork, praxeography, participant observations, interviews, and critical-hermeneutical analysis are all revealing approaches towards the study of the omnipresence of play and games in the digital age.[9] New research perspectives with regards to human beings – their mutual coexistence and approaches to the surrounding world, as well as the influence of media on human behaviour – appear on the academic horizon.

As we have argued, today, the logic of play permeates all the domains of life on an unparalleled scale and feeds back into the everyday. Digitisation seems to go hand in hand with ludification, as digital mass media further encourage playful transformations of everyday practices. Also, digital machines themselves may be perceived as play ensembles, with the interface thought of as a metaphor of a theatre (Laurel 1993). Rule-based systems of play are also moving into the workspace, just as originally non-playful practices, such as working, feed back into play spaces. In those superposition spots, new practices emerge, which until now have been separate, and form waves of the new, floating through the surfaces of being, to express it with the words of the French philosopher Gaston Bachelard (2007: 175). The emerging digital practices are still young and under-explored fields of research. In the years to come, we are about to witness further scholarly impulses and critical studies, preoccupied with the investigation of *ludified everyday cultures*, which may broaden our understanding and contribute to a more informed development of digitised societies.

Further resources

Wikis

Play4Science research project funded by the German Research Society (DFG) (www.play4science.uni-muenchen.de/index.html). It constitutes an informative example of the usage of game mechanics and logics in the collaborative research scenario.

The Higgs Boson Machine Challenge: the website includes project documentation of a scholarly online competition prepared by CERN as ludified means of outsourcing selected research tasks in High-Energy Physics (https://higgsml.lal.in2p3.fr).

Blogs and websites

Blog of an anthropologist and game scholar T.L. Taylor (http://tltaylor.com/teaching/e-sports-and-pro-gaming-literature) with resources, books, project descriptions, and case studies on anthropological research of various gaming cultures and digital domains.

Blog of the Gamification Lab (http://projects.digital-cultures.net/gamification) with an e-pub open-access collected volume *Rethinking Gamification,* project descriptions, and further academic resources.

Online library of the Digital Games Research Association (www.digra.org) with open-access articles focusing on the interdisciplinary game research, including cultural studies and cultural anthropology, amongst others.

A selection of international academic journals

Journal of Gaming and Virtual Worlds (www.intellectbooks.co.uk/journals/view-Journal,id=164).
Journal of Digital Culture & Society (www.transcript-verlag.de/zeitschriften/digital-culture-und-society).

Game Studies (www.gamestudies.org).
Theory, Culture & Society (http://tcs.sagepub.com).
Games and Culture (http://gac.sagepub.com/content/1/1/29.abstract) (www.intellectbooks.co.uk/journals/view-Journal,id=164).
Eludamost: Journal for Computer Game Culture (www.eludamos.org/index.php/eludamos).
Replay: The Polish Journal of Game Studies (www.replay.uni.lodz.pl).
Homo Ludens. The official journal of the Games Research Association of Poland (http://ptbg.org.pl/HomoLudens/).
WASD Bookazine für Gameskultur (https://wasd-magazin.de/about).
GAME. The Italian Journal of Game Studies (www.gamejournal.it).

Notes

1 In original: 'Das gegenwärtige Jahrhundert konnte man in den Geschichtsbüchern nicht besser, als unter dem Titel: Das Freygeister-Journal und Spielsaeculum nennen' (Bernoulli 1769: 387).
2 'Mash-ups', 'memes', and 'GIFs' (Graphics Interchange Format) are all playful forms of replicating, mimicking, blending, and animating, in this case, visual cultural content and then distributing and sharing it online.
3 The oldest discovered board games are the Mancala games, the variations of which are still played in Africa and Asia today. The oldest version of Mancala games dates back to 7200 B.C. and was excavated not far away from Petra in Jordan (Beidha) (Murray 1952).
4 The most prominent ones, amongst many others, are: Aristotle, Gregory Bateson, Roger Caillois, Stewart Culin, Jacques Derrida, René Descartes, Eugen Fink, James Frazer, G.W.F. Fröbel, Erving Goffman, Johan Huizinga, Immanuel Kant, Moritz Lazarus, John Locke, Marcel Mauss, George Herbert Mead, Michel de Montaigne, John von Neumann, Blaise Pascal, Jean Piaget, Friedrich Schiller, Herbert Spencer, Brian Sutton-Smith, Edward Tylor, and Ludwig Wittgenstein.
5 One of the most recognisable games, *World of Warcraft* (2004), had 12 million active players during its peak (Statista).
6 'Raiding' refers to play practices in Massively Multiplayer Online games, which are focused on organised team combat against other teams of players or on completing tasks that otherwise would be too difficult to accomplish alone or in a smaller group.
7 A LAN party refers to a gathering of gamers, who establish a local area network (LAN) between their computers or consoles in order to play multiplayer games.
8 'Gold farming' is a term denoting the practice of playing in order to later sell virtual goods and in-game currency to other players for real money.
9 Praxeography is a recent term introduced into and discussed within German anthropological discourse (Knecht 2012, 2013, Niewöhner et al. 2012).

References

Adamowsky, N., 2000. *Spielfiguren in der virtuellen Welt*. Frankfurt am Main: Campus.
Allen, R., 2014. America's digital army. Games at work and war. *Journal of Gaming and Virtual Worlds*, 6 (2), pp. 179–192.

Aristotle, 1971. *Nikomachische Ethik*. Leipzig: Reclam Verlag.
Bachelard, G., 2007. *Poetik des Raums*. Frankfurt am Main: Fischer.
Bally, G., 1966. *Vom Spielraum der Freiheit. Die Bedeutung des Spiels bei Tier und Mensch*. Basel/Stuttgart: Schwabe.
Bateson, G., 1972. *Steps to an ecology of mind*. Chicago: University of Chicago Press.
Bauer, G., 2006. Mozart, Kavalier und Spieler. In H. Lachmayer (Ed.), *Mozart: Experiment Aufklarung*. Hatje Cantz, pp. 377–388.
Becker, A., 2008. The royal game of Ur. In I. L. Finkel (Ed.), *Ancient board games in perspective: papers from the 1990 British museum colloquium with additional contributions*. London: British Museum Press.
Bernoulli, D., 1769. In *Anonym, Die Kunst die Welt erlaubt mitzunehmen in den verschiedenen Arten der Spiele, so in Gesellschaften höhern Standes, besonders in der Kayserlich=Königlichen Residenz=Stadt Wien üblich sind*, Bd. II, 1st ed. in 1756, Nürnberg.
Blumenberg, H., 1979. *Höhlenausgänge*. Frankfurt am Main: Suhrkamp.
Boellstorff, T., 2008. *Coming of age in second life: an anthropologist explores the virtually human*. Princeton: Princeton University Press.
Boellstorff, T., Nardie, B., Pearce C., and Taylor, T.L. (Eds.), 2012. *Ethnography and virtual worlds: a handbook of method*. Princeton: Princeton University Press.
Bogost, I., 2011a. Persuasive games: exploitationware. *gamasutra*. Available at: www.gamasutra.com/view/feature/6366/persuasive_games_exploitationware.php (accessed 10 January 2016).
Bogost, I., 2011b. Gamification is bullshit! My position statement at the Wharton gamification symposium. *Ian Bogost Blog* [blog], August 8. Available at: www.bogost.com/blog/gamification_is_bullshit.shtml (accessed 10 January 2016).
Brockhaus, 1957. *Spiel*. Der große Brockhaus, Band 11, 16. Auflage. Wiesbaden: F. A. Brockhaus Verlag, pp. 102.
Caillois, R., 1958/2001. *Man, play and games*. Chicago: University of Illinois Press.
Calleja, G., 2011. *In-Game: from immersion to incorporation*. Cambridge: MIT Press.
Chen, M., 2011. *Leet noobs: the life and death of an expert player group in World of Warcraft*. New York: Peter Lang.
Csikszentmihalyi, M., 1990/2008. *Flow: the psychology of optimal experience*. New York: Harper Perennial Modern Classics.
De Mul, J., De Lange, M., Raessens, J., Frissen, V., and Lammes, S. (Eds.), 2015. *Playful identities: the ludification of digital media cultures*. Amsterdam: Amsterdam University Press.
de Souza e Silva, A. and Frith, J., 2012. *Mobile interfaces in public spaces: locational privacy, control, and urban sociability*. New York: Routledge.
de Souza e Silva, A. and Frith, J., 2015. Location-based mobile games: interfaces to urban spaces. In V. Frissen, S. Lammes, M. De Lange, J. De Mul, and J. Raessens (Eds.), *Playful identities: the ludification of digital media cultures*. Amsterdam: Amsterdam University Press.
Deterding, S., Khaled, R., Nacke, R.E., and Dixon, D., 2011. Gamification: toward a definition. Available at: http://gamification-research.org/wp-content/uploads/2011/04/02-Deterding-Khaled-Nacke-Dixon.pdf (accessed 12 December 2015).
Deterding, S. and Walz, S.P. (Eds.), 2015. *The gameful world: approaches, issues, applications*. Cambridge: MIT Press.

Dippel, A., 2014. Shut up, and calculate? – Zur Produktion wissenschaftlicher Tatsachen am CERN. Ethnographische Eindrücke aus der Hochenergiephysik. *A Paper Presented at the Conference of the Interdisciplinary Network for Studies Investigating Science and Technology (INSIST)*, Berlin, Germany, 22–23 October 2014. Berlin Social Science Centre (WZB).

Dippel, A., 2015. The duck conspiracy. Reality, simulation and shamanism at CERN. *A Paper Presented at Art Genève*. Geneva, Switzerland, 31 January 2015.

Dippel, A. and Fizek, S., 2015. Playful laboratories: the significance of games for knowledge production in the digital era. *A Paper Presented at the Digital Games Research Association's Annual Conference (DiGRA)*, Lüneburg, Germany, 14–17 May 2015.

Dippel A. and Fizek, S., 2016 (in print). Laborious playgrounds: citizen science games as new modes of work/play in the digital age. In M. de Lange, J. Raessens, and I. de Vries (Eds.), *The playful citizen: knowledge, creativity, power*. Amsterdam: Amsterdam University Press.

Dragona, D., 2014. Counter-gamification: emerging tatics and practices against the rule of numbers. In M. Fuchs, S. Fizek, N. Schrape, and P. Ruffino (Eds.), *Rethinking gamification*. Lüneburg: meson press.

Egenfeldt-Nielsen, S., Heide Smith, J., and Pajares Tosca, S. (Eds.), 2013. *Understanding video games: the essential introduction*. 2nd ed. London: Routledge.

Fuchs, M., 2014. Predigital precursors of gamification. In M. Fuchs, S. Fizek, N. Schrape, and P. Ruffino (Eds.), *Rethinking gamification*. Lüneburg: Meson press.

Fuchs, M., Fizek, S., Schrape, N., and Ruffino, P. (Eds.), 2014. *Rethinking gamification*. Lüneburg: meson press.

Gartner Inc., 2011. Gartner predicts over 70 percent of global 2000 organisations will have at least one gamified application by 2014. Available at: www.gartner.com/newsroom/id/1844115 (accessed 23 November 2015).

Global Mobile Games Market Report, 2013. Infographic: The global mobile landscape. [online] Available at: www.newzoo.com/infographics/infographic-the-global-mobile-landscape (accessed 10 January 2016).

Huizinga, J., 1938/1992. *Homo ludens: a study of the play-element in culture*. Boston: Beacon Press.

Isabella, S., 2007. Ethnography of online role-playing games: the role of virtual and real contest in the construction of the field. *Forum Qualitative Sozialforschung*. Available at: www.qualitative-research.net/index.php/fqs/article/view/280/615 (accessed 10 October 2015).

Knecht, M., 2012. Ethnographische Praxis im Feld der Wissenschafts-, Medizin- und Technikanthropologie. In S. Beck, J. Niewöhner, and E. Sörensen (Eds.), *STS – eine sozialanthropologische Einführung*. Bielefeld: Transcript, pp. 245–273.

Knecht, M., 2013. Nach Writing Culture, mit Actor-Network: Ethnographie/ Praxeographie im Feld der Wissenschafts- und Technikanthropologie. In S. Hess, M. Schwertl, and H. Moser (Eds.), *Neue Perspektiven volkskundlicher/ethnologischer Methoden*. Berlin: Panama.

Koch, G., 2015. Empirische Kulturanalyse in digitalisierten Lebenswelten. *Zeitschrift für Volkskunde. Beiträge zur Kulturforschung*, 2015 (2). Münster: Waxmann Verlag, pp. 179–200.

Laurel, B., 1993. *Computers as theatre*. Reading: Addison-Wesley Publishing Company.

Malaby, T., 2009. Anthropology of play: the contours of playful experience. *New Literary History*, 40, pp. 205–218.

Marx, K., 1973/1858. *Grundrisse: introduction to the critique of political economy*. Transl. from German with a Foreword by Martin Nicolaus. New York: Random House.

Marx, K., 2015/1887. The fetishism of commodities and the secret thereof. In *Capital: a critique of political economy*. Available at: www.marxists.org/archive/marx/works/1867-c1/ch01.htm#S2 (accessed 10 January 2016).

Mäyrä, F., 2008. *An introduction to game studies*. Games in Culture. London: Sage Publications.

McGonigal, J., 2011. *Reality is broken: why games make us better and how they can change the world*. New York: Penguin Press.

Miller, K., 2008. The accidental carjack: ethnography, Gameworld tourism, and Grand Theft Auto. *Game Studies* 2008/1. Available at: http://gamestudies.org/0801/articles/miller (accessed 10 October 2015).

Murray, H.J.R., 1952. *A history of board-games other than chess*. Oxford: Oxford University Press.

Nakamura, L., 2014. Don't hate the player, hate the game: the racialization of labor in World of Warcraft. In T. Scholz (Ed.), *Digital labour: the Internet as playground and factory*. London: Routledge.

Niewöhner, J., Sørensen, E., and Beck, S., 2012. Science and Technology Studies aus sozial- und kulturanthropologischer Perspektive. In J. Niewöhner, E. Sørensen, S. Beck (Eds.), *Science and Technology Studies. Eine sozialanthropologische Einführung*. Bielefeld: Transcript, pp. 9–40.

O'Donnell, C., 2014. *The developer's dilemma: the secret world of video game creators*. Cambridge: MIT Press.

Radoff, J., 2011. *Game on: energize your business with social media games*. Indianapolis: Wiley.

Raessens, J., 2006. Playful identities, or the ludification of culture. *Games and Culture: A Journal of Interactive Media*, 1 (1), pp. 52–57.

Raessens, J., 2010. *Homo ludens 2.0: the ludic turn in media theory*. Utrecht: Utrecht University Press.

Raessens, J., 2014. The ludification of culture. In M. Fuchs, S. Fizek, N. Schrape, and P. Ruffino (Eds.), *Rethinking gamification*. Lüneburg: meson press.

Rautzenberg, M., 2015. Navigating uncertainty: ludic epistemology in an age of new essentialisms. In M. Fuchs (Ed.), *Diversity of play*. Lüneburg: meson press, pp. 83–106.

Robertson, M., 2010. Can't play, won't play. *Hide and Seek*, October 6. Available at: http://hideandseek.net/2010/10/06/cant-play-wont-play (accessed 10 January 2016).

Scheuerl, H., 1979. *Das Spiel. Untersuchungen über sein Wesen, seine pädagogischen Möglichkeiten und Grenzen*. Weinheim/Basel: Beltz.

Sicart, M., 2014. *Play matters*. Cambridge: MIT Press.

Statista. The statistics portal. Available at: www.statista.com/statistics/276601/number-of-world-of-warcraft-subscribers-by-quarter (accessed 25 January 2016).

Sternstunde Philosophie, '60 Jahre CERN – Harald Lesch über die Rätsel der Physik', TV program with Harald Lesch, SRF Kultur, Zurich, 28 September 2014. Available at: www.srf.ch/sendungen/sternstunde-philosophie/60-jahre-cern-harald-lesch-ueber-die-raetsel-der-physik (accessed 2 February 2016).

Stewart, J.B., 2013. Looking for a lesson in Google's perks. *New York Times*. Available at: www.nytimes.com/2013/03/16/business/at-google-a-place-to-work-and-play.html?_r=0 (accessed 5 September 2015).

Sutton-Smith, B., 1997. *The ambiguity of play.* Cambridge: Harvard University Press.

Taylor, T.L., 2006. *Play between worlds: exploring online gaming culture.* Cambridge: MIT Press.

Taylor, T.L., 2008. Becoming a player: networks, structures, and imagined futures. In Y. Kafai, C. Heeter, J. Denner, and J. Sun (Eds.), *Beyond Barbie and Mortal Kombat: new perspectives on gender, games, and computing.* Cambridge: MIT Press, pp. 50–65.

Taylor, T.L., 2012. *Raising the stakes: e-Sports and the professionalization of computer gaming.* Cambridge: MIT Press.

Taylor, T.L. and Witkowski, E., 2010. This is how we play it: what a mega-LAN can teach us about games. In *Foundations of Digital Games Conference Proceedings.* Monterey.

Tekinbaş Salen, K. and Zimmerman, E., 2003. *Rules of play: game design fundamentals.* Cambridge: MIT Press.

The Bible. Authorized King James Version with Apocrypha. Genesis 3:19. New York: Oxford University Press.

Tomasello, M., 1999. *A natural history of human cognition.* Cambridge: Harvard University Press.

Turner, V., 1969. *The ritual process: structure and anti-structure.* New York: Cornell University Press.

von Neumann, J., 1928. Zur Theorie der Gesellschaftsspiele. *Mathematische Annalen,* 100, pp. 295–300. Berlin: Springer.

von Neumann, J. and Morgenstern, O., 1944. *The theory of games and economic behaviour.* Princeton: Princeton University Press.

Zichermann, G. and Cunningham, C., 2011. *Gamification by design: implementing game mechanics in web and mobile apps.* Sebastopol: O'Reilly Media.

Zichermann, G. and Linder, J., 2013. *The gamification revolution: how leaders leverage game mechanics to crush the competition.* New York: McGraw-Hill.

Zimmermann, E., 2009. Gaming literacy: game design as a model for literacy in the twenty-first century. In M. J. P. Wolf, and B. Perron (Eds.), *The video game theory reader 2.* London: Routledge, pp. 23–32.

Zimmerman, E., 2013. Manifesto for a ludic century. Available at: http://kotaku.com/manifesto-the-21st-century-will-be-defined-by-games-1275355204 (accessed 10 January 2016).

15 Media genealogy
Back to the present of digital cultures

Clemens Apprich and Götz Bachmann

Media genealogy approaches media cultures of the present out of their history. The goal is a critical take on the present. The central metaphor for research is the branching family tree: media genealogy looks for past divergent lines and hidden relationships that point towards the present in critical ways; this includes the dead ends lost to the present. Media genealogy is less concerned with individual media and their respective cultures. In order to 'find out how different kinds of truth games have been formed' (Foucault 1984: 943; trans. authors), media genealogy instead studies the mechanisms and processes that have led to the development of power and truth structures within digital cultures, seeing media as technical apparatuses, but also as arenas for individual and social practices, for ways of life, cultural patterns, knowledge, power, and control. In such an approach, the genealogical method itself becomes a component part of the truth games being examined: it brings into focus not only the history and present of digital cultures but also the positions of the person investigating these cultures.

On the relevance of media genealogy

Many representatives of German 'Medienwissenschaft' – an intellectual tradition strongly associated with the works of Friedrich Kittler (1985, 1986, 1995) and therefore with a very specific interpretation of the works of Michel Foucault – assume that *only* historical research allows us to gain a real understanding of media. The argument goes like this: when we employ an emphatic concept of media, and assume that media form fundamental conditions of all pure forms of knowing and doing, such media are in the present *not* accessible to us, precisely because they represent such fundamental conditions. Consequently, from this kind of media archaeological viewpoint, we are only able to ask what media *were* (Pias 2011): media only become accessible in retrospect, from a time that is defined by new and different media dispositives. Consequently, history or histories of digitisation are situated centuries before the first electronic computers (representative examples are Siegert 2003, Vismann 2008, Krajewski 2011) and computation is analysed by referring back not only to Turing and Shannon but also to

Lull or Leibniz, Boole, or Gödel. As far as tangible technology is concerned, there is less interest in the smartphone than in imagining what one would have seen given the opportunity to take a walk in the house-sized ENIACS, EDVACS, and UNIVACS of the 1940s and 1950s, or into the post offices, court houses, and libraries in the centuries before.

But, even if we do not share such a fundamental, epistemological argument of media's unaccessibilty in the present, a second set of more pragmatic research issues points us to the past when we wish to understand digital technologies. Early texts are more accessible to laypeople: partly because they have been 'declassified' and thus are no longer in the military's locked drawers or internal corporate development labs (Galison 2004), but also because they often expatiate basic assumptions that were later presumed to be obvious. Therefore, if we wish to understand digital cultures, it makes sense to consult Charles Babbage (2009/1832: 153–163) and Ada Lovelace (1842) on the analytical engine and the idea of the algorithm, Alan Turing (1937) on the universal machine, Vannevar Bush (1945) on the organisation of memory, John von Neumann on the architecture of computers (1945), Claude E. Shannon on information transmission (1948), Norbert Wiener on negative feedback (1948), Joseph Licklider on human/machine symbioses (1960), Paul Baran on computer networks (1962), Ivan Sutherland on Sketchpad (1963), Edgar Codd on relational databases (1970), Alan Kay and Adele Goldberg on the Dynabook (1977), Richard Stallmann on software licenses (1985), Tim Berners-Lee on the World Wide Web (1989/1990), Mark Weiser on ubiquitous information technology (1991), or Sergey Brin and Larry Page on the PageRank algorithm (1998), to name only a few canonical texts. Canons can, of course, always be criticised for good reasons[1] – in fact, we shall be doing it here. In spite of this, such technological texts have stood the test of time, not only for their lucid representation of technical facts but also as tools for academic reflection on digital cultures and media. They have influenced our ideas about technology just as deeply as the history of digital technology itself. Accordingly, approaches that rely strongly on media genealogy – even if they were not yet identified as such – began early on to make these texts by engineers fruitful for the critical analysis of digital culture (classical examples in the German context are Coy 1997, Hagen 1997).

Given the productivity of such media-genealogical excavations, it is all the more astonishing that precisely in German empirical research on digitisation of recent years, the canon of historical texts and their analysis seems to be increasingly less present; attempts to systematically question, broaden, or even replace this historical canon are even rarer. Ethnographic media research often comes with the impulse to advance directly to the relevant questions of the present and kindle a critical potential from that present without long detours (Coleman 2010, Knorr 2011, Boellstorff et al. 2012, Horst & Miller 2012, Budka 2013). In fact, at first sight, there are good reasons to turn away from historicity: does a purely historicising approach not all too often misdirect our attention away from the present? Are not many historical investigations into digitisation strangely uncoupled from current debates and critical approaches? Is it not precisely the dynamic aspects of the continual 'becoming-media' (Vogl 2007) in the present lost, when we rely too much on media archaeology as a method? In order to be able to make these questions productive, we will suggest in the following a specific understanding of media genealogy, which ties the historical investigation of digital cultures and their media directly to a critical engagement with the present.

Differentiating media genealogy from media archaeology

Media genealogy does not aim to give media simple, singular dates and place of 'birth' (e.g., Internet, 1969, *Advanced Research Projects Agency*). Instead, it looks at the many origin stories of technological innovation and the negotiations and decisions that accompanied it, as well as the alternative, not-yet-realised possibilities. The complex processes by which technologies come into being are often hidden behind simplistic and intentionally propagated success stories – whether by military institutions or by the marketing departments of big IT companies. The 'invention' of a worldwide computer network, for example, was not a singular act, as the still-dominant narrative of the 'military origins' of the Internet suggests. The 'historical process of differentiation' (Schröter 2004: 356; trans. authors) instead contradicts the assumption that the 'network of networks' purely emerged in the founding context of the *Advanced Research Projects Agency* (ARPA) of the US Department of Defense. Leaving aside the fact that the military itself is not an ahistorical instance, the origin story of the Internet demonstrates a multiplicity of possible contributions. Not only alternative enterprises at the time, such as the research in the National Physical Laboratory in the United Kingdom (Abbate 1999), but also later processes such as the gradual opening of the computer network in the 1980s (especially through *Usenet* and the first *Bulletin Board Systems*) led to the creation of a multilayered network culture in the 1990s, which then in turns allowed an alternative Internet discourse to arise, without which subsequent global networking via the *World Wide Web* cannot be explained (Apprich 2015). The story of how digital technologies became what they are, formulated here in line with Michel Foucault's considerations on a genealogical historical method, confronts the digital beyond of its supposed origin with its historical contingency. It describes, to continue with the example of the Internet, the transfer of different technical networks (e.g., *ARPAnet, NSFnet, Telnet*), the conflict between different protocols (e.g., TCP/IP versus OSI-standard) or the social, political, and economic transformations, which are now concentrated into a stand-alone network discourse.

This kind of multilinear and non-totalising understanding of history, which is less interested in a specific, all-explaining origin than in a multitude of discursive manifestations, is something media genealogy shares with established media archaeology (Parikka 2012). Both represent methodological attempts to conceptualise digital cultures in their origin stories, in order to demarcate themselves from the conventional history of technology. Both of them undermine traditional historicising processes, for which the history of technology and media functions as a sort of teleological intellectual history. But both are also opposed to purely present-oriented research approaches, especially since these often suffer from a striking tendency to forget history when discussing the 'new media' (Huhtamo & Parikka 2011). Both media archaeology and media genealogy rely here on Michel Foucault's methodological concepts: his investigations into the archaeology of knowledge (Foucault 1965, 1970, 1972) and his history of power in the genealogical sense (Foucault 1977b, 1978, 1985, 1986).

However, while archaeological investigation examines collections of historical accounts – i.e., discursive formations about a given time – genealogical investigation focuses on precisely this dynamic process, around the transformation from one discursive formation to another. Genealogy, as formulated by Michel Foucault building on Friedrich Nietzsche, contains a research programme that extends beyond the analysis of collected historical discourses: the analysis focuses not on how something *is* established, but rather how it *became* established or *becomes* established (Foucault 1977a).[2] Media genealogy does not concern itself with the excavation and description of certain media apparatuses (e.g., paper, camera, film projector, radio set, computer) in their respective discourse-historic settings, but rather takes an interest in the set of ideas, practices, and networks that, together, form a strategic power field for the emergence of digital technologies and media. In media genealogical investigations, the media apparatuses or the definition of these apparatuses remain continually in flux. A genealogical perspective promises a way of looking at things that allows us to understand historical processes as an ongoing confrontation of powers and the power relationships that result from them. So, we can also explain the development of the Internet through the meeting of heterogenous and conflicting power relationships, from technological developments over institutional framework conditions up to social and individual interests.

Compare this approach to media archaeology: the latter combines Foucault's methodological approach with borrowings from German media theory as formulated under the influence of Friedrich Kittler (cf. Horn 2007). This theoretical stream is concerned with the material basis of mediality: the elementary function of a technical medium consists in its ability to store information, process it, and transfer it, which is why the development of technical media comes to a 'perfect end' with the computer as discrete universal machine (Kittler 2013: 189f.). All other media can be converted into this universal machine, or, as in the case of the Internet, only emanate from it (Kittler 1996). This, however, raises the question of whether, in Kittler's interpretation of media history, the end was reset at the beginning; the 'invention' of the computer therefore marks an origin that, from now on, will remain imminent to all media processes. Seen from this 'superhistorical perspective' (Nietzsche 1999: 254; trans. authors), the Internet only represents computing in general; it cannot be distinguished from the archive as a 'discrete source' (Ernst 2002: 129; trans. authors). Reducing our focus to the technical structure of media processes in such a way can lead to the exclusion of praxeological and dynamic sides of those processes (Galloway 2012: 18). It tends to freeze the object of its study – in other words, technical media apparatuses – in order to maintain their *a priori*, transcendental status.[3] Contrast that with media genealogy: instead of describing technical media as something 'prior, decisive, determinative' (Winthrop-Young 2008: 122; trans. authors) solely through discourse analysis, it takes an inversive approach. It traces the different descents of digital technologies with the main goal of aiming for the present, always conscious of the myriads of other genealogical lines.[4]

Fred Turner's from cyberculture to counterculture

An example – albeit, no universal blueprint – for the media genealogy approach, as we are suggesting it here, is Fred Turner's 2006 monograph *From Counterculture to Cyberculture*. Turner's theme is the interplay of military, technological, and (pop) cultural discourses in Silicon Valley from the early 1960s to the 1990s. To this end, Turner examines a number of networks and companies in which the organiser, journalist, and advisor Stewart Brand was active from the early 1960s. Brand is famous in part as initiator and publisher of the *Whole Earth Catalog*, a mail-order catalogue for people with alternative lifestyles, as well as for his founding role in the *Whole Earth 'Lectronic Links* or *The WELL*, the 'birthplace of the online communities' (according to *The WELL*'s subtitle), where – beginning in the 1980s – hackers, journalists, academics, deadheads, and representatives of many other societal groups had the initial experiences they later described as 'virtual community', 'electronic frontier', or 'cyberculture'.[5] Turner follows the networks, ideas, and media practices that took shape in the *Whole Earth Catalog* and the *WELL* into the 1990s, where they interconnected in a new way in the techno-libertarian magazine *Wired*[6] – and here, too, Brand played a key role. Turner's focus, then, is a media genealogy of the 'Californian Ideology' (Barbrook & Cameron 1996).

However, Turner does not merely tell a linear story of counterculture to cyberculture, which could either be read from a cultural critical perspective as the story of how the Left sold out its countercultural ideals or from the opposite perspective of how the Californian hippies played a productive role in Silicon Valley.[7] Instead, the history runs from cyberculture v1 to counterculture to cyberculture v2: why – asks Turner, for example – does Norbert Wiener's *Cybernetics* enjoy required reading status in the *Whole Earth Catalog*? On the formal level, too, he finds in the *Whole Earth Catalog* traces of the prototype of modern sales platforms such as *eBay* – with user-generated content and commentaries, with a community that supplies the content it buys, and with a multiplicity of links and networks between dealers and customers. Based on such observations, Turner follows Stewart Brand's life story further back, and arrives, in inverse order, at the technophile artists' communes of the mid-1960s, at his Stanford University studies with the ecologist Paul Ehrlich, and, finally, at a young man who translates love of freedom and consumer orientation into fear of communism and fascination with cybernetics. Fred Turner therefore follows a specific and, in this case, biographical line and arrives at a residuum that was either forgotten or purposely left out. From this newly discovered point, Turner rolls out the subsequent history anew, arguing that, while the politically motivated Left has little to do with the 'Californian ideology', the communalist counterculture in the United States was influenced from its very starting point in the 1960s by the military 'operational research' of World War II.

Like many media genealogists, Turner is a methodological eclecticist: he analyses technical objects and books, follows networks as well as ideas, studies the language of 'trading zones' in the same way as the vocabulary of subcultures, and draws in cultural and social processes along with institutional, economic, and political history. Turner mixes macrological viewpoints – he sits for two years in the library with the goal of accessing *all the lines* – and detailed case studies, 'literally thumbing through it [the *Whole Earth Catalog*] for months, then noting surprises, lots of counting, making connections'.[8] After discovering the connection between the magazine *Wired* and the *Whole Earth Catalog*, he compiles long lists with the names of people who published *Wired*, wrote in *Wired,* or were quoted by *Wired*, as well as lists of topics and ideas that recur in *Wired*. He does a number of long interviews with the central figures and subsequently manages to obtain Brand's personal journals. Turner digs in different places with different tools, always searching for undiscovered lines, which allow us to think differently about the present: 'it was almost like tracking the Mafia ... Partly I had learned that from STS, you know, follow the actor, but mostly I had learned this as a journalist'.

For years, Turner wrestles with the question of whether the thesis of two opposing movements suggested by his material can be verified and what its political implications are for the present. In the monograph, not everything serves to support the goal of formulating this hypothesis; instead, Turner maintains room for alternative interpretations, multifaceted relationships of influence, and multiple genealogical lines.[9] The ambivalent aspects of the book result from this careful reconstruction of divergent scopes of action and horizons of meaning. But they also owe something to the research method: Turner's research project did not initially aim for a geneology of the 'Californian ideology'.[10] Even in the writing process, he tries to avoid theorising his material early on: 'one of my working principles is: "Keep your questions as simple and as open as possible for as long as you can". As soon as you use a term like "network", you start evoking the whole literature around that – and it cripples you'. For him, media genealogy methods are based on a mixture of instinct and planning:

> The analogy I have is like swimming. I jump into the water and I swim, and I swim hard after whatever piece I have. I only see the next 10 ft in front of me, I do not have a plan of the whole yet. I can feel that the whole is there. And one of the tests for a project is: 'if I can lay out the form very early, it's not a good project'.

By 'swimming' in this way, Turner avoids a linear narrative and also avoids exploiting the past in order to support theses that *only* touch the present.

The bases of media genealogy and their challenges

We can now identify three basic traits of media genealogy: (a) the use of multilinear tree structures that branch off towards the top and the bottom with a view to alternative, i.e., decidedly critical and divergent genealogical lines; (b) the collection of multifaceted and reciprocal relationships of influence with the aid of an eclectic theoretical and methodical toolbox; and (c) an intuitive 'swimming' (Turner) in complex contexts of the past, and from there back into the present. Similar ways of thinking can be found in a multitude of related works on digitisation. Philip Mirowski's *Machine Dreams* (2008) is of note, a critical genealogy of the schools that currently dominate the economic sciences. By describing their history since the 1920s from the point of view of the game theory innovations of John von Neumann and his institutional influence as transmitted via the RAND Corporation, Mirowski analyses the economic sciences as 'cyborg sciences' and opens them up to a new kind of critique (Mirowski 2008). Another example is the deconstruction of post-structuralist theory building by Peter Galison in his canonical essay *The Ontology of the Enemy* (1994), in which he investigates its descents in military research during World War II. In its strongest moments, media genealogy leads to entirely new areas of research: Wolfgang Hagen's *Style of Sources* (1997), for example, opened fields that are now known by such names as 'software studies' or 'critical code studies'; while texts such as Jennifer Light's *When Computers Were Women* (1999) have played a similar role for feminist digital research, and Kavita Philip's (2001) raised new questions about 'postcolonial computing.' This disparate collection shows one thing above all: media genealogy opens new perspectives in the most varied disciplines.[11]

Of course, there are also large differences in detail. Many media genealogy studies pay much less attention than Turner to the meticulous tracing of all the narrative lines, working instead with leaps and omissions (Plant 1997). The manner of criticism – in other words, the way the genealogical lines point to the present – can also be constructed in different ways. In a later book, a sort of prequel, Turner himself relies on a different variant of the process (a): whereas *From Counterculture to Cyberculture* follows a 'subversive' line – in other words, a genealogical line that undermines a dominant success story – *The Democratic Surround* (2013) looks for a 'constructive' line in the 1940s and 1950s, in order to recall forgotten traditions of visions of social inclusion through media technologies. Such constructive lines can be an important variant, especially when they do not contribute to hegemonic identity formation, but rather rearticulate lines that have been interrupted or have trailed off like 'undetonated energy from past revolutions' (Freeman 2010: xvi). Media genealogy therefore encompasses a cluster of different variants of processes A–C, always reconstituting itself, not least because it must be continually rethought due to its double placement in the present and past.

With that in mind, we should like to conclude by discussing four challenges. The first consists in the choice of line. Genealogical multilinearity entails the danger that the researcher will get lost in history and cease to see the forest for the trees – or, in this case, the tree for the branches. On the other hand, there is a danger of reduction to one specific line. Peter Galison (2000) describes this problem in an essay about the retrospective evaluation of aircraft accidents, where he shows how the distinction between 'necessary' and 'sufficient' reasons for catastrophes are negotiated: 'necessary' reasons are all those circumstances that are necessary in order for an accident to occur. These are innumerable and hence have the same status as the potentially unlimited lines in genealogy, which branch off more and more with deeper probing. 'Sufficient', in contrast, describes those reasons isolated in investigations of aircraft accidents – similarly to media genealogical investigations – as being 'at fault': a particular material mistake in the case of an aircraft accident, or the networks of ideas and people that influenced Stewart Brand and were influenced by him in turn, according to Turner's analysis. Here, it quickly becomes apparent that, on one hand, the isolation of 'sufficient' reasons from the myriad 'necessary' reasons is contingent in many respects. To that extent, both movements are constitutive for media genealogy: 'swimming' in the sea of 'necessary' conditions, as well as the ability to select boldly among 'sufficient' lines. Whether this selection ultimately leads to the goal of genealogical criticism can only be determined *ex posteriori*.

The problem of isolating 'sufficient' causes from the complexity of all 'necessary' ones exists not only in media genealogy and plane crash analysis, but also in ethnography, since in the latter, too, the analysis is confronted with a large variety of possible, combined explanations, which must be reduced to a few regardless of how much we value complexity. It is all the more astonishing at first glance that media genealogy and ethnology have not yet, or rarely, been brought together.[12] The reason for this could be that both approaches are so similar in their combination of over- and undercomplexity. They not only complement each other, but also arguably must be employed sequentially with one another – which increases the effort substantially and represents the second challenge of media genealogy. In his essay *Burning Man at Google* (2009), Fred Turner offered an ethnographic 'extension' of his *From Counter to Cyberculture* book. Other examples of connecting media genealogy and ethnography are Gabriella Coleman's *Coding Freedom* (2012) and Christopher Kelty's *Two Bits* (2008). In the latter, Kelty traces free software back to the 1970s, as well as in present day developer communities of Boston, Berlin, and Bangalore. Kelty concentrates first on the present, then dives into the history of free software and Internet protocols, finally arriving back in the present.

Kelty describes a 'recursive public' (ibid.: 3f.): a social and technological assemblage that develops great charisma beyond itself precisely through its ability to maintain itself. By functioning as an organic intellectual, Kelty himself becomes part of this recursive public and its self-maintenance. This brings us to the third challenge: as a reflexive process, media genealogy cannot simply describe technological media in a 'positivistic' way. In the sense of 'wirkliche Historie' (Foucault 1977a: 153), media genealogy has to consciously be a part of the historical reality it describes. The observer perspective of the archeologist is replaced by the participant perspective of the genealogist, who knows about her perspective and the associated 'myopia'. Media genealogy studies engage ideally with their object – and from the perspective of 'situated knowledge' (Haraway 1988). If we assume that, along with digital technologies, a 'media unconscious' (Vogl 2004: 374; trans. authors) has taken form, whose implied knowledge becomes visible with the help of media archeology, but cannot necessarily be made legible, we need the same kind of decidedly reflexive knowledge in genealogical studies. This media genealogy reflexivity should not aim merely at the dominant opinions and practices for engaging with media technologies, but rather at the standpoint of the analyst herself. Her own position within the media history being described, her strategic decisions for this or that line, and indeed her own biases about history therefore become objects of study.

A fourth and final challenge concerns the teaching of media genealogy processes at the bachelor's and master's levels. Media genealogy can only really be put into place pedagogically through independent analyses, in which students take on issues and phenomena of the digital present with the aid of the 'historical sense': how can they develop, for example, a contemporary net critique that builds on the experiences of earlier net cultures? Where would they have to begin in order to strip off the ideological layers of 'Big Data'? What would they learn about current online cultures if they explored the history of 'time sharing' in the 1960s? Furthermore, students should be encouraged to formulate problems and identify genealogical lines on their own – even if they still lack a broad perspective, long immersion in the material, or the confidence to decide on this or that line. When students study the problem of 'sharing', engaging with Shannon or Turing will arguably help them little, even though both of them stand for moments that made the computer and therefore digital sharing cultures possible in the first place. Turing or Shannon are in this case necessary, but not sufficient conditions. Developing the accuracy required to find lines that are at the same time sufficient, in the sense of relevant, and divergent is therefore the central challenge not only in research, but also in teaching. If we succeeded in teaching students the kind of 'swimming' Turner speaks of, so that they undertake their own journeys of exploration without the armbands of established interpretive models, much would be accomplished – the history of digital media and cultures would then really 'make sense'.

Notes

1 Obviously, there is a danger here of reducing the history of digital cultures to the inventions of white men (with the exception of Lovelace and Goldberg in this case) – see also Light (1999).
2 The two methods of questioning are not mutually exclusive, but rather should be seen as complementary dimensions of his overarching research plan: 'The archaeological dimension of the analysis made it possible to examine the forms [of problematisation] themselves; its genealogical dimension enabled me to analyse their formation out of the practices and of the modifications undergone by the latter' (Foucault 1985: 11f.).
3 Because media archaeology is not a discipline with its own independent terminology, it is, of course, not possible to paint all the approaches that have until now been designated as 'media archaeological' with a broad brush. So, for example, Siegfried Zielinski's 'anarchaeology' or 'variantology' represents a project of heterogeneous media practices that undermines materialistic points of view about the media and can be considered close to media genealogy (cf. Kluitenberg 2011: 52f.).
4 Admittedly, this does not mean that media archaeological approaches would be superseded by media genealogy, especially given that they are still very much needed for the deep analysis of discursive formations, together with their more or less stable media apparatuses.
5 *The WELL* began as a text-based bulletin board system and became part of a commercial Internet service provider (ISP) in the early 1990s. With the introduction of the World Wide Web, even *The WELL*'s services were eventually integrated into a web-based environment and the ISP was sold to the company *salon.com*. After a string of financial problems, in part due to the dotcom crash, the community collected money in order to buy *The WELL* back from the Salon Media Group in September 2012.
6 *Wired* was probably the most influential US magazine about digital culture in the 1990s and the journalistic mouthpiece of the 'Californian Ideology' (Barbrook & Cameron 1996).
7 The latter history would not be media genealogy as discussed here, but rather an expanded version of precisely the success and origin stories foisted on the world by Stewart Brand himself; in an essay of 1995 with the title *We Owe It All to the Hippies*, which appeared in a special edition of *Time* magazine with the title *Welcome to Cyberspace*, Brand writes: 'Forget antiwar protests, Woodstock, even long hair. The real legacy of the '60s generation is the computer revolution' (Brand 1995). The goal of this approach to media history is not to critique the present, but rather to continue a myth that aims to shore up existing power relationships.
8 Interview with Fred Turner by Götz Bachmann, Armin Beverungen, and Paula Bialski, 21 May 2015. The following quotes refer to this interview.

9 This is especially recognisable in Turner's treatment of the most important figure in the book, Stewart Brand. Throughout the critique, Brand remains, in Turner's representation, always someone who is muddling along through history and achieving impressive things. It is touching that Turner remained friends with Brand after publication. The first review of the book to appear on Amazon. com was by: Stewart Brand.
10 After writing a book on the American cultural history of the Vietnam War (Turner 1996), Turner actually planned to turn his attention to American masculinity. At that time, digital culture was not an area of interest for Turner.
11 For a recent example in the history and theory of computational design and architecture see Llach (2015).
12 One exception is Gabriella Coleman's ethnographic study of the Internet collective *Anonymous*. Although it is not a media genealogy study in the narrower sense, Coleman traces genealogical lines from online phenomena such as trolling, whistleblowing, or hacktivism in order to describe her object (Coleman 2014).

References

Abbate, J., 1999. *Inventing the Internet*. Cambridge: MIT Press.
Apprich, C., 2015. *Vernetzt – Zur Entstehung der Netzwerkgesellschaft*. Bielefeld: Transcript.
Babbage, C., 2009/1832. *On the economy of machinery and manufacture*. Cambridge: Cambridge University Press.
Baran, P., 1962. *On distributed communications networks*. Available at: www.rand.org/about/history/baran-list.html (Accessed 15 May 2016).
Barbrook, R. and Cameron, A., 1996. The Californian ideology. *Science as Culture*, 6 (1), pp. 44–72.
Berners-Lee, T., 1989/1990. *Information management: a proposal*. Available at: www.w3.org/History/1989/proposal.html (Accessed 15 May 2016).
Boellstorff, T., Nardi, B., Pearce, C., and Taylor, T.L. (Eds.), 2012. *Ethnography and virtual worlds: a handbook of method*. Princeton: Princeton University Press.
Brand, S., 1995. *We owe it all to the hippies*. Available at: http://content.time.com/time/magazine/article/0,9171,982602,00.html (Accessed 15 May 2016).
Brin, S. and Page, L., 1998. *The anatomy of a large-scale hypertextual web search engine*. Available at: http://infolab.stanford.edu/~backrub/google.html (Accessed 15 May 2016).
Budka, P., 2013. Digitale Medientechnologien aus kultur- und sozialanthropologischer Perspektive. Überlegungen zu Technologie als materielle Kultur und Fetisch. *Medien und Zeit*, 28 (1), pp. 22–34.
Bush, V., 1945. As we may think: a top U.S. scientist forsees a possible future world in which man-made machines will start to think. *Life*, 19 (11/10), pp. 112–124.
Codd, E.F., 1970. A relational model of data for large shared data banks. *ACM*, 13 (6), pp. 377–387.
Coleman, G., 2010. Ethnographic approaches to digital media. *Annual Review of Anthropology*, 39, pp. 487–505.
Coleman, G., 2012. *Coding freedom: the ethics and aesthetics of hacking*. Princeton: Princeton University Press.
Coleman, G., 2014. *Hacker, hoaxer, whistleblower, spy. The many faces of anonymous*. London: Verso.

Coy, W., 1997. turing@galaxis.com II. In W. Coy, C. Tholen, and M. Warnke (Eds.), *Hyperkult*. Basel: Stroemfeld, pp. 15–32.
Ernst, W., 2002. *Das Rumoren der Archive. Ordnung aus Unordnung*. Berlin: Merve Verlag.
Foucault, M., 1965. *Madness and civilization: a history of insanity in the age of reason*. New York: Random House.
Foucault, M., 1970. *The order of things*. New York: Pantheon.
Foucault, M., 1972. *The archaeology of knowledge*. New York: Pantheon.
Foucault, M., 1977a. Nietzsche, genealogy, history. In M. Foucault and D.F. Bouchard (Eds.), *Language, counter-memory, practice: selected essays and interviews*. Ithaca: Cornell University Press, pp. 139–164.
Foucault, M., 1977b. *Discipline and punish*. New York: Pantheon.
Foucault, M., 1978. *The history of sexuality, Vol. I: An introduction*. New York: Pantheon.
Foucault, M., 1984. Foucault. In D. Huisman (Ed.), *Dictionnaire des philosophes*. Paris: PUF, pp. 942–944.
Foucault, M., 1985. *The history of sexuality, Vol. II: The usage of pleasure*. New York: Vintage.
Foucault, M., 1986. The care of the self. In M. Foucault (Ed.), *The history of sexuality, Vol. III*. New York: Pantheon.
Freeman, E., 2010. *Time binds: queer temporalities, queer histories*. Durham: Duke University Press.
Galison, P., 1994. The ontology of the enemy: Norbert Wiener and the cybernetic vision. *Critical Inquiry*, 21 (1), pp. 228–266.
Galison, P., 2000. An accident of history. In P. Galison and A. Roland (Eds.), *Atmospheric flight in the twentieth century*. Dordrecht: Kluwer, pp. 2–43.
Galison, P., 2004. Removing knowledge. *Critical Inquiry*, 31, pp. 229–240.
Galloway, A.R., 2012. *The interface effect*. Cambridge: Polity Press.
Hagen, W., 1997. Der Stil der Sourcen. Anmerkungen zur Theorie und Geschichte der Programmiersprachen. In W. Coy, C. Tholen, and M. Warnke (Eds.), *Hyperkult*. Basel: Stroemfeld, pp. 33–68.
Haraway, D., 1988. Situated knowledge: the science question in feminism and the privilege of partial perspective. *Feminist Studies*, 14 (3), pp. 575–599.
Horn, E., 2007. There are no media. *Grey Room*, 29, pp. 6–13.
Horst, H.A. and Miller, D., 2012. The digital and the human: A prospectus for the digital anthropology. In H.A. Horst and D. Miller (Eds.), *Digital anthropology*. London: Berg, pp. 3–38.
Huhtamo, E. and Parikka, J., 2011. Introduction: an archaeology of media archaeology. In E. Huhtamo and J. Parikka (Eds.), *Media archaeology: approaches, applications and implications*. Berkeley: University of California Press, pp. 1–21.
Kay, A. and Goldberg, A., 1977. *Personal dynamic media*. Available at: www.vpri.org/pdf/m1977001_dynamedia.pdf (Accessed 15 May 2016).
Kelty, C., 2008. *Two bits: the cultural significance of free software and the Internet*. Durham: Duke University Press.
Kittler, F., 1985. *Aufschreibesysteme 1800/1900*. Munich: Wilhelm Fink.
Kittler, F., 1986. *Grammophon film typewriter*. Berlin: Brinkmann & Bose.
Kittler, F., 1995. *There is no software*. Available at: www.ctheory.net/articles.aspx?id=74 (Accessed 15 May 2016).

Kittler, F., 1996. Das Internet ist eine Emanation: Ein Gespräch mit Friedrich Kittler. In S. Iglhaut, A. Medosch, and F. Rötzer (Eds.), *Stadt am Netz. Ansichten von Telepolis*. Mannheim: Bollmann, pp. 196–203.

Kittler, F., 2013. Die Stadt ist ein Medium. In: F. Kittler and H.U. Gumprecht (Eds.), *Die Wahrheit der technischen Welt. Essays zur Genealogie der Gegenwart.* Frankfurt: Suhrkamp, pp. 181–197.

Kluitenberg, E., 2011. On the archaeology of imaginary media. In E. Huhtamo and J. Parikka (Eds.), *Media archaeology: approaches, applications and implications.* Berkeley: University of California Press, pp. 48–69.

Knorr, A., 2011. *Cyberanthropology.* Wuppertal: Peter Hammer Verlag.

Krajewski, M., 2011. *Paper machines: about cards and catalogs, 1548–1929.* Cambridge: MIT Press.

Licklider, J., 1960. Man-computer symbiosis. *Transactions on Human Factors in Electronics*, HFE-1, pp. 4–11.

Light, J., 1999. When computers were women. *Technology and Culture*, 40 (3), pp. 455–483.

Llach, D.C., 2015. *Builders of the vision: software and the imagination of design.* Oxford: Routledge.

Lovelace, A., 1842. *Notes by the translator of L.F. Menabrea: sketch of the analytical engine, invented by Charles Babbage.* Available at: www.fourmilab.ch/babbage/sketch.html (Accessed 15 May 2016).

Mirowski, P., 2008. *Machine dreams: economics becomes a cyborg science.* Cambridge: Cambridge University Press.

Nietzsche, F., 1999. Unzeitgemäße Betrachtungen. Zweites Stück: Vom Nutzen und Nachteil der Historie für das Leben. In G. Colli and M. Montinari (Eds.), *Friedrich Nietzsche: Sämtliche Werke. Kritische Studienausgabe in 15 Bänden. Vol. 1.* Munich: dtv, pp. 243–334.

Parikka, J., 2012. *What is media archaelogy?* Cambridge: Polity Press.

Philip, K., 2001. Science and technology studies and indigenous knowledge. In N.J. Smelser and P.B Baltes (Eds.), *International Encyclopedia of the Social and Behavioral Sciences.* London: Elsevier.

Pias, C. (Ed.), 2011. *Was waren Medien?* Zürich: diaphanes.

Plant, S., 1997. *Zeros + ones: digital women + the new technoculture.* London: Fourth Estate.

Schröter, J., 2004. Technik und Krieg. Fragen und Überlegungen zur militärischen Herkunft von Computertechnologien am Beispiel des Internets. In H. Segeberg (Ed.), *Die Medien und ihre Technik: Theorien. Modelle. Geschichte.* Marburg: Schüren, pp. 356–370.

Shannon, C.E., 1948. A mathematical theory of communication. In: *Bell System Technical Journal*, 27, pp. 379–423, 623–656.

Siegert, B., 2003. *Passage des Digitalen. Zeichenpraktiken der neuzeitlichen Wissenschaften 1500–1900.* Berlin: Brinkmann und Bose.

Stallman, R., 1985. The GNU Manifesto. *Dr. Dobb's Journal*, 10 (3), pp. 30–35.

Sutherland, I., 2003/1963. *Sketchpad: a man-machine graphical communication system.* Available at: www.cl.cam.ac.uk/techreports/UCAM-CL-TR-574.pdf (Accessed 15 May 2016).

Turing, A., 1937. On computable numbers, with an application to the Entscheidungsproblem. In *Proceedings of the London Mathematical Society*, 42, pp. 230–265.

Turner, F., 1996. *Echoes of combat: trauma, memory, and the vietnam war.* Minneapolis: University of Minnesota Press.

Turner, F., 2006. *From counterculture to cyberculture: Stewart Brand, the whole earth network, and the rise of digital utopianism.* Chicago: University of Chicago Press.

Turner, F., 2009. Burning man at Google: a cultural infrastructure for new media production. *New Media & Society,* 11 (1–2), pp. 73–94.

Turner, F., 2013. *The democratic surround: multimedia and American Liberalism from World War II to the psychedelic sixties.* Chicago: University of Chicago Press.

Vismann, C., 2008. *Files: law and media technology.* Stanford: Stanford University Press.

Vogl, J., 2004. Technologien des Unbewußten. In C. Pias, J. Vogl, L. Engell, O. Fahle, and B. Neitzel (Eds.), *Kursbuch Medienkultur.* Stuttgart: DVA, pp. 373–376.

Vogl, J., 2007. Becoming-media: Galileo's telescope. *Grey Room,* 29, pp. 64–83.

Von Neumann, J., 1945. *First draft of a report on the EDVAC.* Available at: https://sites.google.com/site/michaeldgodfrey/vonneumann/vnedvac.pdf?attredirects=0&d=1 (Accessed 15 May 2016).

Weiser, M., 1991. The computer for the 21st century. *Scientific American,* 265 (3), pp. 94–104.

Wiener, N., 1948. *Cybernetics: or control and communication in the animal and the machine.* New York: John Wiley & Sons.

Winthrop-Young, G., 2008. Von gelobten und verfluchten Medienländern. Kanadischer Gesprächsvorschlag zu einem deutschen Theoriephänomen. *Zeitschrift für Kulturwissenschaften,* 2, pp. 113–127.

Index

activism, activist 97, 99–100, 103, 105–6, 170, 267–72
aesthetics 120, 132, 134, 143, 146, 150
algorithm, algorithmic: computer feature 1, 18, 185–86; computer practice 62, 158–59, 221–22; content analysis 161; learning algorithms 169–73; literature 294; manipulation 214
analogue 1–5, 117, 143, 209, 214, 277–78

bitcoin 254, 271

capital, capitalism, capitalistic 113, 122–23, 252, 255–57, 258–66, 267–68; *see also* labour
collectives 19, 97–100, 102–5
commodity 251–52, 257–61
community, communities: communities of practice 79, 84; concept 55; documentation 41; hacker 102–5; imagined communities 212; Internet 141, 165, 297–98
complex, complexity: archival practices 45–6; assemblages 58–9; Big Data analyses 158, 173; challenges media genealogy 300; digital infrastructures 78, 81; imaginaries 72–3; networks 168–69; ubicomp technologies 199, 202–4, 206, 295; work 281; *see also* infrastructure
computer (universal machine), computing: augmented reality 231–33, 235–36; binary-digital logic 15–17, 19, 214; computerisation 1–6; game 283; hacker 97–100; imaginations 179–80; interface 185–86, 188–91; media genealogy 293–97; operational chains 27; programme functions 147;

transmedia 149; ubiquitous computing 181, 197–200; workplaces 61–3
(re-)configuration 7, 81, 106, 185, 193
consumer, consumption: AR technology 78, 81, 236; consumer expectations 63; consumer versus producer 128; consumption of culture 261; media consumer 140; property 123–24, 266; transfer 191
copy, copyright: artistic production 152–53; copyright reform movement 105; Creative Commons 270; cultural analysis 124–28; different perceptions 123–24; intellectual property 262–65, 269; mash-up 139, 145, 147; original versus copy 117–18; studies 119–23
cyberculture 297, 300–1
cybernetics 1–4, 182, 188, 192, 297–98
cyberspace 53, 57, 59–61, 64, 71, 110, 216

data, database: accumulation 212–16; analogue to data 1–3; Big Data 158–73, 187, 191; collection 203–6; data centre 65–72; geodata 220–22, 226; infrastructure 86; literature 294; manipulation 19; public debates 127; sharing 113–14; traces 197–200; *see also* information
democracy, democratic 122, 124, 189, 252–53, 257, 268
design: algorithms 173; digital design 112; games 276, 278, 281; infrastructure 62, 82–6; interface design 179–83, 185–86, 188–89, 191–93; mash-up culture 143–45; social media platforms 169; spaces 230–31
discourse: archival 43, 47; authenticity 124; Big Data 158–59; copyright 152;

Index

digitisation 13; hacking 109; Internet of Things 184; labour 258; media 252–53; media genealogy 295–97; remix 146; value 118

economy, economic 119, 121–23, 125–26, 152–54, 295; economical 84, 150; economisation 82; new economy 61, 64, 139–41; political economy 251–58, 264, 268–69, 271

engineering, engineers 29–33, 67, 95–6, 102–3, 107–8, 197–98, 294

experimentation 97, 167–68, 237, 269; experiment 71, 104, 183, 239–40, 263, 279, 286

figuration 107, 109–10, 113
finance, financial 86, 107, 152, 237, 252–54, 271
formal 15–17, 25–6, 41–4, 85, 110, 150, 224

genealogy 98, 293–97, 299–302; *see also* medium (media genealogy)
Globalisation 34, 55, 95, 137
Goffman, Erving 88n9, 288n4

habitus 231; habits 21, 44, 72, 82, 107, 164; habitual 181
heritage, cultural heritage 42–3, 127, 239–40
history, historical 33, 41–3, 124–28, 191–92, 265–66, 282–84, 293–302
human, non-human 19–21, 26–32, 83–5, 200–2, 230–35, 278, 285–86

image 81, 84–5, 88, 134–39, 142–43, 230–35, 239; imaginary 96, 276–78; imagination 4–5, 54, 59, 63, 179–80, 184, 212; *see also* visualisation
industry, industries 59–61, 67–71, 110, 128, 198, 205–6, 243; creative industries 251–55, 122–24; industralisation 280, 117–18; media industries 251–53, 257, 283
information: archive 42–4; Big Data analysis 170–73; control 203–5; geoinformational systems 212–15, 218, 220–21; geopolitical turn 71–2; Google campaign 65–6; information age 254–55, 263; information freedom 98–9; information security 113–14; information societies 3–4; information systems 79–86; information transfer 136, 139–40, 296; informational enhancement 230–33, 236, 241–43; interface 181, 183–84; prosumer 148; social media 160, 165–67
information technology (IT, ICT) 13, 29–30, 32–3, 80–1, 109, 239, 255
infrastructure 61–2, 64–5, 78–9, 81–8, 171–72, 216–17, 242–43; *see also* complexity
innovation 86, 113, 120, 147, 254, 265, 295
Internet: archive 40; Big Data 158; capitalism 263; cultural material 124; Darknet 82–3; digital infrastructures 80; imaginaries 64–5, 66–72; internet business 63–4; Internet of Things 71, 87, 179–88, 190–92, 197–98, 200, 241; media genealogy 293–97; medium 261; remix culture 139–41, 144; social web 269–70; surveillance 127; usage 19; *see also* medium
invention 63, 107–8, 110–11, 113–14, 214, 266, 295–96

knowledge, knowing 23–4, 80–1, 83–5, 167–68, 221, 239–40, 301; knowledge society 2–4

labour 68, 113, 122–23, 184, 187, 259–67, 281–84; *see also* capital, work

Marx, Karl 252, 255–61, 264–67, 272, 282
material: copy 117–19; hacker practices 107–9; interface 184–85, 192; map 221; material goods 123–24; material operations 16–19, 22, 26–8; material valuation 126; materiality 62, 65, 71, 200–2, 216, 225–26, 253; play 277; property 267; raw material 255–56, 259; remix 139, 146–47, 152; spaces 230–31, 233
medium 119–22, 135–36, 255, 261, 279, 281, 296; augmented reality 230–38, 239, 242–43; games 276–79, 283; geomedia 209, 215–17, 220–22; mass media 267, 252–55, 257; media, medial 71–2, 80–1; media genealogy 293–95; media studies 151–52; mobile medium 212–13; social media 43, 141, 153, 159–67, 171–72, 269; *see also* Internet, genealogy

Index 309

memory 19–23, 26–8, 40, 43, 45–7, 240, 294
method 128, 134–36, 149–50, 172–73, 198, 277, 293; methodological 85–7, 113, 125, 158, 160–68, 193, 296–98
mind 5, 60, 67, 233–35

network: Actor-Network-Theory 22, 79–80, 168–70, 188, 201; analysing networks 160–61; computer network 294–98; digital network 183–84, 276–77; hacking 109–11; Internet 263; mapping 212–14, 216, 225; networked technologies 64–5, 71–2; neural network 22–4, 59–62; social network 158, 198, 232–33, 279, 286; sociomaterial network 205
norm, normative 25, 61, 79, 188, 241, 256, 262

participation 46–7, 140–42, 171, 198, 224–25, 237–39, 279; *see also* user
perception 42, 84–5, 123, 136, 215, 230–35, 241–43
policy 83, 110, 127, 160, 199, 253, 270
politics 82, 101, 209, 221; geopolitics 71–2, 102; political 45–6, 97, 105–6, 144, 181, 212, 251–55
postcolonial, post–colonialism 55, 299
power 106–13, 186–87, 191–93, 197–98, 212–13, 216–17, 252–53
practice 15–21, 61, 105, 225–26, 276–77, 280, 296; practical 19–20, 82, 108, 118, 214, 230; social practices 4–5, 25, 78, 80–8, 108, 181, 209
programme 17, 21–30, 83, 111, 147, 185–88, 224

remediation 143–45, 242

semantic, semantics 26, 32–3, 125, 151
software 17, 27, 33, 58–9, 62–3, 107–11, 217–24; free software 98–105, 301; software studies 79–81
sound 16, 133–37, 146, 163, 215, 234, 241
Star, Susan Leigh 5, 79, 84
symbol, symbolic 17–18, 29, 100, 181, 213, 225, 230–33

technique: augmented reality 230; cultural techniques 13–22, 34–5, 134, 209–10, 213–14, 225–26; hacker techniques 99, 107–8; mash-up 142, 147, 151; social media 168
tradition 67, 112–14, 120–24, 143–46, 236, 276–77, 293

user 62–5, 80–5, 182–88, 220–22, 240–43, 297–98; user–generated 133–34, 140, 158–61, 164–72, 233–36, 269; *see also* participation

virtual 17–18, 61–5, 216, 230–37, 239–42, 280–81, 285–86
visualisation 67–9, 181, 183–86, 214–15, 217, 220–21, 225–26; visual 132, 142–43, 192, 198, 210, 240–41, 279; *see also* image

Wiener, Norbert 294, 297
work 41–3, 84–6, 152–53, 198–200, 202–6, 258–59, 280–87; *see also* labour